EYEWITNESS TO Science

Edited by
JOHN CAREY

Harvard University Press
Cambridge, Massachusetts

Printed in the United States of America

This is an authorized reprint of the hardcover edition known as *The Faber Book of Science* published by Faber and Faber. For information address Faber and Faber, Inc., 53 Shore Road, Winchester, MA 01890.

First Harvard University Press paperback edition, 1997

Library of Congress Cataloging-in-Publication Data

Faber book of science.
Eyewitness to science / edited by John Carey.
p. cm.
Originally published: The Faber book of science. London ; Boston : Faber and Faber, 1995.
Includes bibliographical references and index.
ISBN 0-674-28755-x (alk. paper)
1. Science—Popular works. I. Carey, John, 1934– . II. Title.
[Q162.F29 1997]
500—dc21 97–8132

Contents

Introduction

The aim of this book is to make science intelligible to non-scientists. Of course, like any anthology, it is meant to be entertaining, intriguing, lendable-to-friends and good-to-read as well, and the first question I asked about any piece I thought of including was, Is this so well written that I want to read it twice? If the answer was no, it was instantly scrapped. But alongside this question I asked, Does this supply, as it goes along, the scientific knowledge you need to understand it? Will it be clear to someone who is not mathematical, and has no extensive scientific education? Even if it was admirable in other ways, failure to qualify on these counts landed it on the reject pile.

Scientists themselves are not always good at judging intelligibility – and why should they be? They are specialists, paid to communicate with fellow specialists. Of course, they have to communicate, too, with industry, the government, grant-giving bodies and other institutions. But they can often assume a level of expertise in these negotiations which is well above that of the general public. Over the last five years I have read many books and articles by scientists, ostensibly for a popular readership, which start out intelligibly and fairly soon hit a quagmire of equations or a thicket of fuse-blowing technicalities, from which no non-scientist could emerge intact. *Relativity: The Special and General Theory. A Popular Exposition*, by Albert Einstein, Ph.D. (1920) is only a particularly distinguished example of a class of 'popular expositions', still being published, that could not conceivably be understood by more than a tiny fraction of any populace.

Fortunately for this anthology, however, popular science has improved immensely in the later twentieth century. Writers like Isaac Asimov, Arthur C. Clarke, Martin Gardner, Freeman Dyson, Carl Sagan, Richard Feynman, Stephen Jay Gould, Peter Medawar, Stephen Hawking, Lewis Wolpert and Richard Dawkins have transformed the genre, combining expert knowledge with an urge to be understood, and bridging the intelligibility gap to delight and instruct huge readerships. In the process, they have created a new kind of late twentieth-century literature, which demands to be recognized as a separate genre, distinct from the old literary forms, and conveying pleasures and triumphs quite distinct from theirs.

True, these writers had predecessors in the nineteenth century – T. H. Huxley, for example, or Charles Darwin himself, who also strove to reach the general reading public. But in the mid-nineteenth century the general reading public was a much smaller and more select thing than it is now. The challenge for a late twentieth-century writer of popular science is different and greater. The books that succeed represent achievements of a remarkable and unprecedented kind. Nor is it clear on what grounds they can be reckoned inferior to novels, poems and other representatives of the older genres. In what respect, for example, is a masterpiece like Richard Dawkins's *The Blind Watchmaker* imaginatively inferior to a distinguished work of fiction such as Martin Amis's *Einstein's Monsters* (or the hundreds of lesser novels that jam the publishers' lists each year)? Both are clearly the products of brilliant minds; both are highly imaginative; and Amis is more excited by scientific ideas than most contemporary writers. Nevertheless, the essential distinction between them seems to be that between knowledge and ignorance. From the viewpoint of late twentieth-century thought, Dawkins's book represents the instructed and Amis's the uninstructed imagination.

Because I wanted the pieces I included to be seriously informative as well as enjoyable, I decided not to allow in science fiction (which would, in any case, need an anthology of its own), or those plentiful anecdotes about scientists' private lives which show how droll or winning they were despite their erudition. The misty precursors of true science – alchemy, astrology – have also been left out, partly because they can now be classified as history not science, and partly because they tend to encourage in the reader an amused and superior response which is not the reaction I am looking for.

For similar reasons I decided, after some hesitation, not to include ancient science (Aristotle, Pliny, etc.). It is true, of course, that this sometimes foreshadows modern science. But even when it does it is often forbiddingly technical, in a way that no amount of jazzing-up in translation can overcome. After a good deal of searching, I concluded that there were virtually no examples of ancient science that would have anything more than curiosity value – if that – for a general reader today. So my anthology starts with the Renaissance, at a point where two sciences, anatomy and astronomy, take decisive steps towards the modern age, and find exponents who can still be read with pleasure.

A final kind of writing I decided (rather quickly) to exclude was the large body of opinionativeness that has gathered around such questions as whether science is a Good or a Bad Thing, and whether we would be better off if we did not know the earth went round the sun. Ignorance and prejudice seem to be the most prolific contributors to this branch of controversy, and I am not anxious to give either house-room.

In the main, then, I have tried to stick to serious science, though serious science softened up for general consumption. Scientists will object quite rightly that I have included technology as well as science. The pieces on the Wright brothers' aeroplane or on Daguerre and the first photograph, for example, would not figure in a strictly scientific anthology. But I included them and others because, for the general reader, science and technology are intimately connected – as, indeed, they are for scientists. Photography and manned flight both became possible because of scientific perceptions, and technology has advanced scientific discovery from the time of Galileo's telescope.

Choosing the passages to include was one thing: arranging them, another. Should I separate out the various sciences – all the biology pieces in one section; all the chemistry in another? Or would a roughly chronological arrangement be better? I decided it would, because jumping from science to science with each item makes for a livelier read, and the chronological framework turns the book into a story – a way of taking in the development of science over the last five centuries. Some of this story-telling is carried on in the introductions to each extract, and sometimes – as, for example, in the sections on Relativity (p. 267) and the Uncertainty Principle (p. 277) – I have drawn together material from several sources, including poets and novelists, to show how a particular scientific discovery did, or did not, enter the blood-stream of the culture.

Broadly speaking science-writing tends towards one of two modes, the mind-stretching and the explanatory. In practice, of course, any particular piece of science-writing will combine the two in various proportions. Still, they seem to be the extremes between which science-writing happens. The mind-stretching, also called the gee-whizz mode, aims to arouse wonder, and corresponds to the Sublime in traditional literary categories. When scientists tell us that if we could place in a row all the capillaries in a single human body they would reach across the Atlantic, or that the average man has 25 billion red blood corpuscles, or that the number of nerve cells in the cerebral cortex of the brain is twice the population of the globe, these are contributions to the mind-stretching mode – which does not mean, of course, that they are not serious and profound in their implications as well. A similarly amazing example, and less flattering to our self-esteem, is the proposition (from an essay by George Wald) that though a planet of the earth's size and temperature is a comparatively rare event in the universe, it is estimated that at least 100,000 planets like the earth exist in our galaxy alone, and since some 100 million galaxies lie within the range of our most powerful telescopes, it follows that throughout observable space we can count on the existence of at least 10 million million planets more or less like ours.

As readers will find, I have included some examples of this mode in my anthology, because the peculiar thrill and spiritual charge of science would not be fairly represented without it. But my preference has been, and is, for the other mode, the explanatory. What I most value in science-writing is the feeling of enlightenment that comes with a piece of evidence being correctly interpreted, or a problem being ingeniously solved, or a scientific principle being exposed and clarified. There are many instances of these three processes in the anthology, but if I had to choose one favourite example of each they would be from Galileo, Darwin and Haldane respectively.

When Galileo looked at the moon through his telescope, he and everyone else thought it was a perfect sphere. He was astonished, he tells us, to see bright points within its darkened part, which gradually increased in size and brightness till they joined up with its bright part. It occurred to him that they were just like mountain tops on earth, which are touched by the sun's morning rays while the lower ground is still in shadow. So he deduced correctly that the moon's surface was not smooth after all, but mountainous. To follow Galileo as he explains his observations step by step (pp. 9–14) is to share an experience of scientific enlightenment that fiction and poetry, for all their powers, cannot give, since they can never be so authentically engaged with actuality and discovery.

Darwin supplies a beautiful example of the second process, the ingenious solution of a problem, when he is faced with the need to explain how species of freshwater plants could spread to remote oceanic islands without being separately created by God (p. 119). It occurs to him that the seeds might be carried on the muddy feet of wading birds that frequent the edges of ponds. But that raises the question of whether pond mud contains seeds in sufficient quantities. So he takes three tablespoonfuls of mud from the edge of his pond in February – enough to fill a breakfast cup – and keeps it covered in his study for six months, pulling up and counting each plant as it grows. Five hundred and thirty-seven plants grow, of many different species, so that Darwin is able to conclude that it would be an 'inexplicable circumstance' if wading birds did not transport the seeds of freshwater plants, as he had suspected. Once again, fiction could not compete with the impact of this, since the force of Darwin's account depends precisely on its not being fiction but fact.

J. B. S. Haldane's famous essay 'On Being the Right Size' (p. 302) superbly exemplifies the third process – the exposition of a scientific principle. Restricting his mathematics to simple arithmetic, and keeping in mind the need for powerful, graphic examples, Haldane is able to demonstrate, unforgettably, by the end of his second paragraph, that the 60-foot-high Giants Pope and Pagan in Bunyan's *Pilgrim's Progress* could never have existed, because they would have broken their thighs every time they walked. The example is, of course, purposefully chosen, for out goes, with Bunyan, the whole world of (as Haldane saw it) religious mumbo-jumbo that Bunyan stood for, and the light of pure reason comes flooding in instead.

But if the explanatory mode is science-writing's breath of life – its armoury, palette and climate – the problem for science-writers is how to explain. How can science be made intelligible to non-scientists? The least hopeful answer is that it cannot. Giving an inkling of what modern science means to readers who cannot manage higher mathematics is, Richard Feynman has proposed, like explaining music to the deaf. This would be a desolating conclusion if Feynman were not himself among the most brilliant of explainers. His success depends upon his genius for making his material human. He saturates his writing with his individual style and personality. But, more than that, he freely imports a kind of animism into his experimental accounts – discussing, for example, how an individual photon 'makes up its mind' which of a number of possible paths to follow.

Ruskin uses animism, too, when – in his masterly tribute to rust – he tells his readers that iron 'breathes', and 'takes oxygen from the atmosphere as eagerly as we do' (p. 110). Miroslav Holub is animistic when (in perhaps the most mind-expanding piece in the whole anthology) he imagines the adrenalin and the stress hormones in the spilt blood of a dead muskrat still sending out their alarms, and the white blood cells still busily trying to perform their accustomed tasks, bewildered by the unusual temperature outside the muskrat's body (p. 489). In fact, Feynman-Ruskin-Holub-type animism is a persistent ally in the popular science-writer's struggle to engage the reader's understanding.

To a scientist, this might seem ridiculous. Lewis Carroll rubbished the whole idea in *The Dynamics of a Particle:*

> It was a lovely Autumn evening, and the glorious effects of chromatic aberration were beginning to show themselves in the atmosphere as the earth revolved away from the great western luminary, when two lines might have been observed wending their weary way across a plane superficies. The elder of the two had by long practice acquired the art, so painful to young and impulsive loci, of lying evenly between his extreme points; but the younger, in her girlish impetuosity, was ever longing to diverge and become a hyperbola or some such romantic and boundless curve . . .

However, it is not clear that animism is as daft as Carroll makes it appear. All science is inevitably drenched in our human presumptions, designs and conceptions. We cannot get outside the human shapes of our brains. Our observation inevitably alters what it observes. This perception is usually associated with Heisenberg (see p. 277). But it was already evident to Francis Bacon at the start of the seventeenth century, who saw that perfect, pure objective science was impossible, not only because we are forced to use language, or some kind of numerical notation, which does not 'naturally' belong to the objects we name or number, but also because we seek patterns, shapes and symmetries in nature which correspond to our own preconceptions, not to anything that is 'really' there. From this viewpoint, to say that iron 'breathes' is no more absurd than to say that it is called 'iron', or that its chemical symbol is Fe. In each case, we add something human to its remote, alien, unknowable nature – a nature that has nothing to do with human thought, and is therefore altered the instant we think about it.

Whatever reservations the reader may have about this line of argument, it remains true that animism is extraordinarily useful to science-writers, as many pieces in this book testify. To preserve a personal element I have also tried, as often as I could, to present scientists talking about themselves at the moment of discovery. Nothing can match the immediacy of such accounts. Francis Jehl's description of the feverish months of trial and error that preceded the development of the world's first electric light bulb in Edison's laboratory (p. 169), or Ronald Ross's memory of the sweltering afternoon in Secunderabad when he saw, through his microscope, the secret of malaria (p. 204), or William Beebe exclaiming at the astonishing blueness of the sea 700 feet beneath the ocean, so intense that it drives even the thought of any other colour out of his head (p. 293) – if I could have found enough of them, I should have been tempted to make my whole anthology out of pieces like this.

Given the boundless human implications of science, it seems strange that poets have not used it more. One of my disappointments in editing this anthology was to find so little poetry – or so little that was not embarrassingly bad. Lifting the embargo on ancient science would have helped a bit, because I could have included some Lucretius – but it would have been too high a price. Among English poets, even Shelley, who knew more about science than most, does not really write scientific poetry. To treat 'The Cloud', say, as a poem about meteorology (though it is that) would be to ignore most of its meaning. Generally speaking, science has had a bad effect on poets, inciting them to bombast (of the 'O thou terrestrial ball' variety) or to drivelling regrets that science has banished 'faery lore'.

Science's dominant position in contemporary culture might surely have been expected to breed some modern scientific poets. Yet most poets remain science-blind. There are a few distinguished exceptions, as the reader will find: John Updike, Lavinia Greenlaw, John Frederick Nims. But neglect is the norm. Why?

Perhaps because it is assumed that the poetic imagination is superior to the scientific, so poets simply need not bother with science. Certainly this used to be a favourite idea. 'I believe the souls of 500 Sir Isaac Newtons would go to the making up of a Shakespeare or a Milton,' pronounced Samuel Taylor Coleridge. Convictions of this kind still linger, especially among those who know nothing about Sir Isaac Newton. Yet Coleridge's credo does not, when you inspect it, mean much. Presumably he relates soul-size to imaginative power – and obviously poets do use their imagination differently from scientists. But there seem no grounds for deciding they use it better – or worse.

The difference can be seen right at the start of the modern scientific era if we glance, for example, at the way Shakespeare and Bacon write about clocks. For Shakespeare a clock is something that tells the time. 'When I do count the clock that tells the time,' one of his sonnets starts. But for Bacon a clock is a machine which, because he understands it scientifically, he can put to various uses. Thinking about weight and gravitation, he wondered if the weight of an object would increase and decrease according to whether it was nearer to or further from the centre of the earth. Obviously you cannot discover this by weighing the object at various heights, because the weights themselves will also have got heavier or lighter, like the object. What you do, Bacon decides, is take two clocks, one worked by weights, the other by a spring. You adjust them so they are running at the same speed, then you take them up a mountain and down a mine. Up the mountain the clock with weights will go slower, because they have become lighter. Down the mine it will go faster.

He was almost right. The clock with weights would go slower up the mountain. But since the earth's weight is not concentrated at its centre, the clock going down the mine would leave progressively more of the earth's mass above it, so it would go slower too. The point, though, is not Bacon's rightness or wrongness, but the way he thinks about clocks compared to Shakespeare. For Shakespeare the idea of a clock has shrunk to something that tells the time. For Bacon, the clock is a machine, which can be engineered in various ways, and which has an experimental potential independent of the time-telling role ordinary language has allocated to it. It seems rather unfair to call Bacon less imaginative than Shakespeare in this instance. The poet remains satisfied with the conventional attributes of clocks, whereas the scientist's exploratory mind takes him to a wholly new function for a clock, which reveals something unexpected about the universe.

Of course this example is grossly slanted in Bacon's favour, and it would be ridiculous to disparage Shakespeare on the strength of it. Shakespeare's sonnet is no less a great poem because it is uninterested in gravitation. I have risked the comparison with Bacon because it shows us already, at the start of the seventeenth century, a scientist needing to rid himself of language's normal constraints (the usual functions language assigns to 'clock'), in order to think. From this historical moment on, scientists increasingly found that they had to develop their own special language, esoteric and forbidding to outsiders, but valuable to scientists because of its freedom from the vast cloud of associations, nuances and ambiguities that ordinary language carries along with it, and on which poets depend.

To poets, the new technical language seemed a sterile sea of jargon, in which the imagination would freeze and drown. John Donne was the first and last English poet *not* to feel like this about scientific language. He was lucky, being born at just the right time (1572), after the beginning of modern science but before its specialized technical vocabularies had really taken off. So for him, scientific language could still be warm, mysterious and sonorous, like poetry. He could think of love, and the scientific methods used for establishing latitude and longitude, as perfectly compatible and mutually enriching subjects:

> How great love is, presence best trial makes
> But absence tries how long this love will be;
> To take a latitude
> Sun, or stars, are fitliest viewed
> At their brightest, but to conclude
> Of longitudes, what other way have we,
> But to mark when, and where, the dark eclipses be?

Not much more than fifty years later, Milton took an altogether different and alienated view of scientists and scientific language, deriding astronomers who:

> Gird the sphere
> With centric and eccentric scribbled o'er,
> Cycle and epicycle, orb in orb.

Comparing the two examples we can see science, in the space of a half-century (the same half-century that saw the foundation of the Royal Society), beginning to become a hated alternative to poetry, barbaric, ugly, offensive to cultured ears. By the early twentieth century the process had developed so far that the Spanish philosopher José Ortega y Gasset, in *The Revolt of the Masses*, could select science (along with democracy) as a key cause of modern 'primitivism and barbarism'. He regretted that 'while there are more scientists than ever before, there are far fewer cultured men.'

Wordsworth, roughly halfway between Donne and us, prophesied that things would not turn out like this. He believed that science should and would become a subject for poetry. In 1800 he wrote:

> If the labours of men of science should ever create any material revolution, direct or indirect in our condition, and in the impressions which we habitually receive, the poet will sleep then no more than at present, he will be ready to follow the steps of the man of science, not only in those general indirect effects, but he will be at his side, carrying sensation into the midst of the objects of the science itself. The remotest discoveries of the chemist, the botanist, or mineralogist, will be as proper objects of the poet's art as any upon which it can be employed.

But Wordsworth was wrong. This has not happened; or not yet. Perhaps, as more scientists follow the trend of the writers I have mentioned, and make science available to general readers, it will permeate the culture and Wordsworth's prophecy will come true. As things are, however, modern poets avoid science, and, it seems, because they feel inferior to it, not (like Coleridge) superior. W. H. Auden expresses the general loss of confidence: 'When I find myself in the company of scientists, I feel like a shabby curate who has strayed by mistake into a drawing room full of dukes.'

Resistance to science among what Ortega y Gasset calls 'cultured men' has sometimes been strengthened by the objection that science is godless and amoral. Both charges need some qualification. It is perfectly possible for a scientist to believe in God, and even to find scientific evidence for God's existence. To sceptics this might suggest a rather nutty combination of laboratory-bore and Jesus-freak. But when a scientist of James Clerk Maxwell's eminence uses molecular structure as an argument for the existence of God (p.167), few will feel qualified to laugh. Of course, atheistical scientists are plentiful too. The zoologist Richard Dawkins has voiced the suspicion that all religions are self-perpetuating mental viruses. But since everything science discovers can, by sufficiently resolute believers, be claimed as religious knowledge, because it must be part of God's design, science cannot be regarded as inherently anti-religious.

On the contrary, its aims seem identical with those of theology, in that they both seek to discover the truth. Science seeks the truth about the physical universe; theology, about God. But these are not essentially distinct objectives, for theologians (or at any rate Christian theologians) believe God created the universe, so may be contacted through it. Admittedly, many scientists insist that science and religion are irreconcilable. The neuropsychologist Richard Gregory has declared: 'The attitudes of science and religion are essentially different, and opposed, as science questions everything rather than accepts traditional beliefs.' This does less than justice to religion's capacity for change. The whole Reformation movement in Europe, for example, was about not accepting traditional beliefs. It might be objected that science depends on evidence, while religion depends on revealed truth, and that this constitutes an insuperable difference. But for the religious, revealed truth is evidence. Theology might, without any paradox, be regarded as a science, committed to persistently questioning and reinterpreting the available evidence about God. True, by calling itself 'theology' it appears to take it for granted that God (*theos*) exists, which, scientifically speaking, is rather a careless usage. However, there is no reason why theological research should not lead the researcher to atheism, and no doubt it often has, just as (as we have seen) scientific research has led some researchers to God.

The real antithesis of science seems to be not theology but politics. Whereas science is a sphere of knowledge, politics is a sphere of opinion. Politics is constructed out of preferences, which it strives to elevate, by the mere multiplication of words, to the status of truths. Politics depends on personalities and rhetoric; social class, race and nationality are elemental to it. All of these are irrelevant to science. Further, politics relies, for its very existence, upon conflict. It presupposes an enemy. It is essentially oppositional, built on warring prejudices. If this oppositional structure were to collapse, politics could not survive. There could be no politics in a world of total consensus. Science, by contrast, is a co-operative not an oppositional venture. Of course, the history of science resounds with ferocious argument and the elaboration and destruction of rival theories. But when consensus is reached science does not collapse, it advances. Another crucial difference is that politics aims to coerce people. It is concerned with the exercise of power. Science has no such designs. It seeks knowledge. The consequence of this difference is that politics can and frequently does use violence (war, genocide, terrorism) to secure its ends. Science cannot. It would be ludicrous to go to war to decide upon the truth or otherwise of the second law of thermodynamics.

Needless to say, the ideal state I have described, in which science is free from and antithetical to politics, is not one that survives in the real world, where politics invades and contaminates science as it does everything else. But the warlike and destructive uses to which science has been put have nothing essentially to do with science: they are the responsibility of politics. Science's apolitical nature is worth stressing, because it helps us to defuse the charge that it is amoral. It allows us to see science's amorality not as a defect but as a condition of its strength and purity. Politics, of course, is inseparable from morality. It battens on morality, or on moralizing, like a tapeworm on the gut. Consequently science could not free itself from politics except by being amoral.

Approaches to life that are, in moral terms, cold, clinical and inhuman, are sometimes labelled 'scientific', but this is a misunderstanding, arising from the simple-minded transference of scientific method to moral attitudes. Science endorses no such transference, and no moral attitudes, cold or otherwise. In different minds, the same set of scientific propositions can prompt quite contrary moral responses. Darwin's theory of evolution, relating humans to apes, seemed – and seems – degrading to many humans. But Bruce Frederick Cummings (p. 262) accepts it with gusto:

> As for me, I am proud of my close kinship with other animals. I take a jealous pride in my Simian ancestry. I like to think that I was once a magnificent hairy fellow living in the trees and that my frame has come down through geological time via sea-jelly and worms and Amphioxus, Fish, Dinosaurs, and Apes. Who would exchange these for the pallid couple in the Garden of Eden?

Scientists themselves may have moral or immoral reasons for pursuing their research. But these leave no mark on their findings, which are right or wrong, to whatever degree, irrespective of their discoverer's motives. David Bodanis (p. 160) may be right to trace a link between Pasteur's loathing of mass humanity and his connection of disease with bacteria. The scientific credentials of the connection are, however, neither strengthened nor weakened by Pasteur's misanthropy.

The last few paragraphs may prompt readers to ask why they should bother to know about science if it cannot help to resolve moral or religious questions. The best answer is that science is, simply, what is known, and the only alternative to it is ignorance. Coleridge (whatever his opinion of Sir Isaac Newton's soul) saw this clearly:

> The first man of science was he who looked into a thing, not to learn whether it could furnish him with food, or shelter, or weapons, or tools, or ornaments, or *play-withs*, but who sought to know it for the gratification of knowing.

As science has grown, so, inevitably, has the ignorance of those who do not know about it. Within the mind of anyone educated exclusively in artistic and literary disciplines, the area of darkness has spread enormously during the later twentieth century, blotting out most of modern knowledge. A new species of educated but benighted being has come into existence – a creature unprecedented in the history of learning, where education has usually aimed to eradicate ignorance. The most highly gifted members of this new species have generally been the most forthright in regretting their deprivation. 'Exclusion from the mode of thought which is habitually said to be the characteristic achievement of the modern age' is, lamented the distinguished American literary critic Lionel Trilling, 'bound to be experienced as a wound to our intellectual self-esteem.'

More recently, however, ignorance of science has acquired a degree of political correctness. The Green movement, blaming science for global pollution, has contributed to this. So has feminism, which has demonized science as the embodiment of the male will-to-power. Even supposing these attacks were justified, however, they would not constitute reasons for relinquishing science, rather the reverse. Countering the pollution that political misdirection of science has caused can only be achieved by scientific means. Even at its most basic level, the monitoring, protection and conservation of endangered plant and animal species is inevitably a scientific endeavour. Nor does the feminist complaint that science is dominated by male aims and attitudes justify the neglect or rejection of science by women. On the contrary, it makes urgently desirable the increased involvement of women in scientific education and research. This is the view put forward by one of the most cogent of the feminist critics, Evelyn Fox Keller, in her book *Reflections on Gender and Science* (1984). Herself a mathematical biophysicist, and a biographer of the Nobel prizewinning geneticist Barbara McClintock, Keller sees scientific knowledge as ideally 'a universal goal', rather than the expression of destructively masculine drives.

A text that has been utilized to reinforce feminist and other disparagements of science is Thomas S. Kuhn's *The Structure of Scientific Revolutions* (1962). This popularized the idea that scientists are not really as rational as they suppose, but follow cultural trends, shifting from one paradigm to another for reasons that have nothing to do with objective truth. A criticism of Kuhn's book often voiced by scientists is that in describing how beliefs came to be held it leaves out of account the question of their truth or falsehood.

The effect of these various devices for discrediting science has been to allow ignorance to appear not merely excusable but righteous. Teachers at British universities will know that most arts students happily forget what little science they learnt in their schooldays. Even if you are prepared for this, however, the extent of their ignorance can come as a shock. Recently, in an Oxford literature seminar, I cited John Donne's lines (quoted on p. 17), where Donne observes that no one at the time he was writing (1612) knew how blood gets from one ventricle of the heart to the other. I asked the class how, in fact, it does. There were about thirty students present, all in their last year of study, all outstandingly intelligent, and none of them knew. One young man ventured haltingly that it might be 'by osmosis'. That the blood circulated round their bodies, they seemed unaware.

The annual hordes competing for places on arts courses in British universities, and the trickle of science applicants, testify to the abandonment of science among the young. Though most academics are wary of saying it straight out, the general consensus seems to be that arts courses are popular because they are easier, and that most arts students would simply not be up to the intellectual demands of a science course. On this issue, Sir Peter Medawar is worth quoting, since he is well qualified to judge, and he disagrees. Commenting on the career of James Watson, the young American who became world famous in 1953 when, with Crick, Wilkins and Franklin, he discovered the molecular structure of DNA, Medawar says:

> In England a schoolboy of Watson's precocity and style of genius would probably have been steered towards literary studies. It just so happens that during the 1950s, the first great age of molecular biology, the English schools of Oxford and particularly of

> Cambridge produced more than a score of graduates of quite outstanding ability – much more brilliant, inventive, articulate and dialectically skilful than most young scientists; right up in the Watson class. But Watson had one towering advantage over all of them: in addition to being extremely clever he had something important to be clever *about*. This is an advantage which scientists enjoy over most other people engaged in intellectual pursuits, and they enjoy it at all levels of capability. To be a first-rate scientist it is not necessary (and certainly not sufficient) to be extremely clever, anyhow in a pyrotechnic sense. One of the great social revolutions brought about by scientific research has been the democratization of learning. Anyone who combines strong common sense with an ordinary degree of imaginativeness can become a creative scientist, and a happy one besides, in so far as happiness depends upon being able to develop to the limit of one's abilities.

Medawar's remarks caused a considerable rumpus, especially his claim that scientists had something to be clever about whereas arts students had not. Surely, he was asked, he did not intend to imply that Shakespeare, Tolstoy, etc. were not proper subjects for cleverness? Less attention was paid to his claim that science could bring happiness, and not just to geniuses but to people of ordinary ability. Yet that was surely the vital part of his message. If young people are to be wooed back to science, it will not be done by telling them that if they continue to spurn it, Britain will face economic decline (true as that may be). But if scientists demonstrate by their writing that Medawar's promises of pleasure and self-fulfilment are true, they will not lack recruits.

The new generation of popular science-writers, whose work I have drawn on in this anthology, are the advance guard of that campaign. If readers ask, as they well might, what I, a professor of literature, think I am up to editing a science anthology, my answer is that I have done it for pleasure, self-fulfilment and (in Coleridge's words) 'the gratification of knowing'.

Prelude: The Misfit from Vinci

A left-handed, vegetarian, homosexual bastard, Leonardo da Vinci (1452–1519) contravened most of the accepted norms of his day. Reared by his peasant grandparents in a remote Tuscan village, he had minimal schooling. He was apprenticed as a painter because his illegitimacy debarred him from respectable professions. (Painting in fifteenth-century Tuscany was regarded not as 'creative art' but as a lowly trade, fit for the sons of peasants and artisans.) Lacking literary culture he was scorned in the highbrow Florence of the Medicis. This turned him towards science and observation. 'Anyone who invokes authors in discussion is not using his intelligence but his memory,' he contended.

He was insatiable for newness, both in art and science. His first known drawing was also the first true landscape drawing in western art. He was the first painter to omit haloes from the heads of figures from scripture and show them in ordinary domestic settings, and he was the first to paint portraits that showed the hands as well as the faces of sitters. His Leda (which does not survive) was the first modern painting inspired by pagan myth. His notebooks, of which over 5,000 pages survive, are all written backwards in mirror writing, and are dense with intricate drawings. They record his observations on geology, optics, acoustics, music, botany, mathematics, anatomy, engineering and hydraulics, together with plans for many inventions, including a bicycle, a tank, a machine gun, a folding bed, a diving suit, a parachute, contact lenses, a water-powered alarm clock, and plastics (made of eggs, glue and vegetable dyes).

It is true that Leonardo was not strictly a scientist, nor always as original as he seems. His war-machines had already been designed by a German engineer, Konrad Keyser; his 'automobile' by an Italian, Martini. Though he came close to formulating some scientific laws, his insights were sporadic and untested by experiment. He thought of looking at the moon through a telescope a century before Galileo (see p. 8), but he did not construct one. He knew no algebra, and made mistakes in simple arithmetic. His man-powered flying machine, designed to flap its wings like a bird, could never have flown. Apart from anything else, it must have weighed about 650 lbs (as against 72 lbs for *Daedalus 88*, the man-powered aircraft which flew 74 miles over the Aegean in 1988).

Despite these reservations his notebooks give an astonishing preview of the new world science was to open. The first of the following extracts, recording two autopsies he carried out in Florence on a very old man and a young child, has been called the first description of arteriosclerosis in the history of medicine. The second anticipates nineteenth-century geology (see p. 71) in deducing from fossil remains that the earth's present land-masses were once covered by sea. (The 'great horse' Leonardo refers to in this extract was his 7-metre-high bronze equestrian statue, planned for Lodovico Sforza in 1493, but never completed.) The third and fourth extracts show the sympathetic observation of birds, which inspired his interest in manned flight. The fifth illustrates Leonardo's irreverent humour and anatomical accuracy.

Autopsies

A few hours before his death, this old man told me that he had lived a hundred years and that he felt no physical pain, only weakness; and thus, seated on a bed in the hospital of Santa Maria Novella [in Florence], without any movement or symptom of distress, he gently passed from life into death. I carried out the autopsy to determine the cause of such a calm death and discovered that it was the result of weakness produced by insufficiency of blood and of the artery supplying the heart and other lower members, which I found to be all withered, shrunken and desiccated. The other postmortem was on a child of two years, and here I discovered the case to be exactly opposite to that of the old man.

Submarine Traces

Why are the bones of great fishes, and oysters and corals and various other shells and sea-snails, found on the high tops of mountains that border the sea, in the same way in which they are found in the depths of the sea? In the mountains of Parma and Piacenza, multitudes of shells and corals filled with worm-holes may be seen still adhering to the rocks, and when I was making the great horse at Milan a large sack of those which had been found in these parts was brought to my workshop by some peasants. The red stone of the mountains of Verona is found with shells all intermingled, which have become part of this stone. And if you should say that these shells have been and still constantly are being created in such places as these by the nature of the locality or by potency of the heavens in these spots, such an opinion cannot exist in brains possessed of any extensive powers of reasoning. Because the years of their growth are numbered upon the outer coverings of their shells; and both small and large ones may be seen; and these would not have grown without feeding, or fed without movement, and here [embedded in rock] they would not have been able to move . . . The peaks of the Apennines once stood up in a sea, in the form of islands surrounded by salt water, and above the plains of Italy where flocks of birds are flying today, fishes were once moving in large shoals.

Birds' Eyes

The eyes of all animals have pupils which have power to increase or diminish of their own accord, according to the greater or lesser light of the sun or other luminary. In birds, however, the difference is greater, and especially with nocturnal birds of the owl species, such as the long-eared, the white and the brown owls; for with these the pupil increases until it almost covers the whole eye, or diminishes to the size of a grain of millet, preserving all the time its round shape. In the horned owl, which is the largest nocturnal bird, the power of vision is so much increased that even in the faintest glimmer of night, which we call darkness, it can see more distinctly than we in the radiance of noon.

Flight

A bird is an instrument working according to a mathematical law, which instrument it is within the capacity of man to reproduce, with all its movements. A bird maintains itself in the air by imperceptible balancing, when near to the mountains or lofty ocean crags. It does this by means of the curves of the winds, which as they strike against these projections, being forced to preserve their first impetus, bend their straight course towards the sky, with divers revolutions, at the beginning of which the birds come to a stop, with their wings open, receiving underneath themselves the continual buffetings of the reflex courses of the winds.

The Penis

It has dealings with human intelligence and sometimes displays an intelligence of its own; where a man may desire it to be stimulated it remains obstinate and follows its own course; and sometimes it moves on its own without permission or any thought by its owner. Whether one is awake or asleep, it does what it pleases; often the man is asleep and it is awake; often the man is awake and it is asleep; or the man would like it to be in action but it refuses; often it desires action and the man forbids it. That is why it seems that this creature often has a life and intelligence separate from that of the man, and it seems that man is wrong to be ashamed of giving it a name or showing it; that which he seeks to cover and hide he ought to expose solemnly like a priest at mass.

Sources: 'Submarine Traces', 'Birds' Eyes' and 'Flight' are from *The Notebooks of Leonardo da Vinci*, Arranged, Rendered into English, and Introduced by Edward MacCurdy, 2 vols, London, Jonathan Cape, 1938. 'Autopsies' and 'The Penis' are from Serge Bramly, *Leonardo: The Artist and the Man*, translated by Sian Reynolds, London, Edward Burlingame Books (an imprint of HarperCollins Publishers), 1991.

Going inside the Body

1543 has a good claim to be the year when modern science began. It saw the publication of Copernicus' *On the Revolutions of the Heavenly Spheres* (see below p. 8) and of Andreas Vesalius' *On the Fabric of the Human Body* (generally known by its Latin title, the *Fabrica*). The text of this book – the foundation of modern anatomy – was accompanied by magnificent illustrations, designed by artists of the school of Titian, and cut on fine pearwood by Venetian block-cutters, which show the arteries, veins, muscles and nerves of the human body.

A well-off Belgian doctor's son, Vesalius (1514–64) had been given the best medical education available, studying at Louvain, Paris and Padua, where he became Professor of Anatomy at the age of 23. His mission was to rescue anatomy from the errors of the ancient Greek physician Galen, who still dominated medicine in the sixteenth century. Galen had had to depend on animal corpses for his knowledge of anatomy, and the prejudice against cutting up human bodies was still strong at the start of Vesalius' career. At Louvain, wishing to construct a human skeleton, he stole the remains of a malefactor from a gibbet outside the city. In order to satisfy his curiosity about the fluid in the pericardium, he contrived to be present when a criminal was quartered alive and (he recalls) carried off for study 'the still-pulsating heart with the lung and the rest of the viscera'. Once he was established in Padua, the magistrates supplied him with corpses fresh from the gallows, and executions were timed to coincide with his anatomy lessons.

Unlike previous professors he did not sit aloof on his throne while a barber surgeon cut up the cadaver, but carried out the dissection himself. The title page of the *Fabrica* – as if to emphasize masculine conquest of 'Mother Nature' – shows him handling the abdominal organs of a naked, cut-open woman, surrounded by tiers of eager male spectators. The woman, Vesalius records, had tried to cheat the gallows by declaring herself pregnant.

By chance an eyewitness account of Vesalius' first public anatomy classes survives, written by a German student, Baldasar Heseler. Held in Bologna in 1540, the classes covered the dissection of three human corpses, but the last class was on a living dog. The question which puzzles the students in this extract had already been answered by Vesalius at the end of his previous lecture, where he pointed out that it was when the heart contracted that it pumped blood into the pulmonary artery – so evidently the students had not been listening.

Finally, he took a dog (which was now the fifth or perhaps the sixth killed in our anatomy). He bound it with ropes to a small beam so that it could not move, similarly he tied his jaws so that it could not bite. Here, Domini, he said, you will see in this living dog the function of the nervi reversivi, and you will hear how the dog will bark as long as these nerves are not injured. I shall cut off one nerve, and half of the voice will disappear, then I shall cut the other nerve, and the voice will no longer be heard. When he had opened the dog, he quickly found the nervi reversivi around the arteries, and all happened as he said. The bark of the dog disappeared when he had by turn cut off the nervi reversivi, and only the breathing remained. But, he said, it can still quite well bite, do not let its jaws free, hold it strongly. Finally, he said, I shall proceed to the heart, so that you shall see its movement, and feel its warmth, and so that you shall here around the ilium feel the pulse with one hand, and with the other the movement of the heart. And please, tell me, what its movement is, whether the arteries are compressed when the heart is dilated, or whether they in the same time also have the same movement as the heart. I saw how the heart of the dog bounded upwards, and when it no longer moved, the dog instantly died. Those mad Italians pulled the dog at all sides so that nobody could really feel these two movements. But some students asked Vesalius what the true fact about these movements was, what he himself thought, whether the arteries followed the movement of the heart, or whether they had a movement different from that of the heart. Vesalius answered: I do not want to give my opinion, please do feel yourselves with your own hands and trust them. He was said always to be so little communicative.

When seventeenth-century poets thought of the human body they still thought of Vesalius' anatomy pictures and executed criminals, as this extract from Andrew Marvell's *Dialogue between the Soul and Body* suggests. Like Vesalius, Marvell considers the heart 'double', formed only of the two ventricles. Vesalius regarded the right atrium as a passageway for the vena cava, and the left as part of the pulmonary vein.

O who shall from this dungeon raise
A soul enslaved so many ways? . . .
A soul hung up, as 'twere, in chains
Of nerves and arteries and veins.
Tortured, besides each other part,
In a vain head and double heart.

Sources: Vesalius translation (slightly altered) from *Andreas Vesalius' First Public Anatomy at Bologna, 1540. An Eyewitness Report* By Baldasar Heseler, ed. Ruben Eriksson, Uppsala and Stockholm, Almquist & Wiksells Boktryckeri AB, 1959.

Galileo and the Telescope

Until the sixteenth century the accepted model of the universe was that developed by the second-century Alexandrian astrologer Ptolemy. According to this, the sun and the planets revolved round the earth. Over the centuries, complex adjustments were added to Ptolemy's system to make it fit astronomical observations.

The Pole Nicolaus Copernicus (1473–1543), a canon of the cathedral church at Frauenberg, and an amateur astronomer, put forward the hypothesis (in his book *On the Revolutions of the Heavenly Spheres*, published in 1543) that the earth moved, and went round the sun, which remained stationary. This contradicted several biblical texts, for example Joshua 10: 12–13, where Joshua commands the sun to stand still, implying that it normally moves. However the Church did not object. Copernicus dedicated his work to Pope Paul III and a cardinal and a bishop were among friends who urged him to publish. His theory was regarded as a harmless mathematical speculation. Most people did not take it seriously. Martin Luther spoke for the general public: 'This fool wishes to reverse the entire science of astronomy, but sacred scripture tells us that Joshua commanded the sun to stand still, and not the earth.'

With the advent of the telescope, however, observation replaced theory, and the old map of the heavens could be shown to be false. The inventor of the telescope is not known, but it was probably an obscure Dutch spectacle-maker living in Middelburg, Hans Lippershey. There is a story that, around 1600, two children were playing with lenses in his shop and found that by holding two together they could magnify the church weathervane. This led him to construct a simple telescope. By 1609 telescopes, under the name of 'Dutch trunks', were being made and sold in several European cities, including Venice, Padua, Paris and London.

Galileo Galilei (1564–1642) was a skilful instrument-maker and Professor of Mathematics at Padua University. To eke out his meagre salary, he kept a small shop in Padua, selling scientific instruments. About May 1609, he heard about telescopes and began constructing them. They were regarded as chiefly useful for observation on land or at sea. But it occurred to him to look at the sky through one. He published the astonishing results in March 1610 in a 24-

page pamphlet called *The Starry Messenger* (*Siderius Nuncius*). It was written in a tersely factual style no scholar had used before, and it fell like a bombshell on the learned world.

About ten months ago a report reached my ears that a Dutchman had constructed a telescope, by the aid of which visible objects, although at a great distance from the eye of the observer, were seen distinctly as if near; and some proofs of its most wonderful performances were reported which some gave credence to, but others contradicted. A few days after, I received confirmation of the report in a letter written from Paris by a noble Frenchman, Jaques Badovere, which finally determined me to give myself up first to inquire into the principle of the telescope, and then to consider the means by which I might compass the invention of a similar instrument, which a little while after I succeeded in doing, through deep study of the theory of Refraction; and I prepared a tube, at first of lead, in the ends of which I fitted two glass lenses, both plane on one side, but on the other side one spherically convex, and the other concave. Then bringing my eye to the concave lens I saw objects satisfactorily large and near, for they appeared one-third of the distance off and nine times larger than when they are seen with the natural eye alone. I shortly afterwards constructed another telescope with more nicety, which magnified objects more than sixty times. At length, by sparing neither labour nor expense, I succeeded in constructing for myself an instrument so superior that objects seen through it appear magnified nearly a thousand times, and more than thirty times nearer than if viewed by the natural powers of sight alone.

It would be altogether a waste of time to enumerate the number and importance of the benefits which this instrument may be expected to confer, when used by land or sea. But without paying attention to its use for terrestrial objects, I betook myself to observations of the heavenly bodies; and first of all, I viewed the Moon as near as if it was scarcely two semi-diameters of the Earth distant. After the Moon, I frequently observed other heavenly bodies, both fixed stars and planets, with incredible delight; and, when I saw their very great number, I began to consider about a method by which I might be able to measure their distances apart, and at length I found one . . .

Now let me review the observations made by me during the two months just past, again inviting the attention of all who are eager for true philosophy to the beginnings which led to the sight of most important phenomena.

Let me speak first of the surface of the Moon, which is turned towards us. For the sake of being understood more easily, I distinguish two parts in it, which I call respectively the brighter and the darker. The brighter part seems to surround and pervade the whole hemisphere; but the darker part, like a sort of cloud, discolours the Moon's surface and makes it appear covered with spots. Now these spots, as they are somewhat dark and of considerable size, are plain to every one, and every age has seen them, wherefore I shall call them *great* or *ancient* spots, to distinguish them from other spots, smaller in size, but so thickly scattered that they sprinkle the whole surface of the Moon, but especially the brighter portion of it. These spots have never been observed by any one before me; and from my observations of them, often repeated, I have been led to that opinion which I have expressed, namely, that I feel sure that the surface of the Moon is not perfectly smooth, free from inequalities and exactly spherical, as a large school of philosophers considers with regard to the Moon and the other heavenly bodies, but that, on the contrary, it is full of inequalities, uneven, full of hollows and protuberances, just like the surface of the Earth itself, which is varied everywhere by lofty mountains and deep valleys.

The appearances from which we may gather these conclusions are of the following nature: – On the fourth or fifth day after new-moon, when the Moon presents itself to us with bright horns, the boundary which divides the part in shadow from the enlightened part does not extend continuously in an ellipse, as would happen in the case of a perfectly spherical body, but it is marked out by an irregular, uneven, and very wavy line . . . for several bright excrescences, as they may be called, extend beyond the boundary of light and shadow into the dark part, and on the other hand pieces of shadow encroach upon the light: – nay, even a great quantity of small blackish spots, altogether separated from the dark part, sprinkle everywhere almost the whole space which is at the time flooded with the Sun's light, with the exception of that part alone which is occupied by the great and ancient spots. I have noticed that the small spots just mentioned have this common characteristic always and in every case, that they have the dark part towards the Sun's position, and on the side away from the Sun they have brighter boundaries, as if they were crowned with shining summits. Now we have an appearance quite similar on the Earth about sunrise, when we behold the valleys, not yet flooded with light, but the mountains surrounding them on the side opposite to the Sun already ablaze with the splendour of his beams; and just as the shadows in the hollows of the Earth diminish in size as the Sun rises higher, so also these spots on the Moon lose their blackness as the illuminated part grows larger and larger. Again, not only are the boundaries of light and shadow in the Moon seen to be uneven and sinuous, but – and this produces still greater astonishment – there appear very many bright points within the darkened portion of the Moon, altogether divided and broken off from the illuminated tract, and separated from it by no inconsiderable interval, which, after a little while, gradually increase in size and brightness, and after an hour or two become joined on to the rest of the bright portion, now become somewhat larger; but in the meantime others, one here and another there, shooting up as if growing, are lighted up within the shaded portion, increase in size, and at last are linked on to the same luminous surface, now still more extended . . . Now, is it not the case on the Earth before sunrise, that while the level plain is still in shadow, the peaks of the most lofty mountains are illuminated by the Sun's rays? After a little while does not the light spread further, while the middle and larger parts of those mountains are becoming illuminated; and at length, when the Sun has risen, do not the illuminated parts of the plains and hills join together? The grandeur, however, of such prominences and depressions in the Moon seems to surpass both in magnitude and extent the ruggedness of the Earth's surface, as I shall hereafter show. And here I cannot refrain from mentioning what a remarkable spectacle I observed while the Moon was rapidly approaching her first quarter . . . A protuberance of the shadow, of great size, indented the illuminated part in the neighbourhood of the lower cusp; and when I had observed this indentation longer, and had seen that it was dark throughout, at length, after about two hours, a bright peak began to arise a little below the middle of the depression; this by degrees increased, and presented a triangular shape, but was as yet quite detached and separated from the illuminated surface. Soon around it three other small points began to shine, until, when the Moon was just about to set, that triangular figure, having now extended and widened, began to be connected with the rest of the illuminated part, and, still girt with the three bright peaks already mentioned, suddenly burst into the indentation of shadow like a vast promontory of light . . .

Galileo goes on to describe the greatly increased number of stars visible through his telescope. The number of stars visible to the naked eye could be counted. But his telescope 'set distinctly before the eyes other stars in myriads which have never been seen before, and which surpass the old, previously known, stars in number more than ten times'. Turning his telescope to the Milky Way, the nature of which had been in dispute for centuries, he was able to establish that it was 'a mass of innumerable stars planted in clusters'. He then went on to his most amazing discovery.

I have now finished my brief account of the observations which I have thus far made with regard to the Moon, the Fixed Stars, and the Galaxy. There remains the matter, which seems to me to deserve to be considered the most important in this work, namely, that I should disclose and publish to the world the occasion of discovering and observing four PLANETS, never seen from the very beginning of the world up to our own times, their positions, and the observations made during the last two months about their movements and their changes of magnitude; and I summon all astronomers to apply themselves to examine and determine their periodic times, which it has not been permitted me to achieve up to this day, owing to the restriction of my time. I give them warning however again, so that they may not approach such an inquiry to no purpose, that they will want a very accurate telescope, and such as I have described in the beginning of this account.

On the 7th day of January in the present year, 1610, in the first hour of the following night, when I was viewing the constellations of the heavens through a telescope, the planet Jupiter presented itself to my view, and as I had prepared for myself a very excellent instrument, I noticed a circumstance which I had never been able to notice before, owing to want of power in my other telescope, namely, that three little stars, small but very bright, were near the planet; and although I believed them to belong to the number of the fixed stars, yet they made me somewhat wonder, because they seemed to be arranged exactly in a straight line, parallel to the ecliptic, and to be brighter than the rest of the stars, equal to them in magnitude. The position of them with reference to one another and to Jupiter was as follows.

On the east side there were two stars, and a single one towards the west. The star which was furthest towards the east, and the western star, appeared rather larger than the third.

I scarcely troubled at all about the distance between them and Jupiter, for, as I have already said, at first I believed them to be fixed stars; but when on January 8th, led by some fatality, I turned again to look at the same part of the heavens, I found a very different state of things, for there were three little stars all west of Jupiter, and nearer together than on the previous night, and they were separated from one another by equal intervals.

At this point, although I had not turned my thoughts at all upon the approximation of the stars to one another, yet my surprise began to be excited, how Jupiter could one day be found to the east of all the aforesaid fixed stars when the day before it had been west of two of them; and forthwith I became afraid lest the planet might have moved differently from the calculation of astronomers, and so had passed those stars by its own proper motion. I therefore waited for the next night with the most intense longing, but I was disappointed of my hope, for the sky was covered with clouds in every direction.

But on January 10th the stars appeared in the following position with regard to Jupiter; there were two only, and both on the east side of Jupiter, the third, as I thought, being hidden by the planet. They were situated just as before, exactly in the same straight line with Jupiter, and along the Zodiac.

When I had seen these phenomena, as I knew that corresponding changes of position could not by any means belong to Jupiter, and as, moreover, I perceived that the stars which I saw had been always the same, for there were no others either in front or behind, within a great distance, along the Zodiac, – at length, changing from doubt into surprise, I discovered that the interchange of position which I saw belonged not to Jupiter, but to the stars to which my attention had been drawn, and I thought therefore that they ought to be observed henceforward with more attention and precision.

Accordingly, on January 11th I saw an arrangement of the following kind, namely, only two stars to the east of Jupiter, the nearer of which was distant from Jupiter three times as far as from the star further to the east; and the star furthest to the east was nearly twice as large as the other one; whereas on the previous night they had appeared nearly of equal magnitude. I therefore concluded, and decided unhesitatingly, that there are three stars in the heavens moving about Jupiter, as Venus and Mercury round the Sun; which at length was established as clear as daylight by numerous other subsequent observations. These observations also established that there are not only three, but four, erratic sidereal bodies performing their revolutions round Jupiter, observations of whose changes of position made with more exactness on succeeding nights the following account will supply . . .

Galileo's discoveries shattered old beliefs. The mountains and craters on the moon showed that the celestial bodies were not perfect spheres, as had always been thought. The thousands of new stars brought home the littleness of our world, lost in space. Afraid of what he would see, the Professor of Philosophy at Padua refused even to look into Galileo's telescope. Further, the four new 'planets' around Jupiter meant that astrology was false, since it had not taken the influence of these bodies into account. The English diplomat Sir Henry Wotton pointed this out when he wrote from Venice to the Earl of Salisbury on 13 March 1610, enclosing a copy of Galileo's book.

Now touching the occurrents of the present, I send herewith unto his Majesty the strangest piece of news (as I may justly call it) that he hath ever yet received from any part of the world; which is the annexed book (come abroad this very day) of the Mathematical Professor at Padua, who by the help of an optical instrument (which both enlargeth and approximateth the object) invented first in Flanders, and bettered by himself, hath discovered four new planets rolling about the sphere of Jupiter, besides many other unknown and lastly, that the moon is not spherical but endued with many prominences, and, which is of all the strangest, illuminated with the solar light by reflection from the body of the earth, as he seemeth to say. So as upon the whole subject he hath first overthrown all former astronomy and next all astrology. For the virtue of these new planets must needs vary the judicial part, and why may there not be yet more?

Galileo became an instant celebrity. He christened the moons of Jupiter 'the Medicean planets' after Grand Duke Cosimo II de' Medici, who became his patron. Invited to Rome in triumph, he was received in audience by Pope Paul V, who refused to let him kneel, and at a grand banquet in his honour his optical instrument was dignified by the Greek name 'telescope' – a title conferred by the Marquis of Monticelli.

Though Galileo had been a convinced Copernican from his early years, he had tactfully said little about the Copernican system in *The Starry Messenger*. But he became less guarded with time, and the Church, awakening to the danger of the new ideas, became less tolerant. In 1632, when he published his Copernican *Dialogue on the Two Chief World Systems*, he was brought to trial before the Inquisition, found guilty, and sentenced to an indefinite term of imprisonment. Under threat of torture, he made a public abjuration.

I, Galileo, son of the late Vincenzo Galilei, Florentine, aged seventy years, arraigned personally before this tribunal and kneeling before you, Most Eminent and Reverend Lord Cardinals Inquisitors-General against heretical pravity throughout the entire Christian commonwealth, having before my eyes and touching with my hands the Holy Gospels, swear that I have always believed, do believe, and by God's help will in the future believe all that is held, preached, and taught by the Holy Catholic and Apostolic Church. But, whereas – after an injunction had been judicially intimated to me by this Holy Office to the effect that I must altogether abandon the false opinion that the Sun is the center of the world and immovable and that the Earth is not the center of the world and moves and that I must not hold, defend, or teach in any way whatsoever, verbally or in writing, the said false doctrine, and after it had been notified to me that the said doctrine was contrary to Holy Scripture – I wrote and printed a book in which I discuss this new doctrine already condemned and adduce arguments of great cogency in its favour without presenting any solution of these, I have been pronounced by the Holy Office to be vehemently suspected of heresy, that is to say, of having held and believed that the Sun is the center of the world and immovable and that the Earth is not the center and moves.

Therefore, desiring to remove from the minds of your Eminences, and of all faithful Christians, this vehement suspicion justly conceived against me, with sincere heart and unfeigned faith I abjure, curse, and detest the aforesaid errors and heresies and generally every other error, heresy, and sect whatsoever contrary to the Holy Church, and I swear that in future I will never again say or assert, verbally or in writing, anything that might furnish occasion for a similar suspicion regarding me; but, should I know any heretic or person suspected of heresy, I will denounce him to this Holy Office.

Confined in a secluded house at Arcetri, near Florence, the old and now blind Galileo was visited, two years before his death, by the young English poet,

John Milton, who recalled the meeting in his classic defence of press freedom *Areopagitica* (1644): 'There it was that I found and visited the famous Galileo, grown old, a prisoner of the Inquisition, for thinking in Astronomy otherwise than the Franciscan and Dominican licensers thought.' In *Paradise Lost* Milton compares the fallen Satan's huge shield, dimly seen amid the murk of Hell, to the strange giant moon that Galileo ('the Tuscan artist') first saw through his telescope from the hills of Fiesole ('Fesole') or from the valley of the Arno ('Valdarno') where Florence stands:

the broad circumference
Hung on his shoulders like the moon, whose orb
Through optic glass the Tuscan artist views
At evening from the top of Fesole,
Or in Valdarno, to descry new lands,
Rivers or mountains in her spotty globe.

However, the universe in Milton's epic is the old earth-centred one, and when Adam asks a visiting angel for an astronomy lesson he is told that God has deliberately put such matters as whether the earth moves round the sun beyond men's grasp:

He his fabric of the heavens
Hath left to their disputes, perhaps to move
His laughter at their quaint opinions wide.

Sources: *The Sidereal Messenger of Galileo Galilei*, ed. and trans. Edward Strafford Carlos, London, Rivingtons, 1880, and *The Life and Letters of Sir Henry Wotton*, ed. Logan Pearsall Smith, Oxford, Clarendon Press, 1907.

William Harvey and the Witches

In 1612 the poet John Donne wrote:

> Knows't thou how blood, which to the heart doth flow,
> Doth from one ventricle to the other go?

This was a rhetorical question – for no one did know. A common idea was that the central division of the heart (the septum) had holes through which the blood passed – though Vesalius had shown it had not. William Harvey (1578–1657), who probably knew Donne, solved this mystery with his discovery of the circulation of the blood. He was already lecturing about this at the College of Physicians in London in 1615, though he did not publish it until 1628. Even then, he records, many medical experts thought his great discovery 'crack-brained', and some, like René Descartes, stuck to the idea that the heart was a kind of furnace rather than, as Harvey had shown, a pump made of muscle.

As private physician to Charles I, Harvey looked after the two royal princes at the Battle of Edgehill, during which he sat under a hedge reading a book. His royal appointment also involved him in the affair of the Lancashire witches, recounted here by his biographer Geoffrey Keynes, which illustrates the gradual advance of science over superstition in the seventeenth century.

It was in 1633 that the events took place in Pendle Forest near Burnley in Lancashire that led to Harvey's being called as a witness in the following year. This remote area in the north-west had been for some years agitated by a series of crimes attributed to witches, gossip leading to fanciful accusations conceived in the fertile brains of imaginative children or even taught them by their elders. The particular story that ultimately concerned Harvey began on 10 February 1633. A boy of 11 named Edmund Robinson made an elaborate deposition before two Justices of the Peace, Richard Shutleworth and John Starkey, at Padiham, alleging that on All Saints Day last (1 November 1632) he was gathering wild plums in Wheatley Lane, when he saw two greyhounds, one brown the other black, running in his direction over the next field. Each dog, he noticed, had a collar which 'did shine like gold', but though each had a string attached there was no one with them. At the same moment he saw a hare, and, thinking to set the dogs off after it, cried 'Loo, loo, loo', but they would not run. This angered him, and tying them by their strings to a bush, he beat them with a stick. Thereupon the black dog stood up in the person of the wife of one Dickenson, and the brown dog as a small boy he did not know. In his fright Robinson made to run away, but was stopped by the woman, who, producing a silver coin from her pocket, offered to give it to him if he would hold his tongue. This he refused, saying, 'Nay, thou art a witch'. She then pulled from her pocket a sort of bridle that jingled, put it on the head of the boy that had been a dog, who then turned into a white horse. Seizing young Robinson, the woman mounted him on the horse in front of her and rode with him to a house called Hoarstones, a locality well known as a gathering place for witches. Many other people then came riding up on horses of various colours to the number of about threescore, and meat was roasted. A young woman tried to make him eat some of this and to drink something out of a glass, but he refused after the first taste of it. He then saw various people go into a neighbouring barn, where six of them kneeled and pulled on ropes fastened to the roof. This brought down smoking flesh, lumps of butter, and milk, which they caught in basins. Then six more people repeated the process, making such fearful faces that he stole out in terror and ran home, where he told his father that he had also seen the woman pricking pictures with thorns. When it was noticed that the boy had escaped, a party of people, several of whom he named, started in pursuit and had nearly caught up with him at a place called Boggard-hole, when two horsemen came up and rescued him. On the same evening Robinson's father sent him to tie up two cows in their stalls, and on the way, in a field called the Ellers, he met another boy who picked a quarrel and made him fight until his ears were made very bloody. Looking down he saw that the aggressor had a cloven foot, which aroused fresh fears. He ran on to find the cows and saw the light of a lantern; thinking it was carried by friends he ran towards it only to find a woman on a bridge, whom he recognized, and turned back to meet again the boy with the cloven foot, who gave him a blow on the back and made him cry. The boy's father in confirmation of the story said he had gone to look for him and found him in a state of terror and crying pitifully, so that he did not recover for nearly a quarter of an hour. In his deposition to the magistrates the boy gave the names of seventeen persons whom he knew as present at Hoarstones and said he could recognize others . . .

The boy was taken round by his father to various churches in the district and identified many more people among the congregations, money being paid for his services. It so happened that at the church of Kildwick, where he was taken, the curate was David Webster, who in 1677 published an important book, *The Displaying of Supposed Witchcraft*, exposing the frauds perpetrated in witch-hunts. Webster related that he asked the boy if he had truly seen and heard the strange things that he described, but two ill-favoured men who were in charge of him forbade the boy to answer, saying that he had already been examined by two Justices. As a result of this nearly thirty people were imprisoned, and a variety of other accusations were hurled at them by their enraged neighbours. A trial took place at Lancaster after the prisoners had been searched for any suspicious marks on their bodies, and seventeen were found guilty on this evidence. Great importance was attached to the discovery of marks on witches' bodies, since it was believed that the devil put his marks on those allied to him, and these places then became callous and insensitive. The law therefore required that the accused should be scrutinized by a jury of the same sex together with one doctor or several. The head was to be shaved and every part of the body handled. Any callous spot that was found was to be pricked with pins, and, if it was insensitive, that was evidence of guilt. Search was also to be made for anything resembling a teat capable of suckling the witch's familiar or imp, which might take the form of a rat, mouse, frog, toad, bird, fly, or spider; sometimes the imps were in the form of larger animals such as a cat or dog. King James in his *Daemonologie* believed firmly in two 'good helps that may be used for their triall: the one is, the finding of the marke and the trying the insensibleness thereof: the other is their fleeting on the water', since those in whom the devil resided were lighter than normal people and so floated when thrown into a pond.

Fortunately for the seventeen Lancashire prisoners found guilty, it was requested that seven of them should be seen by John Bridgeman, Bishop of Chester, in whose diocese they lived. The Bishop went to the gaol, but by then three of them had died and a fourth, Jennet Hargreaves, was very ill. Of the remaining three, two denied all knowledge of witchcraft, but the third, Margaret Johnson, declared herself to have been a witch for six years. She had stated on 9 March 1633 before the same Justices who had examined Edmund Robinson, that in a fit of anger and discontent a devil had appeared to her in the form of a man 'apparrelled in a suite of blake, tied about with silk pointes, whoe offered her, if shee would give him her soule, hee would supply all her wantes, and at her appointment would helpe her to kill and revenge her either of men or beeste, or what she desired'. To this she agreed and the devil bade her call him Memillion, and when she called he would be ready to do her will. She denied being at the meeting at Hoarstones on the particular day described by Robinson, but admitted being there on the next Sunday, when various evil plans were concerted. She further declared 'that such witches as have sharpe boanes are generally for the devil to prick them with which have no papps nor duggs, but raiseth blood from the place pricked with the boane, which witches are more greater and grand witches then they which have papps or duggs'. After further boastings she said that since 'this trouble befell her, her Spiritt hath left her, and shee never saw him since'.

After his examination the Bishop reported the affair to the Secretary of State, Sir John Coke, and so it came to the ears of King Charles. The King was a less credulous man than his father, and he ordered the Lord Privy Seal, Henry Montagu, Earl of Manchester, to write to the Court doctors as follows:

> To Alexander Baker Esq., and Sergiant Clowes his Majesty's Chirurgions.
>
> These shalbe to will and reqire you forth with to make choise of such Midwives as you shall thinke fitt to inspect and search the Boddies of those women that were lately brought by the Sheriff of the Countie of Lancaster indited for witchcraft and to report unto you whether they finde about them any such markes as are pretended; wherein the said midwives are to receave instructions from Mr Dr Harvey his Majesty's Physician and yourselves;
>
> Dated at Whytehall the 29th of June 1634.
>
> H. Manchester

The four prisoners, including Jennet Hargreaves, who had now recovered, had been brought to London and were held at the Ship Tavern in Greenwich. They were now examined by the prescribed jury at Surgeon's Hall in Monkwell Street, and the following report was returned:

> Surgeons Hall in Mugwell Streete London 2[d] July A[o] D[ni] 1634
> We in humble obeyance to your Lordshipps have this day caled unto us the Chirurgeons and Midwyves whose names are hereunder written who have by the directions of Mr. Doctor Harvey (in our presence and his) made diligent searche and Inspection on those women which weare lately brought upp from Lancaster and ffynd as followeth vidz.
>
> On the bodyes of Jennett Hargreaves, Ffrances Dicconson, and Mary Spencer nothinge unnaturalle neyther in their secrets or any other partes of theire bodyes, nor anythinge lyke a teate or marke nor any Signe that any suche thinge haith ever beene.
>
> On the body of Margaret Johnson wee fynd two things maye be called teats the one betweene her secretts and the ffundament on the edge thereof the other on the middle of her left buttocke. The first in shape lyke to the teate of a Bitche, but in our judgements nothinge but the skin of the ffundament drawen out as yt wilbe after the pyles or applicacion of leeches. The seacond is lyke the nipple or teate of a woman's breast but of the same colour with the rest of the skin without any hollowness or yssue for any bloode or juyce to come from thence.
>
> *Midwives*
>
> Margryt Franses — Anna Ashwell
> Aurelia Molins — Ffrancis Palmer
> Amis Willuby — Katheren Manuche Clifton
> Rebecke Layne — Joane Sensions
> Sibell Ffellipps
>
> *Surgeons*
>
> Alexander Read
> W. Clowes — Rich[d] Wateson
> Alex. Baker — Ja. Molins
> Ric. Mapes — Henry Blackley

This statement, bearing every mark of Harvey's precise and logical mind, was not signed by himself, Alexander Read having taken his place. As a result four of the seven witches were pardoned by the King, who had himself seen them. Subsequently the boy Robinson, having been brought to London with his father, was re-examined alone and confessed to being an impostor. His father, he said, and some others had taught him what he was to say with a view to making some money out of the story; in fact at the time of the supposed meeting at Hoarstones he was some distance away gathering plums in another man's orchard.

Source: Geoffrey Keynes, *The Life of William Harvey*, London, Oxford University Press, 1966.

The Hunting Spider

Robert Hooke (1635–1703) was curator of experiments at the Royal Society. An astronomer, physicist and naturalist, he assisted Robert Boyle in constructing the first air pump. His *Micrographia* (1665) contains the earliest illustrations of objects enlarged under the microscope – the crystal structure of snowflakes, a louse, a flea, a weevil, etc. It also contains the first scientific use of the word 'cell', to describe the microscopic honeycomb cavities in cork.

Only about half the world's spiders spread webs to catch prey. The rest hunt or ambush. Hooke's description reflects his close observation of the natural world.

The hunting spider is a small grey spider, prettily bespecked with black spots all over its body, which the microscope discovers to be a kind of feathers, like those on butterflies' wings or the body of the white moth. Its gait is very nimble, by fits, sometimes running and sometimes leaping, like a grasshopper almost, then standing still and setting itself on its hinder legs. It will very nimbly turn its body and look round itself every way. It has six very conspicuous eyes, two looking directly forwards, placed just before; two other, on either side of those, looking forward and sideways; and two other about the middle of the top of its back or head, which look backwards and sidewards. These seemed to be the biggest. The surface of them all was very black, spherical, purely polished, reflecting a very clear and distinct image of all the ambient objects, such as a window, a man's hand, a white paper, or the like.

Hooke discussed hunting spiders with his friend, the English traveller, virtuoso and diarist John Evelyn (1620–1706) who sent him the following description of their behaviour in Italy. Evelyn's brown spider is evidently a different species from Hooke's (which is grey). He identifies it as one of the wolf spiders (*Lupi*). These belong to the family Lycosidae (the family to which the Tarantula and the common wolf spider *Pardosa amentata*, which can often be seen in English gardens sunbathing on rockeries, both belong). They get their name because they chase after their prey like wolves, and there are over 2,500 known species.

Of all the sorts of insects, there is none has afforded me more divertisements than the Venatores, which are a sort of Lupi, that have their dens in the rugged walls and crevices of our houses; a small, brown and delicately spotted kind of spiders, whose hinder legs are longer than the rest.

Such I did frequently observe at Rome, which espying a fly at three or four yards distance, upon the balcony (where I stood) would not make directly to her, but crawl under the rail, till being arrived to the Antipodes, it would steal up, seldom missing its aim; but if it chanced to want anything of being perfectly opposite, would at first peep immediately slide down again, till, taking better notice, it would come the next time exactly upon the fly's back. But if this happened not to be within a competent leap, then would this insect move so softly, as the very shadow of the gnomon [the upright arm of a sundial] seemed not to be more imperceptible, unless the fly moved; and then would the spider move also in the same proportion, keeping that just time with her motion, as if the same soul had animated both those little bodies; and whether it were forwards, backwards, or to either side, without at all turning her body, like a well managed horse: But if the capricious fly took wing, and pitched upon another place behind our huntress, then would the spider whirl its body so nimbly about, as nothing could be imagined more swift; by which means she always kept the head towards her prey, though to appearance as immovable as if it had been a nail driven into the wood, till by that indiscernible progress (being arrived within the sphere of her reach) she made a fatal leap (swift as lightning) upon the fly, catching him in the pole [head], where she never quitted hold till her belly was full, and then carried the remainder home. I have beheld them instructing their young ones how to hunt, which they would sometimes discipline for not well observing. But when any of the old ones did (as sometimes) miss a leap, they would run out of the field, and hide them in their crannies, as ashamed, and haply not be seen abroad for four or five hours after.

Source: Robert Hooke, *Micrographia* (1665).

Early Blood Transfusion

The belief that imbibing blood from another person can restore youth and vigour is very ancient, and there were many early attempts to put it into practice. In 1492 Pope Innocent VIII, when weak and in a coma, was given the blood of three young men, all of whom died. How the blood was administered is not known: probably by mouth.

After Harvey's discovery of the circulation of the blood (see p. 17) the possibility of transferring blood directly from the arteries of the donor to the veins of the recipient through a tube was investigated both in France and in England. On 14 November 1666 the minutes of the Royal Society record that:

> The experiment of transfusing the blood of one dog into another was made before the Society by Mr King and Mr Thomas Coxe, upon a little mastiff and a spaniel, with very good success, the former bleeding to death, and the latter receiving the blood of the other, and emitting so much of his own as to make him capable of receiving the other.

Samuel Pepys, a member of the Society, missed this experiment, but heard about it, and followed the fortunes of the surviving dog, reporting in his diary on 28 November that it was still 'in perfect good health'. The experiment had been masterminded by Robert Boyle, who explored the possible psychological effects of transfusion in a series of questions to the Society – whether a fierce dog could be tamed by receiving blood from a cowardly dog; whether a transfused dog would recognize its master, etc.

The first English blood transfusion into a human being took place on 23 November 1667. The Royal Society tried to procure 'some mad person in the hospital of Bedlam' for the purpose, but the Keeper of Bedlam declined, so the choice fell on Arthur Coga, a 'very freakish and extravagant' Bachelor of Divinity from Cambridge who, being 'indigent', was persuaded by a fee of one guinea to volunteer. The Society's secretary, Henry Oldenburg, recorded the result in a letter to Boyle.

On Thursday next, God willing, a report will be made of the good success of the first trial of transfusion practised on a man, which was by order of the Society, and the approbation of a number of physicians, performed on Saturday last in Arundel House, in the presence of many spectators, among whom were Mr Howard and both his sons, the bishop of Salisbury, four or five physicians, some parliament men, etc., by the management and operation of Dr Lower and Dr King, the latter of whom performed the chief part with great dexterity, and with so much ease to the patient, that he made not the least complaint, nor so much as any grimace during the whole time of the operation; in which the blood of a young sheep, to the quantity of about eight or nine ounces by conjecture, was transmitted into the great vein of the right arm, after the man had let out some six or seven ounces of his own blood. All which was done by the method of Dr King's, which I published in Num. 20 of the Transactions, without any change at all of it, save only in the shape of one of the silver pipes, for more conveniency. Having let out, before the transfusion, into a porringer, so much of the sheep's blood, as would run out in about a minute (which amounted to twelve ounces) to direct us as to the quantity to be transfused into the man, he, when he saw that florid arterial blood in the porringer, was so well pleased with it, that he took some of it upon a knife, and tasted it, and finding it of a good relish, he went the more couragiously to its transmission into his veins, taking a cup or two of sack before, and a glass of wormwood wine and a pipe of tobacco after the operation, which no more disordered him, both by his own confession, and by appearance to all bystanders, than it did any of those that were in the room with him. The pipe being taken out of the man, the blood of the sheep ran a very free stream, to assure the spectators of an uninterrupted course of blood.

The patient found himself very well upon it, his pulse better than before, and so his appetite. His sleep good, his body as soluble as usual, it being observed, that the same day of his operation he had three or four stools, as he used to have before. This morning our president (who by very pressing business could not be present in Arundel House) and I sent to see him pretty early, and found him a bed, very well, as he assured us, and more composed, as his host affirmed, than he had been before.

Coga wrote an account of his operation in Latin, and read it to the Society. Pepys, who was present, concluded that he was 'cracked a little in his head, though he speaks very reasonably and very well'. A second transfusion, this time of 14 ounces of sheep's blood, was given to Coga on 12 December 1667.

Once more, he survived apparently unharmed. However, a patient of the French pioneer of blood transfusion Jean Denis, who taught medicine at Montpellier, died following a transfusion in 1668, and this put a stop to transfusion into humans until the discovery of blood-group antigens and antibodies in 1900 made the practice safer. Blood transfusion was first practised on a large scale in the First World War.

The courtiers and literati persistently ridiculed the Society's experiments, headed by Charles II who 'mightily laughed' (Pepys relates) to hear that the scientists were 'spending time only in weighing of air'. (Boyle's epoch-making experiments on the pressure and volume of gases seem to be what excited the royal mirth on this occasion.) Thomas Shadwell's play *The Virtuoso*, first performed in 1676, presents Sir Nicholas Gimcrack boasting of his exploits in blood transfusion:

I assure you I have transfus'd into a human vein 64 ounces, avoirdupois weight, from one sheep. The emittent sheep died under the operation, but the recipient madman is still alive. He suffer'd some disorder at first, the sheep's blood being heterogeneous, but in a short time it became homogeneous with his own . . . The patient from being maniacal or raging mad became wholly ovine or sheepish: he bleated perpetually and chew'd the cud; he had wool growing on him in great quantities; and a Northamptonshire sheep's tail did soon emerge or arise from his anus or human fundament.

Sources: From Pepys's *Diary* and the *Proceedings of the Royal Society* via Marjorie Hope Nicolson, *Pepys's Diary and the New Science*, Charlottesville, University of Virginia Press, 1965.

Little Animals in Water

Antony van Leeuwenhoek (1632–1723) was the first human being to see living protozoa and bacteria. The son of a basket-maker in Delft, Holland, he received little education, but became a prosperous linen-draper. His friends included the painter Jan Vermeer. Drapers used magnifying glasses to inspect cloth, and van Leeuwenhoek took to grinding his own lenses from glass globules, and constructing microscopes. With these he observed protozoa and bacteria in fresh water, in the bile of various animals, in the human mouth, and in his own excrement. He nearly blinded himself watching the explosion of gunpowder under a microscope. His descriptions, communicated to the Royal Society in London in a series of 190 letters spanning 50 years, are so precise that modern bacteriologists can identify with certainty many of the micro-organisms he saw. In this extract from a letter dated 7 September 1674, he announces his first sighting of 'little animals' in water.

About two hours distant from this Town there lies an inland lake, called the Berkelse Mere, whose bottom in many places is very marshy, or boggy. Its water is in winter very clear, but at the beginning or in the middle of summer it becomes whitish, and there are then little green clouds floating through it; which, according to the saying of the country folk dwelling thereabout, is caused by the dew, which happens to fall at that time, and which they call honey-dew. This water is abounding in fish, which is very good and savoury. Passing just lately over this lake, at a time when the wind blew pretty hard, and seeing the water as above described, I took up a little of it in a glass phial; and examining this water next day, I found floating therein divers earthy particles, and some green streaks, spirally wound serpent-wise, and orderly arranged [identified as the common green alga Spirogyra: the earliest recorded observation of this organism], after the manner of the copper or tin worms, which distillers use to cool their liquors as they distil over. The whole circumference of each of these streaks was about the thickness of a hair of one's head. Other particles had but the beginning of the foresaid streak; but all consisted of very small green globules joined together: and there were very many small green globules as well. Among these there were, besides, very many little animalcules, whereof some were roundish, while others, a bit bigger, consisted of an oval. On these last I saw two little legs near the head, and two little fins at the hindmost end of the body. Others were somewhat longer than an oval, and these were very slow a-moving, and few in number. These animalcules had divers colours, some being whitish and transparent; others with green and very glittering little scales; others again were green in the middle, and before and behind white; others yet were ashen grey. And the motion of most of these animalcules in the water was so swift, and so various upwards, downwards, and round about, that 'twas wonderful to see: and I judge that some of these little creatures were above a thousand times smaller than the smallest ones I have ever yet seen, upon the rind of cheese, in wheaten flour, mould, and the like.

Source: *Antony van Leeuwenhoek and His 'Little Animals'*, ed. trans. and introduced by Clifford Dobell, New York, Russell & Russell Inc., 1958.

An Apple and Colours

Sir Isaac Newton (1642–1727), 'one of the tiny handful of supreme geniuses who have shaped the categories of the human intellect' (in the words of his biographer, Richard S. Westfall), was born into an entirely undistinguished, semi-literate sheep-farming family in rural Lincolnshire. His youthful encounter with an apple is the best known of all scientific stories, and, surprisingly, seems to be true. Dr William Stukeley, who knew Newton well in his old age, records:

On 15 April 1726 I paid a visit to Sir Isaac at his lodgings in Orbels Buildings in Kensington, dined with him, and spent the whole day with him, alone . . .

After dinner, the weather being warm, we went into the garden and drank tea, under the shade of some apple trees, only he and myself. Amidst other discourse, he told me he was just in the same situation as when, formerly, the notion of gravitation came into his mind. It was occasioned by the fall of an apple, as he sat in a contemplative mood. Why should that apple always descend perpendicularly to the ground, thought he to himself? Why should it not go sideways or upwards, but constantly to the earth's centre? Assuredly, the reason is, that the earth draws it. There must be a drawing power in matter: and the sum of the drawing power in the matter of the earth must be in the earth's centre, not in any side of the earth. Therefore does this apple fall perpendicularly, or towards the centre. If matter thus draws matter, it must be in proportion of its quantity. Therefore the apple draws the earth, as well as the earth draws the apple. That there is a power, like that we here call gravity, which extends itself through the universe.

And thus by degrees he began to apply this property of gravitation to the motion of the earth and of the heavenly bodies, to consider their distances, their magnitudes and their periodical revolutions; to find out that this property, conjointly with a progressive motion impressed on them at the beginning, perfectly solved their circular courses; kept the planets from falling upon one another, or dropping all together into one centre; and thus he unfolded the universe. This was the birth of those amazing discoveries, whereby he built philosophy on a solid foundation, to the astonishment of all Europe.

Asked at an earlier stage in his life how he had discovered the law of universal gravitation, Newton had replied 'By thinking on it continually' – a remark that supplements, but does not contradict, Stukeley's apple story.

Newton's law, set out in the *Principia* (1687), states that every particle of matter in the universe attracts every other particle with a force that varies according to its mass and to the inverse square of the distance between them. This remained the accepted explanation of gravity until it was superseded by Einstein's theory of general relativity in 1915 (see p. 267, below).

Newton's other great scientific work was the *Optics*, not published till 1704, but based on experiments he made as a young man at Cambridge to discover the nature of light:

In a very dark chamber, at a round hole, about one third part of an inch broad, made in the shut [shutter] of a window, I placed a glass prism, whereby the beam of the sun's light, which came in at that hole, might be refracted upwards towards the opposite wall of the chamber, and there form a coloured image of the sun . . .

So began Newton's account of his experiments with prisms, which led him to the discovery that ordinary white light is really a mixture of rays of every variety of colour. He found, too, that the ray of each colour bends at a certain definite angle on passing through the prism – red being the least bendable, followed by 'orange, yellow, green, blue, indigo, deep violet'. The richness of his response to colour is evident in his experimental accounts, as here where he is explaining that a ray of a single (or 'homogeneal') colour, shining upon objects, makes them all appear of that colour:

All white, grey, red, yellow, green, blue, violet bodies, as paper, ashes, red lead, orpiment, indigo bice [dark blue], gold, silver, copper, grass, blue flowers, violets, bubbles of water tinged with various colours, peacock's feathers, the tincture of *lignum nephriticum* [a wood imported from Spain, the blue infusions of which were used for kidney-disease], and suchlike, in red homogeneal light appeared totally red, in blue light totally blue, in green light totally green, and so of other colours. In the homogeneal light of any colour they all appeared totally of that same colour, with this only difference, that some of them reflected that light more strongly, others more faintly. I never yet found any body, which by reflecting homogeneal light could sensibly change its colour.

From all which it is manifest that if the sun's light consisted of but one sort of rays, there would be but one colour in the whole world.

Newton's friend Edmond Halley (observer of 'Halley's Comet') had engaged in underwater operations off the Sussex coast in a diving bell, and conversation with him enables Newton to draw imaginative conclusions about underwater colours:

Mr Halley, . . . in diving deep into the sea in a diving vessel, found in a clear sunshine day that, when he was sunk many fathoms deep into the water, the upper part of his hand, on which the sun shone directly through the water and through a small glass window in the vessel, appeared of a red colour, like that of a damask rose, and the water below and the under part of his hand, illuminated by light reflected from the water below, looked green. For thence it may be gathered that the sea-water reflects back the violet and blue-making rays most easily, and lets the red-making rays pass most freely and copiously to great depths. For thereby the sun's direct light at all great depths, by reason of the predominating red-making rays, must appear red.

Newton's theory that white light was not pure but a medley of different colours met with strong opposition. It seemed counter to common sense, which had long associated whiteness with purity and simplicity. Poets, however, responded to the new colour-theory excitedly. The influence of the *Optics* flooded eighteenth-century poetry with colour. Alexander Pope's 'sylphs' – fairy creatures who flit around a young lady's dressing-table in his poem *The Rape of the Lock* – show clear evidence of Newton's prismatic discoveries:

Transparent forms, too fine for mortal sight,
Their fluid bodies half-dissolved in light.
Loose to the wind their airy garments flew,
Thin glittering textures of the filmy dew;
Dipped in the richest tincture of the skies,
Where light disports in ever-mingling dyes,
While every beam new transient colours flings,
Colours that change whene'er they wave their wings

Later, however, in reaction against eighteenth-century Enlightenment values, the Romantic poets condemned Newton for banishing mystery from the universe, and reducing everything to fact and reason. 'Art is the Tree of Life. Science is the Tree of Death,' proclaimed William Blake. John Keats agreed that Newton had 'destroyed all the poetry of the rainbow by reducing it to prismatic colours' – an opinion he versified in *Lamia*:

> There was an awful rainbow once in heaven;
> We know her woof, her texture; she is given
> In the dull catalogue of common things.
> Philosophy will clip an angel's wings,
> Conquer all mysteries by rule and line . . .

These Romantic outbursts suggest a profound ignorance of Newton, who was, in fact, acutely aware of the mystery of the universe. Colour itself, he points out in the *Optics*, is mysterious. What we call 'colour' in an object is merely 'a disposition to reflect this or that sort of rays more copiously than the rest', and the rays themselves are not really 'coloured', but set up a motion that, when it meets our eye, gives us the sensation of colour – 'as sound in a bell or musical string, or other sounding body, is nothing but a trembling motion'. Why particular objects should reflect particular rays, and how they affect the eye to suggest colours, Newton found inexplicable.

In the twentieth century Alfred North Whitehead rephrased this problem, and its implications for poets, in his *Science and the Modern World*:

> There is no light or colour as a fact in external nature. There is merely motion of material. Again, when the light enters your eyes and falls on the retina, there is merely motion of material. Then your nerves are affected and your brain is affected, and again this is merely motion of material . . . The mind, in apprehending, experiences sensations which, properly speaking, are qualities of the mind alone. These sensations are projected by the mind so as to clothe appropriate bodies in external nature. Thus the bodies are perceived as with qualities which in reality do not belong to them, qualities which in fact are purely the offspring of the mind. Thus nature gets credit which should in truth be reserved for ourselves: the rose for its scent; the nightingale for his song; and the sun for his radiance. The poets are entirely mistaken. They should address their lyrics to themselves.

How little Newton's discoveries had diminished his sense of mystery, he himself explained not long before his death:

I don't know what I may seem to the world, but, as to myself, I seem to have been only like a boy playing on the sea shore, and diverting myself in now and then finding a smoother pebble or a prettier shell than ordinary, whilst the great ocean of truth lay all undiscovered before me.

Sources: William Stukeley, *Memoirs of Sir Isaac Newton's Life*, 1752. Sir Isaac Newton, *Optics*, 1704. A. N. Whitehead, *Science and the Modern World*, Cambridge, Cambridge University Press, 1926.

The Little Red Mouse and the Field Cricket

Gilbert White (1720–93) was the first great English field naturalist. A country parson, he spent virtually all his life in the Hampshire village of Selborne, and his letters to two fellow naturalists about the local fauna and flora were published in 1788 as *The Natural History of Selborne*. Among the British species that he was the first to identify was the little red mouse or harvest mouse (*Micromys minutus*.)

November 4, 1767

I have procured some of the mice mentioned in my former letters, a young one and a female with young, both of which I have preserved in brandy. From the colour, shape, size, and manner of nesting, I make no doubt but that the species is nondescript [not previously identified]. They are much smaller and more slender than the *mus domesticus medius* of Ray; and have more of the squirrel or dormouse colour: their belly is white, a straight line along their sides divides the shades of their back and belly. They never enter into houses; are carried into ricks and barns with the sheaves; abound in harvest, and build their nests amidst the straws of the corn above the ground, and sometimes in thistles. They breed as many as eight at a litter, in a little round nest composed of the blades of grass or wheat.

One of these nests I procured this autumn, most artificially platted, and composed of the blades of wheat; perfectly round, and about the size of a cricket-ball; with the aperture so ingeniously closed, that there was no discovering to what part it belonged. It was so compact and well filled, that it would roll across the table without being discomposed, though it contained eight little mice that were naked and blind. As this nest was perfectly full, how could the dam come at her litter respectively so as to administer a teat to each? Perhaps she opens different places for that purpose, adjusting them again when the business is over: but she could not possibly be contained herself in the ball with her young, which moreover would be daily increasing in bulk. This wonderful procreant cradle, an elegant instance of the efforts of instinct, was found in a wheat-field, suspended in the head of a thistle . . .

January 22, 1768

As to the small mice, I have farther to remark, that though they hang their nests for breeding up amidst the straws of the standing corn, above the ground; yet I find that, in the winter, they burrow deep in the earth, and make warm beds of grass: but their grand rendezvous seems to be in corn-ricks, into which they are carried at harvest. A neighbour housed an oat-rick lately, under the thatch of which were assembled near an hundred, most of which were taken; and some I saw. I measured them; and found that, from nose to tail, they were just two inches and a quarter, and their tails just two inches long. Two of them, in a scale, weighed down just one copper halfpenny, which is about a third of an ounce avoirdupois: so that I suppose they are the smallest quadrupeds in this island. A full-grown *mus medius domesticus* weighs, I find, one ounce, lumping weight, which is more than six times as much as the mouse above; and measures from nose to rump four inches and a quarter, and the same in its tail.

Selbourne, Sept. 2, 1774

. . . As my neighbour was housing a rick he observed that his dogs devoured all the little red mice that they could catch, but rejected the common mice: and that his cats ate the common mice, refusing the red . . .

There is a steep abrupt pasture field interspersed with furze close to the back of this village, well known by the name of the Short Lithe, consisting of a rocky dry soil, and inclining to the afternoon sun. This spot abounds with the *gryllus campestris*, or field-cricket; which, though frequent in these parts, is by no means a common insect in many other counties.

As their cheerful summer cry cannot but draw the attention of a naturalist, I have often gone down to examine the œconomy of these *grylli*, and study their mode of life: but they are so shy and cautious that it is no easy matter to get a sight of them; for, feeling a person's footsteps as he advances, they stop short in the midst of their song, and retire backward nimbly into their burrows, where they lurk till all suspicion of danger is over.

At first we attempted to dig them out with a spade, but without any great success; for either we could not get to the bottom of the hole, which often terminated under a great stone; or else, in breaking up the ground, we inadvertently squeezed the poor insect to death. Out of one so bruised we took a multitude of eggs, which were long and narrow, of a yellow colour, and covered with a very tough skin. By this accident we learned to distinguish the male from the female; the former of which is shining black, with a golden stripe across his shoulders; the latter is more dusky, more capacious about the abdomen, and carries a long sword-shaped weapon at her tail, which probably is the instrument with which she deposits her eggs in crannies and safe receptacles.

Where violent methods will not avail, more gentle means will often succeed; and so it proved in the present case; for, though a spade be too boisterous and rough an implement, a pliant stalk of grass, gently insinuated into the caverns, will probe their windings to the bottom, and quickly bring out the inhabitant; and thus the humane inquirer may gratify his curiosity without injuring the object of it. It is remarkable that, though these insects are furnished with long legs behind, and brawny thighs for leaping, like grasshoppers; yet when driven from their holes they show no activity, but crawl along in a shiftless manner, so as easily to be taken: and again, though provided with a curious apparatus of wings, yet they never exert them when there seems to be the greatest occasion. The males only make that shrilling noise perhaps out of rivalry and emulation, as is the case with many animals which exert some sprightly note during their breeding time: it is raised by a brisk friction of one wing against the other. They are solitary beings, living singly male or female, each as it may happen: but there must be a time when the sexes have some intercourse, and then the wings may be useful perhaps during the hours of night. When the males meet they will fight fiercely, as I found by some which I put into the crevices of a dry stone wall, where I should have been glad to have made them settle. For though they seemed distressed by being taken out of their knowledge, yet the first that got possession of the chinks would seize upon any that were obtruded upon them with a vast row of serrated fangs. With their strong jaws, toothed like the shears of a lobster's claws, they perforate and round their curious regular cells, having no fore-claws to dig, like the mole-cricket. When taken in hand I could not but wonder that they never offered to defend themselves, though armed with such formidable weapons. Of such herbs as grow before the mouths of their burrows they eat indiscriminately; and on a little platform, which they make just by, they drop their dung; and never, in the day-time, seem to stir more than two or three inches from home. Sitting in the entrance of their caverns they chirp all night as well as day from the middle of the month of May to the middle of July; and in hot weather, when they are most vigorous, they make the hills echo; and, in the stiller hours of darkness, may be heard to a considerable distance. In the beginning of the season, their notes are more faint and inward; but become louder as the summer advances, and so die away again by degrees.

Sounds do not always give us pleasure according to their sweetness and melody; nor do harsh sounds always displease. We are more apt to be captivated or disgusted with the associations which they promote, than with the notes themselves. Thus the shrilling of the field-cricket, though sharp and stridulous, yet marvellously delights some hearers, filling their minds with a train of summer ideas of everything that is rural, verdurous, and joyous.

About the tenth of March the crickets appear at the mouths of their cells, which they then open and bore, and shape very elegantly. All that ever I have seen at that season were in their pupa state, and had only the rudiments of wings, lying under a skin or coat, which must be cast before the insect can arrive at its perfect state (We have observed that they cast these skins in April, which are then seen lying at the mouths of their holes.) From whence I should suppose that the old ones of last year do not always survive the winter. In August their holes begin to be obliterated, and the insects are seen no more till spring.

Not many summers ago I endeavoured to transplant a colony to the terrace in my garden, by boring deep holes in the sloping turf. The new inhabitants stayed some time, and fed and sung; but wandered away by degrees, and were heard at a farther distance every morning; so that it appears that on this emergency they made use of their wings in attempting to return to the spot from which they were taken.

One of these crickets, when confined in a paper cage and set in the sun, and supplied with plants moistened with water, will feed and thrive, and become so merry and loud as to be irksome in the same room where a person is sitting: if the plants are not wetted it will die.

White is still quoted as an authority on the field cricket in some twentieth-century works on entomology. During the Second World War, as a prisoner of war in Bavaria, R. D. Purchon made a study of the field cricket which confirmed White's observations. Purchon found that the adult crickets die in August, and the young ones continue active until late autumn, when they hibernate in their pupa state.

Source: Gilbert White, *The Natural History of Selbourne*, 1788.

Two Mice Discover Oxygen

Joseph Priestley (1733–1804) was a Unitarian minister and schoolteacher. A keen supporter of the American and French Revolutions, he confessed that he had tended, from an early age, 'to embrace what is generally called the heterodox side of almost every question'.

Introduced to science by Benjamin Franklin, whom he met in London, Priestley made experiments on the 'air' (the current name for a gas) given off by the fermenting liquors in the brewery next door to his house in Leeds. This was carbon dioxide and, dissolving it in water, Priestley invented soda water, for which the Royal Society gave him a medal in 1773.

He then turned his attention to the 'airs' given off when various substances were heated. To examine these he constructed a simple apparatus, described in his *Experiments and Observations on Different Kinds of Air* (1775), consisting of a trough full of mercury, over which glass vessels could be inverted to collect the gas. The substance to be heated was placed in another glass vessel on the surface of the mercury, and Priestley focused the sun's rays on it using a 12-inch lens.

To test his gases Priestley employed mice, which he caught in wire traps, and introduced into the gas-filled vessels. Should the gas be likely to prove noxious, he warned, 'it will be proper (if the operator be desirous of preserving the mice for further use) to keep hold of their tails, that they may be withdrawn as soon as they begin to show signs of uneasiness'.

When Priestley heated red mercuric oxide (which he calls *mercurius calcinatus per se*), using his apparatus, it gave off a colourless gas which, as he describes in the following account, made a candle flame burn brightly. Like other eighteenth-century scientists, Priestley believed that all combustible materials contained an element called 'phlogiston', which was given off when they burned. Air in which things had been burned became less able to support combustion because, it was thought, it was saturated with 'phlogiston'. Accordingly Priestley called his gas in which a candle flame burned brightly 'dephlogisticated air'. In fact it was oxygen.

The contents of this section will furnish a very striking illustration of the truth of a remark, which I have more than once made in my philosophical writings, and which can hardly be too often repeated, as it tends greatly to encourage philosophical investigations viz. that more is owing to what we call *chance*, that is, philosophically speaking, to the observation of events arising from *unknown causes*, then to any proper *design*, or pre-conceived theory in this business.

For my own part, I will frankly acknowledge, that, at the commencement of the experiments recited in this section, I was so far from having formed any hypothesis that led to the discoveries I made in pursuing them, that they would have appeared very improbable to me had I been told of them; and when the decisive facts did at length obtrude themselves upon my notice, it was very slowly, and with great hesitation, that I yielded to the evidence of my senses . . . [Priestley then recounts the construction of the mercury-trough apparatus described above.]

With this apparatus, after a variety of other experiments, an account of which will be found in its proper place, on the 1st of August, 1774, I endeavoured to extract air from *mercurius calcinatus per se*; and I presently found that, by means of the lens, air was expelled from it very readily. Having got about three or four times as much as the bulk of my materials, I admitted water to it, and found that it was not imbibed by it. But what surprised me more than I can well express, was, that a candle burned in this air with a remarkably vigorous flame . . .

I cannot, at this distance of time, recollect what it was that I had in view in making this experiment: but I know I had no expectation of the real issue of it. Having acquired a considerable degree of readiness in making experiments of this kind, a very slight and evanescent motive would be sufficient to induce me to do it. If, however, I had not happened, for some other purpose, to have had a lighted candle before me, I should probably never have made the trial; and the whole train of my future experiments relating to this kind of air might have been prevented . . .

In this case, also, though I did not give sufficient attention to the circumstance at that time, the flame of the candle, besides being larger, burned with more splendour and heat . . . and a piece of red-hot wood sparkled in it, exactly like paper dipped in a solution of nitre, and it consumed very fast . . .

On the 8th of this month [March, 1775] I procured a mouse, and put it into a glass vessel, containing two one-ounce measures of the air from mercurius calcinatus. Had it been common air, a full-grown mouse, as this was, would have lived in it about a quarter of an hour. In this air, however, my mouse lived a full half hour; and though it was taken out seemingly dead, it appeared to have been only exceedingly chilled; for, upon being held to the fire, it presently revived, and appeared not to have received any harm from the experiment.

. . . By this I was confirmed in my conclusion, that the air extracted from mercurius calcinatus, &c. was, *at least, as good* as common air; but I did not certainly conclude that it was any *better*; because, though one mouse would live only a quarter of an hour in a given quantity of air, I knew it was not impossible but that another mouse might live in it half an hour; so little accuracy is there in this method of ascertaining the goodness of air . . .

For my farther satisfaction I procured another mouse, and putting it into less than two ounce-measures of air extracted from mercurius calcinatus and air from red precipitate (which, having found them to be of the same quality, I had mixed together) it lived three quarters of an hour. But not having had the precaution to set the vessel in a warm place, I suspect that the mouse died of cold. However, as it had lived three times as long as it could probably have lived in the same quantity of common air, and I did not expect much accuracy from this kind of test, I did not think it necessary to make any more experiments with mice.

It may hence be inferred, that a quantity of very pure air would agreeably qualify the noxious air of a room in which much company should be confined, and which should be so situated, that it could not be conveniently ventilated; so that from being offensive and unwholesome, it would almost instantly become sweet and wholesome. This air might be brought into the room in casks; or a laboratory might be constructed for generating the air, and throwing it into the room as fast as it should be produced. This pure air would be sufficiently cheap for the purpose of many assemblies, and a very little ingenuity would be sufficient to reduce the scheme into practice . . .

From the greater strength and vivacity of the flame of a candle, in this pure air, it may be conjectured, that it might be peculiarly salutary to the lungs in certain morbid cases, when the common air would not be sufficient to carry off the phlogistic putrid effluvium fast enough. But, perhaps, we may also infer from these experiments, that though pure dephlogisticated air might be very useful as *medicine*, it might not be so proper for us in the usual healthy state of the body: for, as a candle burns out much faster in dephlogisticated than in common air, so we might, as may be said, *live out too fast*, and the animal powers be too soon exhausted in this pure kind of air. A moralist, at least, may say, that the air which nature has provided for us is as good as we deserve.

My reader will not wonder, that, after having ascertained the superior goodness of dephlogisticated air by mice living in it, and the other tests above mentioned, I should have the curiosity to taste it myself. I have gratified that curiosity, by breathing it, drawing it through a glass-syphon, and, by this means, I reduced a large jar full of it to the standard of common air. The feeling of it to my lungs was not sensibly different from that of common air; but I fancied that my breast felt peculiarly light and easy for some time afterwards. Who can tell but that, in time, this pure air may become a fashionable article in luxury. Hitherto only two mice and myself have had the privilege of breathing it . . .

Being at Paris in the October following, and knowing that there were several very eminent chemists in that place . . . I frequently mentioned my surprise at the kind of air which I had got from this preparation to Mr Lavoisier, Mr le Roy, and several other philosophers, who honoured me with their notice in that city; and who, I daresay, cannot fail to recollect the circumstance.

The eminent French chemist Antoin-Laurent Lavoisier (1743–94), to whom Priestley divulged his discovery, understood the theoretical implications of it, as Priestley did not. Lavoisier had already announced, in 1772, that he was 'destined to bring about a revolution in physics and chemistry'. Unlike the older scientists he realized that atmospheric air was not an 'element' but a compound of gases, and he identified Priestley's discovery as the active component of air for which he had been searching. He called it 'oxygen' (Greek: 'acid former'), in the belief that all acids contained it. In 1783 he made public his complete renovation of chemical theory, and Mme Lavoisier ceremonially burned the books of the phlogiston theorists to mark the new era. Unfortunately Lavoisier, who had been a tax-collector under the *ancien régime*, was guillotined at the time of the French Revolution.

Source: Joseph Priestley, *Experiments and Observations on Different Kinds of Air*, London, 1775.

Discovering Uranus

The planet Uranus was discovered by the German-born British astronomer William Herschel (1738–1822). The son of an army musician, Herschel came to England in 1757 to follow a musical career, as teacher, composer and performer, and became organist of a fashionable chapel in Bath. An amateur astronomer, he constructed new and powerful telescopes, grinding the mirrors himself, and it was through one of those that, in 1781, he saw Uranus, the first planet to be discovered since prehistoric times. Fame, and a £200-per-year pension from George III, quickly followed, and Herschel gave up music for full-time astronomy. He developed a theory of the evolution of stars, and was the first to hypothesize that nebulae (misty white patches among the stars, visible through a telescope) were clouds of individual stars, forming separate galaxies.

Uranus takes 84.01 years to orbit the sun. It is an extremely cold planet, and is thought to consist of a rocky core and an ice mantle 8,000 kilometres thick. Nine of its twenty rings were discovered in 1977; the rest were photographed by the Voyager 2 probe in 1986.

Though not a very good poet, Alfred Noyes (1880–1958) was singular in that he wrote a modern epic poem about the progress of science, *The Torch-Bearers*. In the following extract (heavily indebted to Robert Browning's dramatic monologues), Noyes imagines Herschel's thoughts while conducting a concert in Bath.

My periwig's askew, my ruffle stained
With grease from my new telescope!
Ach, to-morrow
How Caroline will be vexed, although she grows
Almost as bad as I, who cannot leave
My workshop for one evening.
I must give
One last recital at St Margaret's,
And then – farewell to music.
Who can lead
Two lives at once?

Yet – it has taught me much,
Thrown curious lights upon our world, to pass
From one life to another. Much that I took
For substance turns to shadow. I shall see
No throngs like this again; wring no more praise
Out of their hearts; forego that instant joy
– Let those who have not known it count it vain –
When human souls at once respond to yours.
Here, on the brink of fortune and of fame,
As men account these things, the moment comes
When I must choose between them and the stars;
And I have chosen.
Handel, good old friend,
We part to-night. Hereafter, I must watch
That other wand, to which the worlds keep time.

What has decided me? That marvellous night
When – ah, how difficult it will be to guide,
With all these wonders whirling through my brain! –
After a Pump-room concert I came home
Hot-foot, out of the fluttering sea of fans,
Coquelicot-ribboned belles and periwigged beaux,
To my Newtonian telescope.
The design
Was his; but more than half the joy my own,
Because it was the work of my own hand,
A new one, with an eye six inches wide,
Better than even the best that Newton made
Then, as I turned it on the *Gemini*,
And the deep stillness of those constant lights,
Castor and Pollux, lucid pilot-stars,
Began to calm the fever of my blood,
I saw, O, first of all mankind I saw
The disk of my new planet gliding there
Beyond our tumults, in that realm of peace.

What will they christen it? Ach – not *Herschel*, no!
Not *Georgium Sidus*, as I once proposed;
Although he scarce could lose it, as he lost
That world in 'seventy-six.

Indeed, so far
From trying to tax it, he has granted me
How much? – two hundred golden pounds a year,
In the great name of science, – half the cost
Of one state-coach, with all those worlds to win! . . .
To-night,
– The music carries me back to it again! –
I see beyond this island universe,
Beyond our sun, and all those other suns
That throng the Milky Way, far, far beyond,
A thousand little wisps, faint nebulae,
Luminous fans and milky streaks of fire;
Some like soft brushes of electric mist
Streaming from one bright point; others that spread
And branch, like growing systems; others discrete,
Keen, ripe, with stars in clusters; others drawn back
By central forces into one dense death,
Thence to be kindled into fire, reborn,
And scattered abroad once more in a delicate spray
Faint as the mist by one bright dewdrop breathed
At dawn, and yet a universe like our own;
Each wisp a universe, a vast galaxy
Wide as our night of stars.
The Milky Way
In which our sun is drowned, to these would seem
Less than to us their faintest drift of haze;
Yet we, who are borne on one dark grain of dust
Around one indistinguishable spark
Of star-mist, lost in one lost feather of light,
Can by the strength of our own thought, ascend
Through universe after universe; trace their growth
Through boundless time, their glory, their decay;
And, on the invisible road of law, more firm
Than granite, range through all their length and breadth,
Their height and depth, past, present, and to come.

Alfred Noyes, *The Torch-Bearers*, London, Sheed & Ward, 1937.

The Big Bang and Vegetable Love

Erasmus Darwin (1731–1802), grandfather of Charles, was a doctor, inventor and poet. He helped to found the Lunar Society of Birmingham, which provided the main intellectual impetus for the Industrial Revolution in England. Among its members – and Erasmus's friends – were Benjamin Franklin, James Watt, of steam-engine fame, and Joseph Priestley (see p. 40). Erasmus's inventions included a speaking-machine, an artificial bird with flapping wings (which remained at the drawing-board stage), a sun-operated device for opening cucumber-frames, and a horizontal windmill, which was used to grind colours at his friend Josiah Wedgwood's pottery.

Over half a century before his grandson's *The Origin of Species*, Erasmus expounded a theory of evolution, declaring that 'all warm-blooded animals have arisen from one living filament', during a time-span of 'millions of ages'. He was the first scientist to analyse plant nutrition and photosynthesis, and to explain the process of cloud formation.

He took up poetry-writing in his fifties, and his two-part poem *The Botanic Garden* anticipates the 'big-bang' theory of the universe. The first event in the cosmos, according to Erasmus's account, is an explosion, sparked off by God saying 'Let there be light', whereupon:

> . . the mass starts into a million suns;
> Earths round each sun with quick explosions burst,
> And second planets issue from the first.

Erasmus defends his explosion theory in one of the poem's many 'Philosophical Notes':

It may be objected that if the stars had been projected from a Chaos by explosions, that they must have returned again into it from the known laws of gravitation; this however would not happen, if the whole Chaos, like grains of gunpowder, was exploded at the same time, and dispersed through infinite space at once, or in quick succession, in every possible direction.

One of the 'second planets' to 'issue from the first' in Erasmus's account is the moon, which separates from the earth leaving a hole now occupied by the South Pacific. The Goddess of Botany, accompanied by various Gnomes, Sylphs and Nymphs, is a witness of these cosmic disturbances, and she reminds the Gnomes of the alarm they felt at the moon's emergence:

> Gnomes! how you shrieked! when through the troubled air
> Roared the fierce din of elemental war;
> When rose the continents, and sunk the main,
> And Earth's huge sphere exploding burst in twain.
> Gnomes! how you gazed! when from her wounded side,
> Where now the South Sea heaves its waste of tide,
> Rose on swift wheels the Moon's refulgent car,
> Circling the solar orb, a sister star,
> Dimpled with vales, with shining hills embossed,
> And rolled round Earth her airless realms of frost.

The notion that the moon originated by fission from the earth, which has found some supporters in the twentieth century, became known as the 'Darwinian theory', not because of Erasmus but because of his great-grandson Sir George Darwin, who worked out a mathematical basis for the idea.

The second part of Erasmus's poem, *The Loves of the Plants* (1789), ministered to the craze for botany in the 1770s and 1780s. Captain Cook's famous voyage in the *Endeavour* had brought back to England, via Botany Bay, 1,300 hitherto unknown species of plants, thanks to the labours of young Joseph Banks, the botanist who accompanied Cook. The founding of the Royal Botanical Gardens at Kew, celebrated in Erasmus's poem, was a monument to this new enthusiasm for greenery. The great Swedish botanist Carl Linnaeus (1707–78) had introduced the system of modern plant classification in the middle years of the eighteenth century, and Erasmus translated some of his works. *The Loves of the Plants* personifies 90 different species, and recounts their sex-lives, paying strict attention to Linnaeus's botanical descriptions:

> Sweet blooms Genista in the myrtle shade,
> And *ten* fond brothers woo the haughty maid.
> *Two* knights before thy fragrant altar bend,
> Adored Melissa! and *two* squires attend.
> Meadia's soft chains *five* suppliant beaux confess,
> And hand in hand the laughing belle address;

Alike to all, she bows with wanton air,
Rolls her dark eye, and waves her golden hair.

What this means, as Erasmus's notes explain, is that the flower of the Broom (Genista) has ten males (stamens) and one female (pistil); the Balm (Melissa) has four males and one female, with two of the males standing higher than the other two; and the American Cowslip (Meadia) has five males and one female, with the males' anthers touching one another.

Erasmus's romances become more complicated as the plants' male and female organs increase in number. Lychnis, for example (Ragged Robin), has ten males and five females, the males and females being found on different plants, often at some distance from each other. 'When the females arrive at their maturity', Erasmus recounts, 'they rise above the petals as if looking abroad for their distant husbands. The scarlet ones contribute much to the beauty of our meadows in May and June.' Versified, this becomes:

Five sister-nymphs to join Diana's train
With thee, fair Lychnis! vow – but vow in vain;
Beneath one roof resides the virgin band,
Flies the fond swain, and scorns his offered hand;
But when soft hours on breezy pinions move,
And smiling May attunes her lute to love,
Each wanton beauty, tricked in all her grace
Shakes the bright dew-drops from her blushing face;
In gay undress displays her rival charms,
And calls her wondering lovers to her arms.

Erasmus's notes abound with curious botanical information. Of Madder, a plant that yields a red dye, he records: 'If mixed with the food of young pigs or chickens, it colours their bones red. If they are fed alternate fortnights with a mixture of madder and with their usual food alone, their bones will consist of concentric circles of white and red'; of Menispermum (a climbing tropical plant), Erasmus notes that its berries, dropped into water, make fish drunk.

The poem correctly predicts several technological developments – among them submarines, which will exploit Priestley's discovery of oxygen.

Led by the Sage, lo! Britain's sons shall guide
Huge Sea-Balloons beneath the tossing tide;
The diving castles, roofed with spheric glass,
Ribbed with strong oak, and barred with bolts of brass,

Buoyed with pure air shall endless tracks pursue,
And Priestley's hand the vital flood renew.

Rather surprisingly, Erasmus was an extremely popular and influential poet. Young Wordsworth imitated him. Coleridge called him 'the first *literary* character in Europe', and his great fantasy-poems, *Kubla Khan* and *The Ancient Mariner*, borrow scenes and phrases from Erasmus. Shelley, his keenest disciple, took from him the idea of combining science and poetry in famous lyrics like 'The Cloud' and 'The Sensitive Plant', and followed his lead in attacking superstition, tyrants, slavery, war and alcohol.

Source: Erasmus Darwin, *The Botanic Garden*, 1789–91.

Taming the Speckled Monster

Smallpox is a killer disease that has been compared in virulence to the Black Death. Until the eighteenth century epidemics were frequent. Survivors were often blinded or disfigured. A mode of partial immunization common in China, India and the near East was to inject some of the pus from a smallpox vesicle into the body of a healthy person. The English bluestocking Lady Mary Wortley Montagu (who had herself been scarred by a smallpox attack two years earlier) discovered this in 1717 while resident in Adrianople, where her husband was British ambassador, and wrote to her friend Sarah Chiswell with the news.

A propos of Distempers, I am going to tell you a thing that I am sure will make you wish your selfe here. The Small Pox so fatal and so general amongst us is here entirely harmless by the invention of engrafting (which is the term they gave it). There is a set of old Women who make it their business to perform the Operation. Every Autumn in the month of September, when the great Heat is abated, people send to one another to know if any of their family has a mind to have the small pox. They make partys for this purpose, and when they are met (commonly 15 or 16 together) the old Woman comes with a nutshell full of the matter of the best sort of small-pox and asks what veins you please to have open'd. She immediately rips open that you offer to her with a large needle (which gives you no more pain than a common scratch) and puts into the vein as much venom as can lye upon the head of her needle, and after binds up the little wound with a hollow bit of shell, and in this manner opens 4 or 5 veins. The Grecians have commonly the superstition of opening one in the Middle of the forehead, in each arm and on the breast to mark the sign of the cross, but this has a very ill Effect, all these wounds leaving little Scars, and is not done by those that are not superstitious, who chuse to have them in the legs or that part of the arm that is conceal'd. The children or young patients play together all the rest of the day and are in perfect health till the 8th. Then the fever begins to seize 'em and they keep their beds 2 days, very seldom 3. They have very rarely above 20 or 30 in their faces, which never mark, and in 8 days time they are as well as before their illness. Where they are wounded there remains running sores during the Distemper, which I don't doubt is a great releife to it. Every year thousands undergo this Operation, and the French Ambassador says pleasantly that they take the Small pox here by way of diversion as they take the Waters in other Countrys. There is no example of any one that has dy'd in it, and you may believe I am very well satisfy'd of the safety of the Experiment since I intend to try it on my dear little Son. I am Patriot enough to take pains to bring this useful invention into fashion in England, and I should not fail to write to some of our Doctors very particularly about it if I knew any one of 'em that I thought had Virtue enough to destroy such a considerable branch of their Revenue for the good of Mankind, but that Distemper is too beneficial to them not to expose to all their Resentment the hardy wight that should undertake to put an end to it. Perhaps if I live to return I may, however, have courrage to war with 'em.

Lady Mary was as good as her word. She had her own son and daughter inoculated and, on her return to England, did all in her power to encourage the practice. When a smallpox epidemic hit London in 1721, she urged Princess Caroline to inoculate the royal children. As a preliminary safeguard, six condemned criminals in Newgate were allowed to volunteer for the operation, with freedom as their reward should they survive it – as they did. The operation was then performed on the pauper children of St James's parish, again successfully. Persuaded by these experiments, the Princess had two of her daughters inoculated, and the practice instantly became fashionable, though opposed by some clergymen who denounced it as a defiance of God's will.

Lady Mary's estimate of its safety was, however, over-hopeful. Roughly 1 in 50 died of inoculation, and inoculated patients tended to spread the disease. Salvation came via an obscure country doctor Edward Jenner (1749–1823), working in the Cotswolds. He was familiar with old wives' tales to the effect that the unsightly but harmless disease of cow pox, which milkmaids caught from cows' udders, gave protection against smallpox, and he hit on the idea that cow pox might be artificially induced. The human guinea pigs he used were a young dairymaid, Sarah Neimes, with a fresh cow pox lesion on her finger, and a boy, James Phipps, a labourer's son. Jenner announced the results in his epoch-making paper *An Inquiry into the Causes and Effects of the Variolae Vaccinae, Known by the Name of Cow-Pox* (1798).

The more accurately to observe the progress of the infection I selected a healthy boy, about eight years old, for the purpose of inoculating for the cow-pox. The matter was taken from a sore on the hand of a Dairymaid, who was infected by her master's cows, and it was inserted on the 14th day of May, 1796, into the arm of the boy by means of two superficial incisions, barely penetrating the cutis [skin], each about an inch long.

On the seventh day he complained of uneasiness in the axilla [armpit] and on the ninth he became a little chilly, lost his appetite, and had a slight headache. During the whole of this day he was perceptibly indisposed, and spent the night with some degree of restlessness, but on the day following he was perfectly well . . .

In order to ascertain whether the boy, after feeling so slight an affection of the system from the cow-pox virus, was secure from the contagion of the smallpox, he was inoculated the 1st of July following with variolous [smallpox] matter, immediately taken from a pustule. Several slight punctures and incisions were made on both his arms, and the matter was carefully inserted, but no disease followed. The same appearances were observable on the arms as we commonly see when a patient has had variolous matter applied, after having either the cow-pox or smallpox. Several months afterwards he was again inoculated with variolous matter, but no sensible effect was produced on the constitution.

Jenner had invented the Latin name *Variolae vaccinae* (meaning 'smallpox of the cow') for cow pox. The English word 'vaccination' was not invented until 1800, by Jenner's disciple Richard Denning. Following the publication of his paper, vaccination swiftly spread through Europe and America, earning him worldwide fame. He became, in his own lifetime, the acclaimed saviour of thousands of Germans, Spaniards, Italians and Russians. Ironically, the English were slow to follow suit. Whereas early nineteenth-century Vienna, where an intelligent vaccination programme was enforced, became virtually smallpox-free, in London 'the speckled monster' (as Jenner called it) still claimed 1,700 lives each year.

Sources: Edward Jenner, *Inquiry* (1798) and *The Complete Letters of Lady Mary Wortley Montagu*, ed. Robert Halsband, Oxford, Clarendon Press, 1965–7, volume I, pp. 338–9.

The Menace of Population

Thomas Malthus (1766–1834), mathematician and clergyman, published his *Essay on the Principle of Population* in 1798, arousing a storm of abuse and controversy. By applying scientific thought to the question of population, which no one had done before, he contrived to show that it was impossible – despite the dreams of Utopian philosophers – for the whole of mankind to live in happiness and plenty. Idealists and reformers were enraged by Malthus's demonstration that social welfare, if it consisted of cash handouts to the poor, did more harm than good. Co-founder of Communism, Friedrich Engels denounced 'this vile, infamous theory, this revolting blasphemy against nature and mankind'.

Malthus's case – that food supply increases only arithmetically (1, 2, 3, 4, 5, etc.), whereas population increases geometrically (1, 2, 4, 8, 16, 32, etc.) – has been modified by developments in agricultural technology. Also, his contention that 'misery' (war, disease, starvation, etc.) and 'vice' (abortion, prostitution, etc.) are the only possible checks to population fails to take account of contraception, which was not publicly advocated in England until the 1820s. But despite its flaws Malthus's theory did not exaggerate the prodigious effects of unchecked population increase. In 1956 Professor W. A. Lewis calculated that if the world population were to double every 25 years (a rate of increase currently observable in some parts of Africa and Asia), it would reach 173,500 thousand million by the year 2330, 'at which time there would be standing room only, since this is the number of square yards on the land surface of the earth'.

Population, when unchecked, increases in a geometrical ratio. Subsistence increases only in an arithmetical ratio. A slight acquaintance with numbers will shew the immensity of the first power in comparison of the second.

By that law of our nature which makes food necessary to the life of man, the effects of these two unequal powers must be kept equal.

This implies a strong and constantly operating check on population from the difficulty of subsistence. This difficulty must fall somewhere and must necessarily be severely felt by a large portion of mankind.

Through the animal and vegetable kingdoms, nature has scattered the seeds of life abroad with the most profuse and liberal hand. She has been comparatively sparing in the room and the nourishment necessary to rear them. The germs of existence contained in this spot of earth, with ample food, and ample room to expand in, would fill millions of worlds in the course of a few thousand years. Necessity, that imperious all pervading law of nature, restrains them within the prescribed bounds. The race of plants and the race of animals shrink under this great restrictive law. And the race of man cannot, by any efforts of reason, escape from it. Among plants and animals its effects are waste of seed, sickness, and premature death. Among mankind, misery and vice. The former, misery, is an absolutely necessary consequence of it. Vice is a highly probable consequence, and we therefore see it abundantly prevail, but it ought not, perhaps, to be called an absolutely necessary consequence. The ordeal of virtue is to resist all temptation to evil.

This natural inequality of the two powers of population and of production in the earth, and that great law of our nature which must constantly keep their effects equal, form the great difficulty that to me appears insurmountable in the way to the perfectibility of society. All other arguments are of slight and subordinate consideration in comparison of this. I see no way by which man can escape from the weight of this law which pervades all animated nature. No fancied equality, no agrarian regulations in their utmost extent, could remove the pressure of it even for a single century. And it appears, therefore, to be decisive against the possible existence of a society, all the members of which should live in ease, happiness, and comparative leisure; and feel no anxiety about providing the means of subsistence for themselves and families.

Consequently, if the premises are just, the argument is conclusive against the perfectibility of the mass of mankind . . . The poor laws of England tend to depress the general condition of the poor in two ways. Their first obvious tendency is to increase population without increasing the food for its support. A poor man may marry with little or no prospect of being able to support a family in independence. They may be said therefore in some measure to create the poor which they maintain, and as the provisions of the country must, in consequence of the increased population, be distributed to every man in smaller proportions, it is evident that the labour of those who are not supported by parish assistance will purchase a smaller quantity of provisions than before and consequently more of them must be driven to ask for support.

Secondly, the quantity of provisions consumed in workhouses upon a part of the society that cannot in general be considered as the most valuable part diminishes the shares that would otherwise belong to more industrious and more worthy members, and thus in the same manner forces more to become dependent. If the poor in the workhouses were to live better than they now do, this new distribution of the money of the society would tend more conspicuously to depress the condition of those out of the workhouses by occasioning a rise in the price of provisions.

Fortunately for England, a spirit of independence still remains among the peasantry. The poor laws are strongly calculated to eradicate this spirit. They have succeeded in part, but had they succeeded as completely as might have been expected their pernicious tendency would not have been so long concealed.

Hard as it may appear in individual instances, dependent poverty ought to be held disgraceful. Such a stimulus seems to be absolutely necessary to promote the happiness of the great mass of mankind, and every general attempt to weaken this stimulus, however benevolent its apparent intention, will always defeat its own purpose. If men are induced to marry from a prospect of parish provision, with little or no chance of maintaining their families in independence, they are not only unjustly tempted to bring unhappiness and dependence upon themselves and children, but they are tempted, without knowing it, to injure all in the same class with themselves. A labourer who marries without being able to support a family may in some respects be considered as an enemy to all his fellow-labourers . . .

The evil is perhaps gone too far to be remedied, but I feel little doubt in my own mind that if the poor laws had never existed, though there might have been a few more instances of very severe distress, yet that the aggregate mass of happiness among the common people would have been much greater than it is at present.

Malthus's work triggered the theory of evolution. Alfred Russel Wallace recalls in *My Life: A Record of Events and Opinions* (1905) that in January 1858 he had just arrived at Ternate in the Moluccas to collect butterflies and beetles:

I was suffering from a sharp attack of intermittent fever, and every day during the cold and succeeding hot fits had to lie down for several hours, during which time I had nothing to do but to think over any subject then particularly interesting to me. One day something brought to my recollection Malthus's 'Principles of Population', which I had read twelve years before. I thought of his clear exposition of 'the positive checks to increase' – disease, accidents, war and famine – which keep down the population of savage races to so much lower an average than that of more civilized peoples. It then occurred to me that these causes or their equivalents are continually acting in the case of animals also; and as animals usually breed much more rapidly than does mankind, the destruction every year from these causes must be enormous in order to keep down the numbers of each species . . . as otherwise the world would have been densely crowded with those that breed more quickly . . . Why do some die and some live? And the answer was clearly, that on the whole the best fitted live. From the effects of disease the most healthy escaped; from enemies, the strongest, the swiftest or the most cunning; from famine, the best hunters or those with the best digestion; and so on. Then it suddenly flashed upon me that this self-acting process would necessarily *improve the race*, because in every generation the inferior would inevitably be killed off and the superior would remain – that is, *the fittest would survive* . . . I awaited anxiously for the termination of my fit so that I might at once make notes for a paper on the subject.

He wrote the paper during the following two evenings and sent it to Darwin by the next post. Alarmed to find that a rival had reached the same conclusions as himself, Darwin was spurred to publish the *Origin of Species*, in which he states that 'The struggle for existence is the doctrine of Malthus applied with manifold force to the whole animal and vegetable kingdom'.

Sources: Thomas Malthus, *Essay on the Principle of Population*, 1798; Alfred Russel Wallace, *My Life*, London, Chapman & Hall, 1905.

How the Giraffe Got its Neck

Jean-Baptiste Lamarck (1744–1829), who invented the word 'biology', did not intend to be a biologist. His father destined him for the Church, but he quit the seminary at Amiens at the age of 16 and joined the French army on the eve of the Battle of Fissingshausen. By the end of the next day all the officers in his company had been killed, and he was commissioned for his gallantry. When he was 22, however, he hurt his neck during some horseplay and had to leave the army. His scientific interests were stimulated by Jean-Jacques Rousseau, with whom he went on botanical excursions. His three-volume work on French flora (1778) brought him fame and the job of keeper of the herbarium at the Paris Jardin du Roi (renamed the Jardin des Plantes at the Revolution). Appointed professor of 'insects and worms' at the Museum of Natural History, he reformed the study of invertebrates. His *Zoological Philosophy* (1809) propounded a theory of evolution half a century before Darwin's *Origin of Species*. It argues that animals, birds and fishes exercise willpower to adapt themselves to their living conditions. They strengthen some organs by use, and weaken others by under-use, and pass on these acquired characteristics to the offspring.

The bird which is drawn to the water by its need of finding there the prey on which it lives, separates the digits of its feet in trying to strike the water and move about on the surface. The skin which unites these digits at their base acquires the habit of being stretched by these continually repeated separations of the digits; thus in course of time there are formed large webs which unite the digits of ducks, geese, etc., as we actually find them. In the same way efforts to swim, that is to push against the water so as to move about in it, have stretched the membranes between the digits of frogs, sea-tortoises, the otter, beaver, etc.

On the other hand, a bird which is accustomed to perch on trees and which springs from individuals all of whom had acquired this habit, necessarily has longer digits on its feet and differently shaped from those of the aquatic animals that I have just named. Its claws in time become lengthened, sharpened and curved into hooks, to clasp the branches on which the animal so often rests.

We find in the same way that the bird of the water-side which does not like swimming and yet is in need of going to the water's edge to secure its prey, is continually liable to sink in the mud. Now this bird tries to act in such a way that its body should not be immersed in the liquid, and hence makes its best efforts to stretch and lengthen its legs. The long-established habit acquired by this bird and all its race of continually stretching and lengthening its legs, results in the individuals of this race becoming raised as though on stilts, and gradually obtaining long, bare legs, denuded of feathers up to the thighs and often higher still.

We note again that this same bird wants to fish without wetting its body, and is thus obliged to make continual efforts to lengthen its neck. Now these habitual efforts in this individual and its race must have resulted in course of time in a remarkable lengthening, as indeed we actually find in the long necks of all water-side birds . . .

It is interesting to observe the result of habit in the peculiar shape and size of the giraffe (*Camelo-pardalis*): this animal, the largest of the mammals, is known to live in the interior of Africa in places where the soil is nearly always arid and barren, so that it is obliged to browse on the leaves of trees and to make constant efforts to reach them. From this habit long maintained in all its race, it has resulted that the animal's fore-legs have become longer than its hind legs, and that its neck is lengthened to such a degree that the giraffe, without standing up on its hind legs, attains a height of six metres (nearly 20 feet).

After Lamarck came Darwin (see p. 114), who attributed evolution not to the inheritance of acquired characteristics, but to 'Natural Selection', i.e. random genetic mutation plus the survival of those mutations that were better fitted to their environment than others.

Lamarckism has been discredited by most twentieth-century geneticists, but it has great attractions for progressives and social reformers, who wish to believe in the perfectibility of mankind. Promoted by the agriculturalist Trofim Lysenko (1898–1976), it remained official Soviet scientific doctrine until the 1960s.

It appealed, too, to George Bernard Shaw. Brainpower and energy, such as his own, could, he believed, bring about a New Jerusalem through 'Creative Evolution'. As a Neo-Lamarckian he denounced Darwin and 'Circumstantial' (i.e. Natural) Selection in the Preface to *Back to Methuselah* (1921). They were both 'ghastly and damnable', he declared, because they emptied the universe of 'beauty and intelligence, of strength and purpose, of honour and aspiration' and reduced it to a 'universal struggle for hogwash'. Shaw knew nothing of science, and his Preface is a good example of how the combined powers of ignorance, rhetoric and common sense approach a scientific problem.

Lamarck really held as his fundamental proposition that living organisms changed because they wanted to. As he stated it, the great factor in Evolution is use and disuse. If you have no eyes, and want to see, and keep trying to see, you will finally get eyes. If, like a mole or a subterranean fish, you have eyes and don't want to see, you will lose your eyes. If you like eating the tender tops of trees enough to make you concentrate all your energies on the stretching of your neck, you will finally get a long neck, like the giraffe. This seems absurd to inconsiderate people at the first blush; but it is within the personal experience of all of us that it is just by this process that a child tumbling about the floor becomes a boy walking erect; and that a man sprawling on the road with a bruised chin, or supine on the ice with a bashed occiput, becomes a bicyclist and a skater. The process is not continuous, as it would be if mere practice had anything to do with it; for though you may improve at each bicycling lesson *during* the lesson, when you begin your next lesson you do not begin at the point at which you left off: you relapse apparently to the beginning. Finally, you succeed quite suddenly, and do not relapse again. More miraculous still, you at once exercise the new power unconsciously. Although you are adapting your front wheel to your balance so elaborately and actively that the accidental locking of your handle bars for a second will throw you off; though five minutes before you could not do it at all, yet now you do it as unconsciously as you grow your finger nails. You have a new faculty, and must have created some new bodily tissue as its organ. And you have done it solely by willing. For here there can be no question of Circumstantial Selection, or the survival of the fittest. The man who is learning how to ride a bicycle has no advantage over the non-cyclist in the struggle for existence: quite the contrary. He has acquired a new habit, an automatic unconscious habit, solely because he wanted to, and kept trying until it was added unto him.

But when your son tries to skate or bicycle in his turn, he does not pick up the accomplishment where you left it, any more than he is born six feet high with a beard and a tall hat. The set-back that occurred between your lessons occurs again. The race learns exactly as the individual learns. Your son relapses, not to the very beginning, but to a point which no mortal method of measurement can distinguish from the beginning. Now this is odd; for certain other habits of yours, equally acquired (to the Evolutionist, of course, all habits are acquired), equally unconscious, equally automatic, are transmitted without any perceptible relapse. For instance, the very first act of your son when he enters the world as a separate individual is to yell with indignation: that yell which Shakespear thought the most tragic and piteous of all sounds. In the act of yelling he begins to breathe: another habit, and not even a necessary one, as the object of breathing can be achieved in other ways, as by deep sea fishes. He circulates his blood by pumping it with his heart. He demands a meal, and proceeds at once to perform the most elaborate chemical operations on the food he swallows. He manufactures teeth; discards them; and replaces them with fresh ones. Compared to these habitual feats, walking, standing upright, and bicycling are the merest trifles; yet it is only by going through the wanting, trying process that he can stand, walk, or cycle, whereas in the other and far more difficult and complex habits he not only does not consciously want nor consciously try, but actually consciously objects very strongly. Take that early habit of cutting the teeth: would he do that if he could help it? Take that later habit of decaying and eliminating himself by death – equally an acquired habit, remember – how he abhors it! Yet the habit has become so rooted, so automatic, that he must do it in spite of himself, even to his own destruction.

We have here a routine which, given time enough for it to operate, will finally produce the most elaborate forms of organized life on Lamarckian lines without the intervention of Circumstantial Selection at all. If you can turn a pedestrian into a cyclist, and a cyclist into a pianist or violinist, without the intervention of Circumstantial Selection, you can turn an amœba into a man, or a man into a superman, without it. All of which is rank heresy to the Neo-Darwinian, who imagines that if you stop Circumstantial Selection, you not only stop development but inaugurate a rapid and disastrous degeneration.

Let us fix the Lamarckian evolutionary process well in our minds. You are alive; and you want to be more alive. You want an extension of consciousness and of power. You want, consequently, additional organs, or additional uses of your existing organs: that is, additional habits. You get them because you want them badly enough to keep trying for them until they come. Nobody knows how: nobody knows why: all we know is that the thing actually takes place. We relapse miserably from effort to effort until the old organ is modified or the new one created, when suddenly the impossible becomes possible and the habit is formed. The moment we form it we want to get rid of the consciousness of it so as to economize our consciousness for fresh conquests of life; as all consciousness means preoccupation and obstruction. If we had to think about breathing or digesting or circulating our blood we should have no attention to spare for anything else, as we find to our cost when anything goes wrong with these operations. We want to be unconscious of them just as we wanted to acquire them; and we finally win what we want. But we win unconsciousness of our habits at the cost of losing our control of them; and we also build one habit and its corresponding functional modification of our organs on another, and so become dependent on our old habits. Consequently we have to persist in them even when they hurt us. We cannot stop breathing to avoid an attack of asthma, or to escape drowning. We can lose a habit and discard an organ when we no longer need them, just as we acquired them; but this process is slow and broken by relapses; and relics of the organ and the habit long survive its utility. And if other and still indispensable habits and modifications have been built on the ones we wish to discard, we must provide a new foundation for them before we demolish the old one. This is also a slow process and a very curious one.

The relapses between the efforts to acquire a habit are important because, as we have seen, they recur not only from effort to effort in the case of the individual, but from generation to generation in the case of the race. This relapsing from generation to generation is an invariable characteristic of the evolutionary process. For instance, Raphael, though descended from eight uninterrupted generations of painters, had to learn to paint apparently as if no Sanzio had ever handled a brush before. But he had also to learn to breathe, and digest, and circulate his blood. Although his father and mother were fully grown adults when he was conceived, he was not conceived or even born fully grown: he had to go back and begin as a speck of protoplasm, and to struggle through an embryonic lifetime, during part of which he was indistinguishable from an embryonic dog, and had neither a skull nor a backbone. When he at last acquired these articles, he was for some time doubtful whether he was a bird or a fish. He had to compress untold centuries of development into nine months before he was human enough to break loose as an independent being. And even then he was still so incomplete that his parents might well have exclaimed, 'Good Heavens! have you learnt nothing from our experience that you come into the world in this ridiculously elementary state? Why can't you talk and walk and paint and behave decently?' To that question Baby Raphael had no answer. All he could have said was that this is how evolution or transformation happens. The time may come when the same force that compressed the development of millions of years into nine months may pack many more millions into even a shorter space; so that Raphaels may be born painters as they are now born breathers and blood circulators. But they will still begin as specks of protoplasm, and acquire the faculty of painting in their mother's womb at quite a late stage of their embryonic life. They must recapitulate the history of mankind in their own persons, however briefly they may condense it.

Nothing was so astonishing and significant in the discoveries of the embryologists, nor anything so absurdly little appreciated, as this recapitulation, as it is now called: this power of hurrying up into months a process which was once so long and tedious that the mere contemplation of it is unendurable by men whose span of life is three-score-and-ten. It widened human possibilities to the extent of enabling us to hope that the most prolonged and difficult operations of our minds may yet become instantaneous, or, as we call it, instinctive.

A more recent celebration of Lamarck is by the American poet Richard Wilbur, who distinguishes between the senses (sight, hearing etc.), created by the real world, and the unreal worlds we imagine, following cosmic lyres/liars:

Lamarck Elaborated

'The environment creates the organ'

The Greeks were wrong who said our eyes have rays;
Not from these sockets or these sparkling poles
Comes the illumination of our days.
It was the sun that bored these two blue holes.

It was the song of doves begot the ear
And not the ear that first conceived of sound:
That organ bloomed in vibrant atmosphere,
As music conjured Ilium from the ground.

The yielding water, the repugnant stone,
The poisoned berry and the flaring rose
Attired in sense the tactless finger-bone
And set the taste-buds and inspired the nose.

Out of our vivid ambiance came unsought
All sense but that most formidably dim.
The shell of balance rolls in seas of thought.
It was the mind that taught the head to swim.

Newtonian numbers set to cosmic lyres
Whelmed us in whirling worlds we could not know,
And by the imagined floods of our desires
The voice of Sirens gave us vertigo.

Sources: J. B. Lamarck, *Zoological Philosophy*, trans. Hugh Elliot, London, Macmillan, 1914. George Bernard Shaw, Preface to *Back to Methuselah* (Copyright Society of Authors, 84 Drayton Gardens, SW10). Richard Wilbur, *Things of This World*, New York, Harcourt Brace, 1956.

Medical Studies, Paris 1821

The composer Hector Berlioz (1803–69) had a brief spell as a medical student, until a visit to the Opera drew him irresistibly towards music. This is from his *Memoirs*.

On arriving in Paris in 1821 with my fellow-student Alphonse Robert, I gave myself up wholly to studying for the career which had been thrust upon me, and loyally kept the promise I had given my father on leaving. It was soon put to a somewhat severe test when Robert, having announced one morning that he had bought a 'subject' (a corpse), took me for the first time to the dissecting-room at the Hospice de la Pitié. At the sight of that terrible charnel-house – the fragments of limbs, the grinning faces and gaping skulls, the bloody quagmire underfoot and the atrocious smell it gave off, the swarms of sparrows wrangling over scraps of lung, the rats in their corner gnawing the bleeding vertebrae – such a feeling of revulsion possessed me that I leapt through the window of the dissecting-room and fled for home as though Death and all his hideous train were at my heels. The shock of that first impression lasted for twenty-four hours. I did not want to hear another word about anatomy, dissection or medicine, and I meditated a hundred mad schemes of escape from the future that hung over me.

Robert lavished his eloquence in a vain attempt to argue away my disgust and demonstrate the absurdity of my plans. In the end he got me to agree to make another effort. For the second time I accompanied him to the hospital and we entered the house of the dead. How strange! The objects which before had filled me with extreme horror had absolutely no effect upon me now. I felt nothing but a cold distaste; I was already as hardened to the scene as any seasoned medical student. The crisis was past. I found I actually enjoyed groping about in a poor fellow's chest and feeding the winged inhabitants of the delightful place their ration of lung. 'Hallo!' Robert cried, laughing, 'you're getting civilized. "Thou giv'st the little birds their daily bread."' '"An o'er all nature's realm my bounty spread,"' I retorted, tossing a shoulder-blade to a great rat staring at me with famished eyes.

Source: *The Memoirs of Hector Berlioz, Member of the French Institute, including his travels in Italy, Germany, Russia and England, 1803–1865*, trans. and ed. David Cairns, London, Cardinal, 1990.

The Man with a Lid on his Stomach

On 6 June 1822, at an army station in Michigan, an 18-year-old French Canadian, Alexis St Martin, was accidentally shot at close range. A US army surgeon, William Beaumont (1785–1853), who was called to the scene, describes the wound:

The charge, consisting of powder and duck shot, was received in the left side of the youth, he being at a distance of not more than one yard from the muzzle of the gun. The contents entered posteriorly, and in an oblique direction, forward and inward, literally blowing off integuments and muscles of the size of a man's hand, fracturing and carrying away the anterior half of the sixth rib, fracturing the fifth, lacerating the lower portion of the left lobe of the lungs, the diaphragm, and perforating the stomach.

The whole mass of materials forced from the musket, together with fragments of clothing and pieces of fractured ribs, were driven into the muscles and cavity of the chest.

I saw him in twenty-five or thirty minutes after the accident occurred, and, on examination, found a portion of the lung, as large as a Turkey's egg, protruding through the external wound, lacerated and burnt; and immediately below this, another protrusion, which, on further examination, proved to be a portion of the stomach, lacerated through all its coats, and pouring out the food he had taken for his breakfast, through an orifice large enough to admit the fore finger.

In attempting to return the protruded portion of the lung, I was prevented by a sharp point of the fractured rib, over which it had caught by its membranes; but by raising it with my finger, and clipping off the point of the rib, I was able to return it into its proper cavity, though it could not be retained there, on account of the incessant efforts to cough.

The projecting portion of the stomach was nearly as large as that of the lung. It passed through the lacerated diaphragm and external wound, mingling the food with the bloody mucus blown from the lungs.

Beaumont cleaned and dressed the wounds, after 'replacing the stomach and lungs as far as practicable', but he did not expect his patient to live. St Martin, however, was a tough young man, and he pulled through. At first he could not retain any food in his stomach, because of the hole in it, but, Beaumont reports, 'firm dressings were applied and the contents of the stomach retained'.

Despite all Beaumont's efforts, the hole in St Martin's stomach would not heal completely. But gradually 'a small fold or doubling of the coats of the stomach' grew over and filled the aperture, forming a kind of valve which could be opened by hand. Beaumont, who nursed, fed and clothed St Martin for the first two years of his recovery, realized that he had at his disposal a walking laboratory for the experimental study of human digestion. He could place foodstuffs directly into St Martin's stomach through the hole, and remove them at intervals to observe how much had been digested. As he later acknowledged:

I had opportunities for the examination of the interior of the stomach, and its secretions, which has never before been so fully offered to any one. This most important organ, its secretions and its operations, have been submitted to my observation in a very extraordinary manner, in a state of perfect health, and for years in succession.

The first experiment was carried out on 1 August, 1825:

At 12 o'clock I introduced through the perforation, into the stomach, the following articles of diet, suspended by a silk string, and fastened at proper distances, so as to pass in without pain – viz.: – a piece of high seasoned *a la mode beef*; a piece of *raw, salted, fat pork*; a piece of *raw, salted, lean beef*; a piece of *boiled, salted beef*; a piece of *stale bread*; and a bunch of *raw, sliced cabbage*; each piece weighing about two drachms; the lad continuing his usual employment about the house.

At 1 o'clock, p.m., withdrew and examined them – found the *cabbage* and *bread* about half digested: the pieces of *meat* unchanged. Returned them into the stomach.

At 2 o'clock, p.m., withdrew them again – found the *cabbage, bread, pork*, and *boiled beef*, all cleanly digested, and gone from the string; the other pieces of meat but very little affected. Returned them into the stomach again.

Over the next eight years Beaumont performed hundreds of similar experiments, gathering a vast amount of precise information about the speed or difficulty with which the stomach digests different types of food. He also established by direct observation the harmful effects of mental disturbance on digestion. During this whole period St Martin led an active, vigorous life, married and fathered four children, and became a sergeant in the US army. His comrades joked about 'the man with a lid on his stomach'. However, he was able to make considerable sums by touring and showing his internal organs to interested doctors.

Besides watching food in St Martin's stomach, Beaumont was able to extract gastric juice from it through a tube:

The usual method of extracting the gastric juice, for experiment, is by placing the subject on his right side, depressing the valve within the aperture, introducing a gum-elastic tube, of the size of a large quill, five or six inches into the stomach, and then turning him on the left side, until the orifice becomes dependent. In health, and when free from food, the stomach is *usually* entirely empty, and contracted upon itself. On introducing the tube, the fluid soon begins to flow, first by drops, then in an interrupted, and sometimes in a short continuous stream. Moving the tube about, up and down, or backwards and forwards, increases the discharge. The quantity of fluid ordinarily obtained is from four drachms to one and a half or two ounces, varying with the circumstances and condition of the stomach. Its extraction is generally attended by that peculiar sensation at the pit of the stomach, termed sinking, with some degree of faintness, which renders it necessary to stop the operation. The usual time of extracting the juice is early in the morning, before he has eaten, when the stomach is empty and clean.

Obtaining gastric juice in this manner, Beaumont was able to answer a question that had puzzled medical science – namely, how the stomach digests. Previous accounts had suggested that the stomach worked like a fermenting vat or a mill or a cooking vessel. Beaumont showed that gastric juice, placed in a glass vessel, would dissolve foodstuffs at just the same rate and in just the same way as they were dissolved inside St Martin's stomach. He deduced that gastric juice was a chemical agent, and he rightly identified its important acid component as hydrochloric (which he calls 'muriatic'):

I think I am warranted, from the result of all the experiments, in saying, that the gastric juice, so far from being 'inert as water,' as some authors assert, is the most general solvent in nature, of alimentary matter – even the hardest bone cannot withstand its action. It is capable, *even out of the stomach*, of effecting perfect digestion, with the aid of due and uniform degrees of heat (100° Fahrenheit), and gentle agitation . . . We must, I think, regard this fluid as a chemical agent, and its operation as a chemical action. It is certainly every way analogous to it; and I can see no more objection to accounting for the change effected on the food, on the supposition of a chemical process, than I do in accounting for the various and diversified modifications of matter, which are operated on in the same way. The decay of the dead body is a chemical operation, separating it into its elementary principles – and why not the solution of aliment in the stomach . . .

Pure gastric juice, when taken directly out of the stomach of a healthy adult, unmixed with any other fluid, save a portion of the mucus of the stomach, with which it is most commonly, and perhaps always combined, is a clear, transparent fluid; inodorous; a little saltish; and very perceptibly acid. Its taste, when applied to the tongue, is similar to thin mucilaginous water, slightly acidulated with muriatic acid. It is readily diffusible in water, wine or spirits; slightly effervesces with alkalis; and is an effectual solvent of the *materia alimentaria* [food]. It possesses the property of coagulating albumen, in an eminent degree; is powerfully antiseptic, checking the putrefaction of meat; and effectually restorative of healthy action, when applied to old, fœtid sores, and foul, ulcerating surfaces.

Beaumont's publication of his findings in 1833 made him famous throughout the medical world.

Source: William Beaumont, *Experiments and Observations on the Gastric Juice and the Physiology of Digestion*, Burlington, Chauncey Goodrich, 1833.

Those Dreadful Hammers: Lyell and the New Geology

Geology revolutionized thought and feeling in the early nineteenth century. Its effects spread far beyond the scientific community, destroying established truths, and forcing ordinary men and women to realize that they, and everything they thought of as time and history, were a mere blip in the unimaginable millions of years of the earth's existence. Faced with these mind-blanking immensities, many found their religious faith ebbing away. Orthodox, Bible-based estimates of the earth's age, such as that of Archbishop Ussher (who fixed the date of the creation of the world as 23 October 4004 BC), now seemed ridiculously inadequate. 'If only the Geologists would let me alone, I could do very well,' lamented John Ruskin in 1851, 'but those dreadful Hammers! I hear the clink of them at the end of every cadence of the Bible verses.'

The manifesto of the new science was Charles Lyell's *Principles of Geology* (1830–3). What this book set out to shatter was the assumption that the earth – its oceans, land masses, and geological strata – had remained much the same since the Creation, or since an age of vast volcanic upheavals which, it was imagined, had taken place very early in its history. Lyell argued that, on the contrary, the surface of the earth is continuously changing. The agents that have changed it in the past are still active today though since they work very slowly we tend to overlook them. They are essentially two – water (rivers, tides, etc.) and subterranean fire (causing earthquakes and volcanoes). They work in opposite ways – water wearing down, and subterranean fire elevating the earth's surface:

We know that one earthquake may raise the coast of Chile for a hundred miles to an average height of about five feet. A repetition of two thousand shocks of equal violence might produce a mountain chain one hundred miles long and ten thousand feet high. Now should . . . one of these conclusions happen in a century, it would be consistent with the order of events experienced by the Chileans from the earliest times.

On this reckoning 200,000 years – a brief period in geological time – could produce a mountain range where flat land, or sea, had been before. It is this agency, Lyell argued, that has created the continents:

There is scarcely any land hitherto examined in Europe, North Asia, or North America, which has not been raised from the bosom of the deep, since the origins of the carboniferous rocks . . . If we were to submerge again all the marine strata [i.e. rock layers containing underwater remains], from the transition limestone to the most recent shelly beds, the summits of some primary mountains would alone remain above the waters.

Great as such a change might seem to us it is, Lyell points out, quite normal in geological terms.

However constant we believe the relative proportion of sea and land to continue, we know that there is annually some small variation in their respective geographical positions, and that in every century the land is in some parts raised, and in others depressed by earthquakes, and so likewise is the bed of the sea. By these and other ceaseless changes, the configuration of the earth's surface has been remodelled again and again since it was the habitation of organic beings, and the bed of the ocean has been lifted up to the height of some of the loftiest mountains. The imagination is apt to take alarm, when called upon to admit the formation of such irregularities of the crust of the earth, after it had become the habitation of living creatures; but if time be allowed, the operation need not subvert the ordinary repose of nature, and the result is insignificant, if we consider how slightly the highest mountain chains cause our globe to differ from a perfect sphere. Chimborazo [a mountain in Ecuador], although it rises to more than 21,000 feet above the surface of the sea, would only be represented on an artificial globe, of about six feet in diameter, by a grain of sand less than one-twentieth of an inch in thickness. The superficial inequalities of the earth, then, may be deemed minute in quantity, and their distribution at any particular epoch must be regarded in geology as temporary peculiarities.

Rivers, carrying and depositing silt, are also, Lyell points out, working perpetually to change the configuration of land and sea, sometimes at an astonishing rate.

One of the most extraordinary statements is that of Major Rennell, in his excellent paper on the Delta of the Ganges. 'A glass of water', he says, 'taken out of the river when at its height, yields about one part in four of mud. No wonder, then, that the subsiding waters should quickly form a stratum of earth, or that the delta should encroach on the sea!' The same hydrographer computed with much care the number of cubic feet of water discharged by the Ganges into the sea, and estimated the mean quantity through the whole year to be eighty thousand cubic feet in a second. When the river is most swollen, and its velocity much accelerated, the quantity is four hundred and five thousand cubic feet in a second . . . We are somewhat staggered by the results to which we must arrive if we compare the proportion of mud, as given by Rennell, with his computation of the quantity of water discharged, which latter is probably very correct. If it were true that the Ganges, in the flood season, contained one part in four of mud, we should then be obliged to suppose that there passes down, every four days, a quantity of mud equal in volume to the water which is discharged in the course of twenty-four hours. If the mud be assumed to be equal to one-half the specific gravity of granite (it would, however, be more), the weight of matter *daily* carried down in the flood season, would be about equal to seventy-four times the weight of the Great Pyramid of Egypt. Even if it could be proved that the turbid waters of the Ganges contain one part *in a hundred* of mud, which is affirmed to be the case in regard to the Rhine, we should be brought to the extraordinary conclusion that there passes down, in every two days, into the Bay of Bengal, a mass about equal in weight and bulk to the Great Pyramid . . . We may confidently affirm that when the aggregate amount of solid matter transported by rivers in a given number of centuries from a large continent, shall be reduced to arithmetical computation, the result will appear most astonishing to those who are not in the habit of reflecting how many of the mightiest operations in nature are effected insensibly, without noise or disorder.

In later editions of the *Principles* Lyell modified his estimate of mud in the Ganges – but this did not affect his main argument.

Since we normally see only what is happening on the surface of the earth, we are, he stresses, badly placed as geological observers. We do not see new strata – such as Ganges mud – being laid down on the ocean floor. Nor can our eyes penetrate to the subterranean rivers and reservoirs of liquid rock far beneath the earth's surface. Hence we tend to assume that rocks such as granite are older than the 'sedimentary' rocks, composed of underwater deposits, and belong to some 'primeval' state of nature. Lyell's argument, however, is that granite, and other 'primeval' rocks, are constantly being formed out of sedimentary rocks by subterranean fire, and if we lived in the depths of the earth we should assume that they were the new rocks and the sedimentary ones the old.

If we may be allowed so far to indulge the imagination, as to suppose a being entirely confined to the nether world – some 'dusky melancholy sprite', like Umbriel [a gnome in Alexander Pope's poem *The Rape of the Lock*], who could 'flit on sooty pinions to the central earth', but who was never permitted to 'sully the fair face of light' and emerge into the regions of water and air; and if this being should busy himself in investigating the structure of the globe, he might frame theories the exact converse of those adopted by human philosophers. He might infer that the stratified rocks, containing shells and other organic remains, were the oldest of created things, belonging to some original and nascent state of the planet. 'Of these masses', he might say, 'whether they consist of loose, incoherent sand, soft clay, or solid rock, none have been formed in modern times. Every year some parts of them are broken and shattered by earthquakes, or melted up by volcanic fire; and when they cool down slowly from a state of fusion, they assume a crystalline form perfectly distinct from those inexplicable rocks which are so regularly bedded, and contain stones full of curious impressions and fantastic markings. This process cannot have been carried on for an indefinite time, for in that case all the stratified rocks would ere this have been fused and crystallised. It is therefore probable that the whole planet once consisted of these curiously-bedded formations, at a time when the volcanic fire had not yet been brought into activity. Since that period there seems to have been a gradual development of heat, and this augmentation we may expect to continue till the whole globe shall be in a state of fluidity and incandescence.'

Such might be the system of the Gnome at the very same time that the followers of Leibnitz [1646–1716, a German philosopher], reasoning on what they saw on the outer surface, would be teaching the doctrine of gradual refrigeration, and averring that the earth had begun its career as a fiery comet, and would hereafter become an icy mass . . . Man observes the annual decomposition of crystalline and igneous rocks, and may sometimes see their conversion into stratified deposits; but he cannot witness the reconversion of the sedimentary into the crystalline by subterranean fire. He is in the habit of regarding all the sedimentary rocks as more recent than the unstratified, for the same reason that we may suppose him to fall into the opposite error if he saw the origin of the igneous class only.

Of the many instances Lyell gives of the immense periods of time needful for geological processes, none is more striking than his comment on the rocks known as marls, which were produced as sediments on the floors of freshwater lakes during the Eocene period (which extended from about 54 million to about 38 million years ago).

The entire thickness of these marls is unknown, but it certainly exceeds, in some places, 700 feet. They are for the most part either light-green or white, and usually calcareous. They are thinly foliated, a character which frequently arises from the innumerable thin plates or scales of that small animal called *cypris*, a genus which comprises several species, of which some are recent, and may be seen swimming rapidly through the waters of our stagnant pools and ditches. This animal resides within two small valves like those of a bivalve shell, and it moults its integuments annually . . . Countless myriads of the shells of *cypris* were shed in the Eocene lakes, so as to give rise to divisions in the marl as thin as paper, and that too in stratified masses several hundred feet thick. A more convincing proof . . . of the slow and gradual process by which the lake was filled up with fine mud cannot be desired.

But the most dramatic implications of Lyell's theory relate to the future not the past. The perpetual interchange of sea and land on the earth's surface must, he predicts, go on, since the agents that caused it are still in operation. The northern hemisphere was once a vast ocean, dotted with islands, like the archipelagoes of the South Pacific, and it will return to this state again.

The existence of enormous seas of fresh water, such as the North American lakes, the largest of which is elevated more than six hundred feet above the level of the ocean, and is in parts twelve hundred feet deep, is alone sufficient to assure us, that the time will come, however distant, when a deluge will lay waste a considerable part of the American continent . . . Notwithstanding, therefore, that we have not witnessed within the last three thousand years the devastation by deluge of a large continent, yet, as we may predict the future occurrence of such catastrophes, we are authorized to regard them as part of the present order of Nature.

Redistribution of the land masses will, Lyell points out, cause radical changes in climate. Ages of intense heat and cold will succeed each other in the future, as they have in the past – the summers and winters of the geological 'great year'. As the climate of our hemisphere changes, so will its vegetation, and the animal life it supports.

Then might those genera of animals return, of which the memorials are preserved in the ancient rocks of our continents. The huge iguanodon might reappear in the woods, and the icthyosaur in the sea, while the pterodactyl might flit again through umbrageous groves of tree ferns. Coral reefs might be prolonged beyond the arctic circle, where the whale and the narwal now abound. Turtles might deposit eggs in the sand of the sea beach, where now the walrus sleeps, and where the seal is drifted on the ice-floe.

When the poet Alfred Tennyson read Lyell's *Principles*, he felt, like many Victorians, dismay. Lyell's demonstration of the temporariness of the familiar world shocked him. In his great poem *In Memoriam*, the classic expression of Victorian angst, he keeps Lyell's book in mind. Simple geological observation had allowed Lyell to assert that:

Millions of our race are now supported by lands situated where deep seas once prevailed in earlier ages. In many districts not yet occupied by man, land animals and forests now abound where the anchor once sank into the oozy bottom.

In Tennyson, this becomes:

> There rolls the deep where grew the tree.
> O earth, what changes hast thou seen!
> There where the long street roars, hath been
> The stillness of the central sea.

> The hills are shadows, and they flow
> From form to form, and nothing stands;
> They melt like mist, the solid lands,
> Like clouds they shape themselves and go.

Equally appalling was Lyell's assurance that:

> Amidst the vicissitudes of the earth's surface, species cannot be immortal, but must perish, one after the other, like the individuals which compose them. There is no possibility of escaping from this conclusion.

This wrung an anguished cry from Tennyson:

> Are God and Nature then at strife
> That Nature lends such evil dreams?
> So careful of the type she seems,
> So careless of the single life . . .
>
> 'So careful of the type?' but no
> From scarped cliff and quarried stone
> She cries, 'A thousand types are gone:
> I care for nothing, all shall go.'

Lyell's own response was calmer and more chilling. He foresaw not merely the extinction of the human race, but the gradual obliteration of every single trace of its existence.

> We may anticipate with confidence that many edifices and implements of human workmanship, and the skeletons of men, and casts of the human form, will continue to exist when the great part of the present mountains, continents, and seas have disappeared. Assuming the future duration of the planet to be indefinitely protracted, we can foresee no limit to the perpetuation of some of the memorials of man, which are continually entombed in the bowels of the earth or in the bed of the ocean, unless we carry forward our views to a period sufficient to allow the various causes of change, both igneous and aqueous, to remodel more than once the entire crust of the earth. *One* complete revolution will be inadequate to efface every monument of

our existence . . . Yet it is no less true that none of the works of a mortal can be eternal . . . Even when they have been included in rocky strata, when they have been made to enter as it were into the solid framework of the globe itself, they must nevertheless eventually perish; for every year some portion of the earth's crust is shattered by earthquakes or melted by volcanic fire, or ground to dust by the moving waters on the surface.

Source: Charles Lyell, *Principles of Geology*, London, John Murray, 1830–3.

The Discovery of Worrying

Changes in early nineteenth-century mentality, related to the progress of science, may help to explain a linguistic development noted by Adam Phillips in his book *On Kissing, Tickling and Being Bored* (1993). Phillips is Principal Child Psychotherapist at Charing Cross Hospital.

The history of the word *worrying* is itself revealing. Deriving from the Old English *wyrgan*, meaning 'to kill by strangulation', it was originally a hunting term, describing what dogs did to their prey as they caught it. The *Oxford English Dictionary* has, among several meanings from the fourteenth to the early nineteenth century: 'To swallow greedily or to devour . . . to choke a person or animal with a mouthful of food . . . to seize by the throat with the teeth and tear or lacerate; to kill or injure by biting or shaking. Said, e.g., of dogs or wolves attacking sheep, or of hounds when they seize their prey.' Johnson's *Dictionary* of 1755 has for *worry*: 'To tear or mangle as a beast tears its prey. To harass or persecute brutally.' A worrier for Johnson is someone who persecutes others, 'one who worries or torments them'. Two things are immediately striking in all of this. First there is the original violence of the term, the way it signifies the vicious but successful outcome of pursuing an object of desire. This sense of brutal foreplay is picked up in Dryden's wonderful lines in *All for Love*: 'And then he grew familiar with her hand / Squeezed it, and worry'd it with ravenous kisses.' Worrying, then, is devouring, a peculiarly intense, ravenous form of eating. The second striking thing is that worrying, until the nineteenth century, is something one does to somebody or something else. In other words, at a certain point in history worrying became something that people could do to themselves. Using, appropriately enough, an analogy from hunting, worrying becomes a consuming, or rather self-consuming, passion. What was once thought of as animal becomes human, indeed all too human. What was once done by the mouths of the rapacious, the desirous, is now done, often with a relentless weariness, by the minds of the troubled.

It is not until the early nineteenth century, a time of significant social transformation, that we get the psychological sense of worrying as something that goes on inside someone, what the *Oxford English Dictionary* calls 'denoting a state of mind', giving as illustration a quotation from Hazlitt's *Table Talk*: 'Small pains are . . . more within our reach: we can fret and worry ourselves about them.' Domestic agitation replaces any sense of quest in Hazlitt's essay 'On Great and Little Things'. By the 1850s we find many of Dickens's characters worrying or 'worriting'. Where once wild or not-so-wild animals had worried their prey, we find Dickens's people worrying their lives away about love and money and social status. From, perhaps, the middle of the nineteenth century people began to prey on themselves in a new kind of way. Worry begins to catch on as a description of a new state of mind. It is now impossible to imagine a life without worry. In little more than a century worrying has become what we call a fact of life, as integral to our lives, as apparently ahistorical, as any of our most familar feelings. So in Philip Roth's recent fictional autobiography, *The Facts*, it is surprising to find the word made interesting again in the narrator's description of his hardworking Jewish father: 'Despite a raw emotional nature that makes him prey to intractable worry, his life has been distinguished by the power of resurgence.' The pun on *prey* suggests the devotion of that generation of American Jews to a new God. But the narrator also implies that his father's nature and history make him subject to his own persecution in the form of relentless worrying, and also that something about his life is reflected in the quality of his worry, its intractability, its obstinate persistence. A new kind of heroic resilience is required to deal with the worries of everyday life.

Source: Adam Phillips, *On Kissing, Tickling and Being Bored,* Cambridge, Mass., Harvard University Press, 1993.

Pictures for the Million

The camera obscura (or 'dark room') began to be used as a drawing-aid by Italian artists in the Renaissance. They found that light entering a room through a pinhole in a window-shutter formed an inverted image, on the opposite wall, of the scene outside. In 1568 a professor at Padua, Daniel Barbaro, substituted a lens for the pinhole, explaining that it produced a more brilliant image.

Close all the shutters and doors until no light enters the room except through the lens, and opposite hold a sheet of paper, which you move forward and backward until the scene appears in sharper detail. There on the paper you will see the whole view as it really is, with its distances, its colours and shadows and motion, the clouds, the waters twinkling, the birds flying. By holding the paper steady you can trace the whole perspective with a pen, shade it, and delicately colour it from nature.

The camera obscura was made portable by fitting the lens into one side of a closed box and covering the opposite side with frosted glass. By the eighteenth century portable camera obscuras were standard artists' equipment. All that is necessary to make a camera obscura into a camera is to put a sheet of light-sensitive material at the back of the box, which will fix the image. Owing to ignorance of chemistry, however, this step took nearly three centuries.

Around 1817 a French inventor Nicéphore Niepce found that he could fix the camera image using a plate coated in bitumen that hardened on exposure to light. After taking the plate from the camera he washed off the unhardened bitumen in oil of lavender, so developing the picture. A murky view of a farmyard, photographed (with an eight-hour exposure) from his upstairs window, still survives (now in Austin, Texas).

In Paris Niepce met the theatrical scene-painter Louis Daguerre, who shared his interest in the potential of photography. They became partners, and Daguerre continued to experiment independently after Niepce's death in 1829. To fix the camera image, he found that he could use a sheet of silver-plated copper, which he put over a box containing iodine. The iodine fumes reacted with the silver, forming light-sensitive silver iodide. When Daguerre exposed this in his camera, the light reduced the silver-oxide to silver, in proportion to its intensity. He then put the exposed plate over a box containing heated mercury, which gave off fumes that amalgamated with the silver, so that the image became visible. Finally he washed the plate in a strong solution of common salt, rendering the unexposed silver iodide insensitive to further light.

The earliest surviving 'daguerrotype' – a still-life of plaster casts and a wicker bottle – dates from 1837. The English country gentleman William Henry Fox Talbot had developed a similar process independently, and may have anticipated Daguerre. A photograph of a latticed window at his house, Lacock Abbey, dates from August 1835. 'This I believe to be the first instance on record of a house having painted its own portrait,' he claimed. However, he then laid aside photography to pursue classical studies.

Daguerre kept his process secret, but exhibited examples of his photographs. They caused a sensation. Samuel F. B. Morse, the American painter and inventor, who was in Paris at the time, invited Daguerre to a demonstration of his electric telegraph, and was invited, in return, to see some daguerrotypes. His reactions were published in the New York *Observer.*

The exquisite minuteness of the delineation cannot be conceived. No painting or engraving ever approached it. For example: in a view up the street, a distant sign would be perceived, and the eye could just discern that there were lines of letters upon it, but so minute as not to be read with the naked eye. By the assistance of a powerful lens, which magnified 50 times, applied to the delineation, every letter was clearly and distinctly legible, and so also were the minutest breaks and lines in the walls of the buildings, and the pavements of the streets. The effect of the lens upon the picture was in a great degree like that of the telescope in nature.

Objects moving are not impressed. The Boulevard, so constantly filled with the moving throng of pedestrians and carriages, was perfectly solitary, except for an individual who was having his boots brushed. His feet were compelled, of course, to be stationary for some time, one being on the box of the boot black, and the other on the ground. Consequently his boots and legs were well defined, but he is without body or head, because these were in motion.

Daguerre proposed to sell his secret to the highest bidder. In the event it was bought by the French state. The agreement was signed by Louis Philippe on 7 August 1839, granting Daguerre 6,000 francs a year for life. Twelve days later, at a special joint meeting of the Academy of Sciences and the Academy of Fine Arts, the technical details were made public. An eyewitness, Marc Antonine Gaudin, has left this account.

> The Palace of the Institute was stormed by a swarm of the curious at the memorable sitting on August 19, 1839, when the process was at long last divulged. Although I came two hours beforehand, like many others I was barred from the hall. I was on the watch with the crowd for everything that happened outside. At one moment an excited man comes out; he is surrounded, he is questioned, and he answers with a know-it-all air that bitumen of Judea and lavender oil is the secret. Questions are multiplied, but as he knows nothing more, we are reduced to talking about bitumen of Judea and lavender oil. Soon the crowd surrounds a newcomer, more startled than the last. He tells us with no further comment that it is iodine and mercury . . . Finally the sitting is over, the secret is divulged.
>
> A few days later, opticians' shops were crowded with amateurs panting for daguerrotype apparatus, and everywhere cameras were trained on buildings. Everyone wanted to record the view from his window, and he was lucky who at first trial got a silhouette of roof tops against the sky. He went into ecstasies over chimneys, counted over and over roof tiles and chimney bricks – in a word the technique was so new that even the poorest plate gave him indescribable joy.

Technical improvements followed, and portrait studios opened up all over the western world. By 1853 New York alone had 86 studios. The enormous demand for family pictures was due partly to the high nineteenth-century mortality rates, especially among children.

> Secure the shadow ere the substance fade,
> Let Nature imitate what Nature made

ran the advertising slogan. An important advance came in the late 1870s with the introduction of highly light-sensitive gelatin emulsion, which allowed fraction-of-a-second exposures and made action photographs possible. The pioneer was the Surrey-born American immigrant Eadward Muybridge (a name adopted in the belief that it was the Anglo-Saxon original of his real name, Edward James Muggeridge). He invented one of the first high-speed shutters, and took a series of photos of a galloping horse by arranging a row of cameras beside the race track, operated by tapes which the horse ran through. This series, published in England and America in 1879, caused consternation, since it showed that all previous ideas of how a horse moves were incorrect. Muybridge's photos proved that the horse's feet are all off the ground at once at one stage in the gallop, but only when they are bunched together under the belly. None of the photos showed the 'hobbyhorse attitude', with front legs stretched forwards and hind legs stretched back, traditional in painting. In 1880, using a device he named the zoogyroscope or zoopraxiscope, Muybridge projected his horse pictures in quick succession on to a screen at the California School of Fine Arts, thus inaugurating motion pictures.

In the 1880s hand-held box cameras became common, the best-known being the Kodak, invented by George Eastman, a New York photographic-plate maker. The name, he explained, was quite new, not derived from any existing word. He had chosen it because it was short, memorable and 'incapable of being misspelled so as to destroy its identity'. This consideration, vital in New York's polyglot immigrant population, points towards the twentieth-century transition from verbal languages to the universal language of pictures. The cheap, ready-loaded Kodak brought photography, said Eastman:

> within the reach of every human being who desires to preserve a record of what he sees. Such a photographic notebook is an enduring record of many things seen only once in a lifetime, and enables the fortunate possessor to go back, by the light of his own fireside, to scenes which would otherwise fade from memory and be lost.

Source: *The History of Photography: From 1839 to the Present*, completely revised and enlarged, by Beaumont Newhall, London, Secker & Warburg, 1982. The Barbaro, Gaudin and Eastman quotes are from Newhall, the Barbaro and Gaudin being (presumably) translated by him. The Morse is also quoted by Newhall, but is from the New York *Observer* for 1839.

The Battle of the Ants

On 4 July 1845, the American poet and essayist Henry David Thoreau (1817–62) moved to a cabin he had built himself beside Walden Pond near Concord, Massachusetts. He lived alone there for two years, thinking, observing nature, and growing his own food (beans). His masterpiece *Walden; or Life in the Woods* (1854) from which this extract is taken, records his solitary happiness. The Battle of Concord (1775), to which he refers, naming some of the combatants, was one of the opening actions of the American War of Independence.

One day when I went to my wood-pile, or rather my pile of stumps, I observed two large ants, the one red, the other much larger, nearly half an inch long, and black, fiercely contending with one another. Having once got hold they never let go, but struggled and wrestled and rolled on the chips incessantly. Looking farther, I was surprised to find that the chips were covered with such combatants, that it was not a *duellum*, but a *bellum*, a war between two races of ants, the red always pitted against the black, and frequently two red ones to one black. The legions of these Myrmidons covered all the hills and vales in my wood-yard, and the ground was already strewn with the dead and dying, both red and black. It was the only battle which I have ever witnessed, the only battle-field I ever trod while the battle was raging; internecine war; the red republicans on the one hand, and the black imperialists on the other. On every side they were engaged in deadly combat, yet without any noise that I could hear, and human soldiers never fought so resolutely. I watched a couple that were fast locked in each other's embraces, in a little sunny valley amid the chips, now at noon-day prepared to fight till the sun went down, or life went out. The smaller red champion had fastened himself like a vice to his adversary's front, and through all the tumblings on that field never for an instant ceased to gnaw at one of his feelers near the root, having already caused the other to go by the board; while the stronger black one dashed him from side to side, and as I saw on looking nearer, had already divested him of several of his members. They fought with more pertinacity than bull-dogs. Neither manifested the least disposition to retreat. It was evident that their battle-cry was Conquer or die. In the meanwhile there came along a single red ant on the hill-side of this valley, evidently full of excitement, who either had despatched his foe, or had not yet taken part in the battle; probably the latter, for he had lost none of his limbs; whose mother had charged him to return with his shield or upon it. Or perchance he was some Achilles, who had nourished his wrath apart, and had now come to avenge or rescue his Patroclus. He saw this unequal combat from afar – for the blacks were nearly twice the size of the red – he drew near with rapid pace till he stood on his guard within half an inch of the combatants; then, watching his opportunity, he sprang upon the black warrior, and commenced his operations near the root of his right fore-leg, leaving the foe to select among his own members; and so there were three united for life, as if a new kind of attraction had been invented which put all other locks and cements to shame. I should not have wondered by this time to find that they had their respective musical bands stationed on some eminent chip, and playing their national airs the while, to excite the slow and cheer the dying combatants. I was myself excited somewhat even as if they had been men. The more you think of it, the less the difference. And certainly there is not a fight recorded in Concord history, at least, if in the history of America, that will bear a moment's comparison with this, whether for the numbers engaged in it, or for the patriotism and heroism displayed. For numbers and for carnage it was an Austerlitz or Dresden. Concord Fight! Two killed on the patriots' side, and Luther Blanchard wounded! Why here every ant was a Buttrick, – 'Fire! for God's sake, fire!' – and thousands shared the fate of Davis and Hosmer. There was not one hireling there. I have no doubt that it was a principle they fought for, as much as our ancestors, and not to avoid a threepenny tax on their tea; and the results of this battle will be as important and memorable to those whom it concerns as those of the battle of Bunker Hill, at least.

I took up the chip on which the three I have particularly described were struggling, carried it into my house, and placed it under a tumbler on my window-sill, in order to see the issue. Holding a microscope to the first-mentioned red ant, I saw that, though he was assiduously gnawing at the near fore-leg of his enemy, having severed his remaining feeler, his own breast was all torn away, exposing what vitals he had there to the jaws of the black warrior, whose breast-plate was apparently too thick for him to pierce; and the dark carbuncles of the sufferer's eyes shone with ferocity such as war only could excite. They struggled half-an-hour longer under the tumbler, and when I looked again the black soldier had severed the heads of his foes from their bodies, and the still living heads were hanging on either side of him like ghastly trophies at his saddle-bow, still apparently as firmly fastened as ever, and he was endeavouring with feeble struggles, being without feelers and with only the remnant of a leg, and I know not how many other wounds, to divest himself of them; which at length, after half-an-hour more, he accomplished. I raised the glass, and he went off over the window-sill in that crippled state. Whether he finally survived that combat, and spent the remainder of his days in some Hotel des Invalides, I do not know; but I thought that his industry would not be worth much thereafter. I never learned which party was victorious, nor the cause of the war: but I felt for the rest of that day as if I had had my feelings excited and harrowed by witnessing the struggle, the ferocity and carnage, of a human battle before my door.

Source: Henry David Thoreau, *Walden; or, Life in the Woods*, Boston, Mass. Ticknor and Fields, 1854.

On a Candle

Michael Faraday (1791–1867) was a blacksmith's son who taught himself science after training as a bookbinder, and was apprenticed, aged 21, to Sir Humphry Davy. A pioneer of electromagnetism, he constructed the first electric motor and the first dynamo. He started the Christmas lectures for children at the Royal Institution in Albemarle Street, and his famous series 'On the Chemical History of a Candle' was first given in 1849. A model of lecturing technique, the series included experiments that produced bangs, flashes, soap bubbles filled with hydrogen floating roofwards, and other spectacular effects. To illustrate the expansion of water when frozen, for example, Faraday placed two vessels made of half-inch-thick iron, and filled with water, in a freezing solution, then went on lecturing until the vessels exploded.

Faraday aimed to show that 'there is not a law under which any part of this universe is governed' which the burning of a candle, and the simple experiments leading from it, could not illustrate. Charles Dickens, keen, like Faraday, to bring science to a wider audience, borrowed the notes for the lectures, and in 1850 his family magazine *Household Words* carried a semi-fictionalized version of them in which a rather priggish nephew, who has attended Faraday's course, explains the wonders of chemistry to his appreciative uncle.

This extract represents the climax of the course – the conclusion of the sixth and last lecture.

Now I must take you to a very interesting part of our subject – to the relation between the combustion of a candle and that living kind of combustion which goes on within us. In every one of us there is a living process of combustion going on very similar to that of a candle, and I must try to make that plain to you. For it is not merely true in a poetical sense – the relation of the life of man to a taper; and if you will follow, I think I can make this clear . . .

We consume food: the food goes through that strange set of vessels and organs within us, and is brought into various parts of the system, into the digestive parts especially; and alternately the portion which is so changed is carried through our lungs by one set of vessels, while the air that we inhale and exhale is drawn into and thrown out of the lungs by another set of vessels, so that the air and the food come close together, separated only by an exceedingly thin surface: the air can thus act upon the blood by this process, producing precisely the same results in kind as we have seen in the case of the candle. The candle combines with parts of the air, forming carbonic acid, and evolves heat; so in the lungs there is this curious, wonderful change taking place. The air entering, combines with the carbon (not carbon in a free state, but, as in this case, placed ready for action at the moment), and makes carbonic acid, and is so thrown out into the atmosphere, and thus this singular result takes place; we may thus look upon the food as fuel. Let me take that piece of sugar, which will serve my purpose. It is a compound of carbon, hydrogen, and oxygen, similar to a candle, as containing the same elements, though not in the same proportion; the proportions being as shown in this table:

SUGAR

Carbon	72	
Hydrogen	11	99
Oxygen	88	

This is, indeed, a very curious thing, which you can well remember, for the oxygen and hydrogen are in exactly the proportions which form water, so that sugar may be said to be compounded of 72 parts of carbon and 99 parts of water; and it is the carbon in the sugar that combines with the oxygen carried in by the air in the process of respiration, so making us like candles; producing these actions, warmth, and far more wonderful results besides, for the sustenance of the system, by a most beautiful and simple process. To make this still more striking, I will take a little sugar; or to hasten the experiment I will use some syrup, which contains about three-fourths of sugar and a little water. If I put a little oil of vitriol on it, it takes away the water, and leaves the carbon in a black mass. [The Lecturer mixed the two together.] You see how the carbon is coming out, and before long we shall have a solid mass of charcoal, all of which has come out of sugar. Sugar, as you know, is food, and here we have absolutely a solid lump of carbon where you would not have expected it. And if I make arrangements so as to oxidize the carbon of sugar, we shall have a much more striking result. Here is sugar, and I have here an oxidizer –

a quicker one than the atmosphere; and so we shall oxidize this fuel by a process different from respiration in its form, though not different in its kind. It is the combustion of the carbon by the contact of oxygen which the body has supplied to it. If I set this into action at once, you will see combustion produced. Just what occurs in my lungs – taking in oxygen from another source, namely, the atmosphere, takes place here by a more rapid process.

You will be astonished when I tell you what this curious play of carbon amounts to. A candle will burn some four, five, six, or seven hours. What then must be the daily amount of carbon going up into the air in the way of carbonic acid! What a quantity of carbon must go from each of us in respiration! What a wonderful change of carbon must take place under these circumstances of combustion or respiration! A man in twenty-four hours converts as much as seven ounces of carbon into carbonic acid; a milch cow will convert seventy ounces, and a horse seventy-nine ounces, solely by the act of respiration. That is, the horse in twenty-four hours burns seventy-nine ounces of charcoal, or carbon, in his organs of respiration to supply his natural warmth in that time. All the warm-blooded animals get their warmth in this way, by the conversion of carbon, not in a free state, but in a state of combustion. And what an extraordinary notion this gives us of the alterations going on in our atmosphere. As much as 5,000,000 pounds, or 548 tons, of carbonic acid is formed by respiration in London alone in twenty-four hours. And where does all this go? Up into the air. If the carbon had been like the lead which I showed you, or the iron which, in burning, produces a solid substance, what would happen? Combustion could not go on. As charcoal burns it becomes a vapour and passes off into the atmosphere, which is the great vehicle, the great carrier for conveying it away to other places. Then what becomes of it? Wonderful is it to find that the change produced by respiration, which seems so injurious to us (for we cannot breathe air twice over), is the very life and support of plants and vegetables that grow upon the surface of the earth. It is the same also under the surface in the great bodies of water; for fishes and other animals respire upon the same principle, though not exactly by contact with the open air.

Such fish as I have here [pointing to a globe of gold-fish] respire by the oxygen which is dissolved from the air by the water, and from carbonic acid, and they all move about to produce the one great work of making the animal and vegetable kingdoms subservient to each other. And all the plants growing upon the surface of the earth, like that which I have brought here to serve as an illustration, absorb carbon; these leaves are taking up their carbon from the atmosphere to which we have given it in the form of carbonic acid, and they are growing and prospering. Give them a pure air like ours, and they could not live in it; give them carbon with other matters, and they live and rejoice. This piece of wood gets all its carbon, as the trees and plants get theirs, from the atmosphere, which, as we have seen, carries away what is bad for us and at the same time good for them – what is disease to the one being health to the other. So are we made dependent not merely upon our fellow-creatures, but upon our fellow-existers, all Nature being tied together by the laws that make one part conduce to the good of another.

There is another little point which I must mention before we draw to a close – a point which concerns the whole of these operations, and most curious and beautiful it is to see it clustering upon and associated with the bodies that concern us – oxygen, hydrogen, and carbon, in different states of their existence. I showed you just now some powdered lead, which I set burning; and you saw that the moment the fuel was brought to the air it acted, even before it got out of the bottle – the moment the air crept in it acted. Now, there is a case of chemical affinity by which all our operations proceed. When we breathe, the same operation is going on within us. When we burn a candle, the attraction of the different parts one to the other is going on. Here it is going on in this case of the lead, and it is a beautiful instance of chemical affinity. If the products of combustion rose off from the surface, the lead would take fire, and go on burning to the end; but you remember that we have this difference between charcoal and lead – that, while the lead can start into action at once if there be access of air to it, the carbon will remain days, weeks, months, or years. The manuscripts of Herculaneum were written with carbonaceous ink, and there they have been for 1,800 years or more, not having been at all changed by the atmosphere, though coming in contact with it under various circumstances. Now, what is the circumstance which makes the lead and carbon differ in this respect? It is a striking thing to see that the matter which is appointed to serve the purpose of fuel *waits* in its action; it does not start off burning, like the lead and many other things that I could show you, but which I have not encumbered the table with; but it waits for action. This waiting is a curious and wonderful thing. Candles – those Japanese candles, for instance – do not start into action at once like the lead or iron (for iron finely divided does the same thing as lead), but there they wait for years, perhaps for ages, without undergoing any alteration. I have here a supply of coal-gas. The jet is giving forth the gas, but you see it does not take fire – it comes out into the air, but it waits till it is hot enough before it burns. If I make it hot enough, it takes fire. If I blow it out, the gas that is issuing forth waits till the light is applied to it again. It is curious to see how different substances wait – how some will wait till the temperature is raised a little, and others till it is raised a good deal. I have here a little gunpowder and some gun-cotton; even these things differ in the conditions under which they will burn. The gunpowder is composed of carbon and other substances, making it highly combustible; and the gun-cotton is another combustible preparation. They are both waiting, but they will start into activity at different degrees of heat, or under different conditions. By applying a heated wire to them, we shall see which will start first [touching the gun-cotton with the hot iron]. You see the gun-cotton has gone off, but not even the hottest part of the wire is now hot enough to fire the gunpowder. How beautifully that shows you the difference in the degree in which bodies act in this way! In the one case the substance will wait any time until the associated bodies are made active by heat; but, in the other, as in the process of respiration, it waits no time. In the lungs, as soon as the air enters, it unites with the carbon; even in the lowest temperature which the body can bear short of being frozen, the action begins at once, producing the carbonic acid of respiration; and so all things go on fitly and properly. Thus you see the analogy between respiration and combustion is rendered still more beautiful and striking. Indeed, all I can say to you at the end of these lectures (for we must come to an end at one time or other) is to express a wish that you may, in your generation, be fit to compare to a candle; that you may, like it, shine as lights to those about you; that, in all your actions, you may justify the beauty of the taper by making your deeds honourable and effectual in the discharge of your duty to your fellow-men.

Source: Michael Faraday, *A Course of Six Lectures on the Chemical History of a Candle*, London, Chatto & Windus, 1861.

Heat Death

The concept of entropy (the dissipation of available energy) was the brainchild of the German physicist Rudolf Clausius (1822–88) who, in 1850, formulated the second law of thermodynamics (that heat cannot of itself pass from a colder to a hotter body). In any closed system, Clausius pointed out, there is an inevitable waste or loss of energy (entropy), and the amount of energy available for work will decrease as entropy increases. Regarding the universe as a closed system, he predicted that its entropy would eventually be maximized, i.e. that it would run down, reaching a state of equilibrium and uniform temperature, with no further energy available for doing work – a condition known as 'the heat death of the universe'. The American novelist and poet John Updike (b.1932) protests against this conclusion in his 'Ode to Entropy':

Some day – can it be believed? –
in the year 10^{70} or so,
single electrons and positrons will orbit
one another to form atoms bonded
across regions of space
greater than the present observable universe.
'Heat death' will prevail.
The stars long since will have burnt their hydrogen
and turned to iron.
Even the black holes will have decayed.
Entropy!
thou seal on extinction,
thou curse on Creation.
All change distributes energy,
spills what cannot be gathered again.
Each meal, each smile,
each foot-race to the well by Jack and Jill
scatters treasure, lets fall

gold straws once woven from the resurgent dust.
The night sky blazes with Byzantine waste.
The bird's throbbling is expenditure,
and the tide's soughing,
and the tungsten filament illumining my hand.

A ramp has been built into probability
the universe cannot re-ascend.
For our small span,
the sun has fuel, the moon lifts the lulling sea,
the highway shudders with stolen hydrocarbons.
How measure these inequalities
so massive and luminous
in which one's self is secreted
like a jewel mislaid in mountains of garbage?
Or like that bright infant Prince William,
with his whorled nostrils and blank blue eyes,
to whom empire and all its estates are already assigned.
Does its final diffusion
deny a miracle?
Those future voids are scrims of the mind,
pedagogic as blackboards.

Did you know
that four-fifths of the body's intake goes merely
to maintain our temperature of 98.6°?
Or that Karl Barth, addressing prisoners, said
the prayer for stronger faith is the one prayer
that has never been denied?
Death exists nowhere in nature, not
in the minds of birds or the consciousness of flowers,
not even in the numb brain of the wildebeest calf
gone under to the grinning crocodile, nowhere
in the mesh of woods or the tons of sea, only
in our forebodings, our formulae.
There is still enough energy in one overlooked star
to power all the heavens madmen have ever proposed.

Source: John Updike, *Facing Nature: Poems*, London, André Deutsch, 1986.

Adam's Navel

Philip Henry Gosse, the man who said God had put fossils in the rocks to deceive geologists, is a laughing stock of popular science. His reputation is defended here by one of the foremost modern science writers, Stephen Jay Gould. A research biologist, Gould has devoted many years to the study of the Bahamian land snail *Cerion*. 'I love *Cerion*', he has declared, 'with all my heart and intellect.' However, it is the clarity and originality of his scientific writing that has brought him the widest fame. Four collections of his monthly columns in *Natural History* magazine have been published. This is from the fourth, *Hen's Teeth and Horse's Toes* (1990).

The ample fig leaf served our artistic forefathers well as a botanical shield against indecent exposure for Adam and Eve, our naked parents in the primeval bliss and innocence of Eden. Yet, in many ancient paintings, foliage hides more than Adam's genitalia; a wandering vine covers his navel as well. If modesty enjoined the genital shroud, a very different motive – mystery – placed a plant over his belly. In a theological debate more portentous than the old argument about angels on pinpoints, many earnest people of faith had wondered whether Adam had a navel.

He was, after all, not born of a woman and required no remnant of his nonexistent umbilical cord. Yet, in creating a prototype, would not God make his first man like all the rest to follow? Would God, in other words, not create with the appearance of preexistence? In the absence of definite guidance to resolve this vexatious issue, and not wishing to incur anyone's wrath, many painters literally hedged and covered Adam's belly.

A few centuries later, as the nascent science of geology gathered evidence for the earth's enormous antiquity, some advocates of biblical literalism revived this old argument for our entire planet. The strata and their entombed fossils surely seem to represent a sequential record of countless years, but wouldn't God create his earth with the appearance of preexistence? Why should we not believe that he created strata and fossils to give modern life a harmonious order by granting it a sensible (if illusory) past? As God provided Adam with a navel to stress continuity with future men, so too did he endow a pristine world with the appearance of an ordered history. Thus, the earth might be but a few thousand years old, as Genesis literally affirmed, and still record an apparent tale of untold eons.

This argument, so often cited as a premier example of reason at its most perfectly and preciously ridiculous, was most seriously and comprehensively set forth by the British naturalist Philip Henry Gosse in 1857. Gosse paid proper homage to historical context in choosing a title for his volume. He named it *Omphalos* (Greek for navel), in Adam's honor, and added as a subtitle: *An Attempt to Untie the Geological Knot.*

Since *Omphalos* is such spectacular nonsense, readers may rightly ask why I choose to discuss it at all. I do so, first of all, because its author was such a serious and fascinating man, not a hopeless crank or malcontent. Any honest passion merits our attention, if only for the oldest of stated reasons – Terence's celebrated *Homo sum: humani nihil a me alienum puto* (I am human, and am therefore indifferent to nothing done by humans).

Philip Henry Gosse (1810–88) was the David Attenborough of his day, Britain's finest popular narrator of nature's fascination. He wrote a dozen books on plants and animals, lectured widely to popular audiences, and published several technical papers on marine invertebrates. He was also, in an age given to strong religious feeling as a mode for expressing human passions denied vent elsewhere, an extreme and committed fundamentalist of the Plymouth Brethren sect. Although his *History of the British Sea-Anemones* and other assorted ramblings in natural history are no longer read, Gosse retains some notoriety as the elder figure in that classical work of late Victorian self-analysis and personal exposé, his son Edmund's wonderful account of a young boy's struggle against a crushing religious extremism imposed by a caring and beloved parent – *Father and Son.*

My second reason for considering *Omphalos* invokes the same theme surrounding so many of these essays about nature's small oddities: Exceptions do prove rules (prove, that is, in the sense of probe or test, not affirm). If you want to understand what ordinary folks do, one thoughtful deviant will teach you more than ten thousand solid citizens. When we grasp why *Omphalos* is so unacceptable (and not, by the way, for the reason usually cited), we will understand better how science and useful logic proceed. In any case, as an exercise in the anthropology of knowledge, *Omphalos* has no parallel – for its surpassing strangeness arose in the mind of a stolid Englishman, whose general character and cultural setting we can grasp as akin to our own, while the exotic systems of alien cultures are terra incognita both for their content and their context.

To understand *Omphalos*, we must begin with a paradox. The argument that strata and fossils were created all at once with the earth, and only present an illusion of elapsed time, might be easier to appreciate if its author had been an urban armchair theologian with no feelings or affection for nature's works. But how could a keen naturalist, who had spent days, nay months, on geological excursions, and who had studied fossils hour after hour, learning their distinctions and memorizing their names, possibly be content with the prospect that these objects of his devoted attention had never existed – were, indeed, a kind of grand joke perpetrated upon us by the Lord of All?

Philip Henry Gosse was the finest descriptive naturalist of his day. His son wrote: 'As a collector of facts and marshaller of observations, he had not a rival in that age.' The problem lies with the usual caricature of *Omphalos* as an argument that God, in fashioning the earth, had consciously and elaborately lied either to test our faith or simply to indulge in some inscrutable fit of arcane humor. Gosse, so fiercely committed both to his fossils and his God, advanced an opposing interpretation that commanded us to study geology with diligence and to respect all its facts even though they had no existence in real time. When we understand why a dedicated empiricist could embrace the argument of *Omphalos* ('creation with the appearance of preexistence'), only then can we understand its deeper fallacies.

Gosse began his argument with a central, but dubious, premise: All natural processes, he declared, move endlessly round in a circle: egg to chicken to egg, oak to acorn to oak.

> This, then, is the order of all organic nature. When once we are in any portion of the course, we find ourselves running in a circular groove, as endless as the course of a blind horse in a mill . . . [In premechanized mills, horses wore blinders or, sad to say, were actually blinded, so that they would continue to walk a circular

> course and not attempt to move straight forward, as horses relying on visual cues tend to do.] This is not the law of some particular species, but of all: it pervades all classes of animals, all classes of plants, from the queenly palm down to the protococcus, from the monad up to man: the life of every organic being is whirling in a ceaseless circle, to which one knows not how to assign any commencement. . . . The cow is as inevitable as a sequence of the embryo, as the embryo is of the cow.

When God creates, and Gosse entertained not the slightest doubt that all species arose by divine fiat with no subsequent evolution, he must break (or 'erupt,' as Gosse wrote) somewhere into this ideal circle. Wherever God enters the circle (or 'places his wafer of creation,' as Gosse stated in metaphor), his initial product must bear traces of previous stages in the circle, even if these stages had no existence in real time. If God chooses to create humans as adults, their hair and nails (not to mention their navels) testify to previous growth that never occurred. Even if he decides to create us as a simple fertilized ovum, this initial form implies a phantom mother's womb and two nonexistent parents to pass along the fruit of inheritance.

> Creation can be nothing else than a series of irruptions into circles . . . Supposing the irruption to have been made at what part of the circle we please, and varying this condition indefinitely at will, – we cannot avoid the conclusion that each organism was from the first marked with the records of a previous being. But since creation and previous history are inconsistent with each other; as the very idea of the creation of an organism excludes the idea of pre-existence of that organism, or of any part of it; it follows, that such records are false, so far as they testify to time.

Gosse then invented a terminology to contrast the two parts of a circle before and after an act of creation. He labeled as 'prochronic,' or occurring outside of time, those appearances of preexistence actually fashioned by God at the moment of creation but seeming to mark earlier stages in the circle of life. Subsequent events occurring after creation, and unfolding in conventional time, he called 'diachronic.' Adam's navel was prochronic, the 930 years of his earthly life diachronic.

Gosse devoted more than 300 pages, some 90 per cent of his text, to a simple list of examples for the following small part of his complete argument – if species arise by sudden creation at any point in their life cycle, their initial form must present illusory (prochronic) appearances of preexistence. Let me choose just one among his numerous illustrations, both to characterize his style of argument and to present his gloriously purple prose. If God created vertebrates as adults, Gosse claimed, their teeth imply a prochronic past in patterns of wear and replacement.

Gosse leads us on an imaginary tour of life just an hour after its creation in the wilderness. He pauses at the sea-shore and scans the distant waves:

> I see yonder a . . . terrific tyrant of the sea . . . It is the grisly shark. How stealthily he glides along. . . . Let us go and look into his mouth. . . . Is not this an awful array of knives and lancets? Is not this a case of surgical instruments enough to make you shudder? What would be the amputation of your leg to this row of triangular scalpels?

Yet the teeth grow in spirals, one behind the next, each waiting to take its turn as those in current use wear down and drop out:

> It follows, therefore, that the teeth which we now see erect and threatening, are the successors of former ones that have passed away, and that they were once dormant like those we see behind them. . . . Hence we are compelled by the phenomena to infer a long past existence to this animal, which yet has been called into being within an hour.

Should we try to argue that teeth in current use are the first members of their spiral, implying no predecessors after all, Gosse replies that their state of wear indicates a prochronic past. Should we propose that these initial teeth might be unmarred in a newly created shark, Gosse moves on to another example.

> Away to a broader river. Here wallows and riots the huge hippopotamus. What can we make of his dentition?

All modern adult hippos possess strongly worn and beveled canines and incisors, a clear sign of active use throughout a long life. May we not, however, as for our shark, argue that a newly created hippo might have sharp and pristine front teeth? Gosse argues correctly that no hippo could work properly with teeth in such a state. A created adult hippo must contain worn teeth as witnesses of a prochronic past:

> The polished surfaces of the teeth, worn away by mutual action, afford striking evidence of the lapse of time. Some one may possibly object . . . 'What right have you to assume that these teeth were worn away at the moment of its creation, admitting the animal to have been created adult. May they not have been entire?' I reply, Impossible: the Hippopotamus's teeth would have been perfectly useless to him, except in the ground-down condition: nay, the unworn canines would have effectually prevented his jaws from closing, necessitating the keeping of the mouth wide open until the attrition was performed; long before which, of course, he would have starved. . . . The degree of attrition is merely a question of time. . . . How distinct an evidence of past action, and yet, in the case of the created individual, how illusory!

This could go on forever (it nearly does in the book), but just one more dental example. Gosse, continuing upward on the topographic trajectory of his imaginary journey, reaches an inland wood and meets *Babirussa*, the famous Asian pig with upper canines growing out and arching back, almost piercing the skull:

> In the thickets of this nutmeg grove beside us there is a Babiroussa; let us examine him. Here he is, almost submerged in this tepid pool. Gentle swine with the circular tusk, please to open your pretty mouth!

The pig, created by God but an hour ago, obliges, thus displaying his worn molars and, particularly, the arching canines themselves, a product of long and continuous growth.

I find this part of Gosse's argument quite satisfactory as a solution, within the boundaries of his assumptions, to that classical dilemma of reasoning (comparable in importance to angels on pinpoints and Adam's navel): 'Which came first, the chicken or the egg?' Gosse's answer: 'Either, at God's pleasure, with prochronic traces of the other.' But arguments are only as good as their premises, and Gosse's inspired nonsense fails because an alternative assumption, now accepted as undoubtedly correct, renders the question irrelevant – namely, evolution itself. Gosse's circles do not spin around eternally; each life cycle traces an ancestry back to inorganic chemicals in a primeval ocean. If organisms arose by acts of creation *ab nihilo*, then Gosse's argument about prochronic traces must be respected. But if organisms evolved to their current state, *Omphalos* collapses to massive irrelevance. Gosse understood this threat perfectly well and chose to meet it by abrupt dismissal. Evolution, he allowed, discredited his system, but only a fool could accept such patent nonsense and idolatry (Gosse wrote *Omphalos* two years before Darwin published the *Origin of Species*).

> If any choose to maintain, as many do, that species were gradually brought to their present maturity from humbler forms . . . he is welcome to his hypothesis, but I have nothing to do with it. These pages will not touch him.

But Gosse then faced a second and larger difficulty: the prochronic argument may work for organisms and their life cycles, but how can it be applied to the entire earth and its fossil record – for Gosse intended *Omphalos* as a treatise to reconcile the earth with biblical chronology, 'an attempt to untie the geological knot.' His statements about prochronic parts in organisms are only meant as collateral support for the primary geological argument. And Gosse's geological claim fails precisely because it rests upon such dubious analogy with what he recognized (since he gave it so much more space) as a much stronger argument about modern organisms.

Gosse tried valiantly to advance for the entire earth the same two premises that made his prochronic argument work for organisms. But an unwilling world rebelled against such forced reasoning and *Omphalos* collapsed under its own weight of illogic. Gosse first tried to argue that all geological processes, like organic life cycles, move in circles:

> The problem, then, to be solved before we can certainly determine the question of analogy between the globe and the organism, is this – Is the life-history of the globe a cycle? If it is (and there are many reasons why this is probable), then I am sure prochronism must have been evident at its creation, since there is no point in a circle which does not imply previous points.

But Gosse could never document any inevitable geological cyclicity, and his argument drowned in a sea of rhetoric and biblical allusion from Ecclesiastes: 'All the rivers run into the sea; yet the sea *is* not full.

Unto the place from whence the rivers come, thither they return again.'

Secondly, to make fossils prochronic, Gosse had to establish an analogy so riddled with holes that it would make the most ardent mental tester shudder – embryo is to adult as fossil is to modern organism. One might admit that chickens require previous eggs, but why should a modern reptile (especially for an antievolutionist like Gosse) be necessarily linked to a previous dinosaur as part of a cosmic cycle? A python surely does not imply the ineluctable entombment of an illusory *Triceratops* into prochronic strata.

With this epitome of Gosse's argument, we can resolve the paradox posed at the outset. Gosse could accept strata and fossils as illusory and still advocate their study because he did not regard the prochronic part of a cycle as any less 'true' or informative than its conventional diachronic segment. God decreed two kinds of existence – one constructed all at once with the appearance of elapsed time, the other progressing sequentially. Both dovetail harmoniously to form uninterrupted circles that, in their order and majesty, give us insight into God's thoughts and plans.

The prochronic part is neither a joke nor a test of faith; it represents God's obedience to his own logic, given his decision to order creation in circles. As thoughts in God's mind, solidified in stone by creation *ab nihilo*, strata and fossils are just as true as if they recorded the products of conventional time. A geologist should study them with as much care and zeal, for we learn God's ways from both his prochronic and his diachronic objects. The geological time scale is no more meaningful as a yardstick than as a map of God's thoughts.

> The acceptance of the principles presented in this volume . . . would not, in the least degree, affect the study of scientific geology. The character and order of the strata; . . . the successive floras and faunas; and all the other phenomena, would be facts still. They would still be, as now, legitimate subjects of examination and inquiry . . . We might still speak of the inconceivably long duration of the processes in question, provided we understand ideal instead of actual time – that the duration was projected in the mind of God, and not really existent.

Thus, Gosse offered *Omphalos* to practicing scientists as a helpful resolution of potential religious conflicts, not a challenge to their procedures or the relevance of their information.

His son Edmund wrote of the great hopes that Gosse held for *Omphalos:*

> Never was a book cast upon the waters with greater anticipations of success than was this curious, this obstinate, this fanatical volume. My father lived in a fever of suspense, waiting for the tremendous issue. This '*Omphalos*' of his, he thought, was to bring all the turmoil of scientific speculation to a close, fling geology into the arms of Scripture, and make the lion eat grass with the lamb.

Yet readers greeted *Omphalos* with disbelief, ridicule, or worse, stunned silence. Edmund Gosse continued:

> He offered it, with a glowing gesture, to atheists and Christians alike. This was to be the universal panacea; this the system of intellectual therapeutics which could not but heal all the maladies of the age. But, alas! atheists and Christians alike looked at it and laughed, and threw it away.

Although Gosse reconciled himself to a God who would create such a minutely detailed, illusory past, this notion was anathema to most of his countrymen. The British are a practical, empirical people, 'a nation of shopkeepers' in Adam Smith's famous phrase; they tend to respect the facts of nature at face value and rarely favor the complex systems of nonobvious interpretation so popular in much of continental thought. Prochronism was simply too much to swallow. The Reverend Charles Kingsley, an intellectual leader of unquestionable devotion to both God and science, spoke for a consensus in stating that he could not 'give up the painful and slow conclusion of five and twenty years' study of geology, and believe that God has written on the rocks one enormous and superfluous lie.'

And so it has gone for the argument of *Omphalos* ever since. Gosse did not invent it, and a few creationists ever since have revived it from time to time. But it has never been welcome or popular because it violates our intuitive notion of divine benevolence as free of devious behavior – for while Gosse saw divine brilliance in the idea of prochronism, most people cannot shuck their seat-of-the-pants feeling that it smacks of plain old unfairness. Our modern American creationists reject it vehemently as imputing a dubious moral character to God and opt instead for the even more ridiculous notion that our miles of fossiliferous strata are all products of Noah's flood and can therefore be telescoped into the literal time scale of Genesis.

But what is so desperately wrong with *Omphalos*? Only this really (and perhaps paradoxically): that we can devise no way to find out whether it is wrong – or, for that matter, right. *Omphalos* is the classical example of an utterly untestable notion, for the world will look exactly the same in all its intricate detail whether fossils and strata are prochronic or products of an extended history. When we realize that *Omphalos* must be rejected for this methodological absurdity, not for any demonstrated factual inaccuracy, then we will understand science as a way of knowing, and *Omphalos* will serve its purpose as an intellectual foil or prod.

Science is a procedure for testing and rejecting hypotheses, not a compendium of certain knowledge. Claims that can be proved incorrect lie within its domain (as false statements to be sure, but as proposals that meet the primary methodological criterion of testability). But theories that cannot be tested in principle are not part of science. Science is doing, not clever cogitation; we reject *Omphalos* as useless, not wrong.

Gosse's deep error lay in his failure to appreciate this essential character of scientific reasoning. He hammered his own coffin nails by continually emphasizing that *Omphalos* made no practical difference – that the world would look exactly the same with a prochronic or diachronic past. (Gosse thought that this admission would make his argument acceptable to conventional geologists; he never realized that it could only lead them to reject his entire scheme as irrelevant.) 'I do not know,' he wrote, 'that a single conclusion, now accepted, would need to be given up, except that of actual chronology.'

Gosse emphasized that we cannot know where God placed his wafer of creation into the cosmic circle because prochronic objects, created *ab nihilo*, look exactly like diachronic products of actual time. To those who argued that coprolites (fossil excrement) prove the existence of active, feeding animals in a real geological past, Gosse replied that as God would create adults with feces in their intestines, so too would he place petrified turds into his created strata. (I am not making up this example for comic effect; you will find it on page 353 of *Omphalos*.) Thus, with these words, Gosse sealed his fate and placed himself outside the pale of science:

> Now, again I repeat, there is no imaginable difference to sense between the prochronic and the diachronic development. Every argument by which the physiologist can prove to demonstration that yonder cow was once a foetus in the uterus of its dam, will apply with exactly the same power to show that the newly created cow was an embryo, some years before its creation . . . There is, and can be, nothing in the phenomena to indicate a commencement there, any more than anywhere else, or indeed, anywhere at all. The commencement, as a fact, I must learn from testimony; I have no means whatever of inferring it from phenomena.

Gosse was emotionally crushed by the failure of *Omphalos*. During the long winter evenings of his discontent, in the January cold of 1858, he sat by the fire with his eight-year-old son, trying to ward off bitter thoughts by discussing the grisly details of past and current murders. Young Edmund heard of Mrs Manning, who buried her victim in quicklime and was hanged in black satin; of Burke and Hare, the Scottish ghouls; and of the 'carpetbag mystery,' a sackful of neatly butchered human parts hung from a pier on Waterloo Bridge. This may not have been the most appropriate subject for an impressionable lad (Edmund was, by his own memory, 'nearly frozen into stone with horror'), yet I take some comfort in the thought that Philip Henry Gosse, smitten with the pain of rejection for his untestable theory, could take refuge in something so unambiguously factual, so utterly concrete.

Source: Stephen Jay Gould, *Hen's Teeth and Horse's Toes*, New York and London, W. W. Norton, 1983.

Submarine Gardens of Eden: Devon, 1858–9

Edmund Gosse (1849–1928) was eight years old when, following his mother's death from cancer, he and his father Philip (see p. 95) moved to the village of Marychurch in Devon, where Philip worked on his *History of the British Sea Anemones and Corals* (1860). It was Philip Gosse's bestseller *The Aquarium* (a word he coined) that had started the mid-Victorian passion for sea-shore collecting, which Edmund complains of in this extract from *Father and Son*.

It was down on the shore, tramping along the pebbled terraces of the beach, clambering over the great blocks of fallen conglomerate which broke the white curve with rufous promontories that jutted into the sea, or, finally, bending over those shallow tidal pools in the limestone rocks which were our proper hunting-ground, – it was in such circumstances as these that my Father became most easy, most happy, most human. That hard look across his brows, which it wearied me to see, the look that came from sleepless anxiety of conscience, faded away, and left the dark countenance still always stern indeed, but serene and unupbraiding. Those pools were our mirrors, in which, reflected in the dark hyaline and framed by the sleek and shining fronds of oar-weed there used to appear the shapes of a middle-aged man and a funny little boy, equally eager, and, I almost find the presumption to say, equally well prepared for business.

If anyone goes down to those shores now, if man or boy seeks to follow in our traces, let him realize at once, before he takes the trouble to roll up his sleeves, that his zeal will end in labour lost. There is nothing, now, where in our days there was so much. Then the rocks between tide and tide were submarine gardens of a beauty that seemed often to be fabulous, and was positively delusive, since, if we delicately lifted the weed-curtains of a windless pool, though we might for a moment see its sides and floor paven with living blossoms, ivory-white, rosy-red, orange and amethyst, yet all that panoply would melt away, furled into the hollow rock, if we so much as dropped a pebble in to disturb the magic dream.

Half a century ago, in many parts of the coast of Devonshire and Cornwall, where the limestone at the water's edge is wrought into crevices and hollows, the tide-line was, like Keats' Grecian vase, 'a still unravished bride of quietness'. These cups and basins were always full, whether the tide was high or low, and the only way in which they were affected was that twice in the twenty-four hours they were replenished by cold streams from the great sea, and then twice were left brimming to be vivified by the temperate movement of the upper air. They were living flower-beds, so exquisite in their perfection, that my Father, in spite of his scientific requirements, used not seldom to pause before he began to rifle them, ejaculating that it was indeed a pity to disturb such congregated beauty. The antiquity of these rock-pools, and the infinite succession of the soft and radiant forms, sea-anemones, sea-weeds, shells, fishes, which had inhabited them, undisturbed since the creation of the world, used to occupy my Father's fancy. We burst in, he used to say, where no one had ever thought of intruding before; and if the Garden of Eden had been situate in Devonshire, Adam and Eve, stepping lightly down to bathe in the rainbow-coloured spray, would have seen the identical sights that we now saw – the great prawns gliding like transparent launches, anthea waving in the twilight its thick white waxen tentacles, and the fronds of the dulse faintly streaming on the water, like huge red banners in some reverted atmosphere.

All this is long over, and done with. The ring of living beauty drawn about our shores was a very thin and fragile one. It had existed all those centuries solely in consequence of the indifference, the blissful ignorance of man. These rock-basins, fringed by corallines, filled with still water almost as pellucid as the upper air itself, thronged with beautiful sensitive forms of life, – they exist no longer, they are all profaned, and emptied and vulgarized. An army of 'collectors' has passed over them, and ravaged every corner of them. The fairy paradise has been violated, the exquisite product of centuries of natural selection has been crushed under the rough paw of well-meaning, idle-minded curiosity. That my Father, himself so reverent, so conservative, had by the popularity of his books acquired the direct responsibility for a calamity that he had never anticipated became clear enough to himself before many years had passed, and cost him great chagrin. No one will see again on the shore of England what I saw in my early childhood, the submarine vision of dark rocks, speckled and starred with an infinite variety of colour, and streamed over by silken flags of royal crimson and purple.

In reviving these impressions, I am unable to give any exact chronological sequence to them. These particular adventures began early in 1858, they reached their greatest intensity in the summer of 1859, and they did not altogether cease, so far as my Father was concerned, until nearly twenty years later. But it was while he was composing what, as I am told by scientific men of today, continues to be his most valuable contribution to knowledge, his *History of the British Sea-Anemones and Corals*, that we worked together on the shore for a definite purpose, and the last instalment of that still-classic volume was ready for press by the close of 1859.

The way in which my Father worked, in his most desperate escapades, was to wade breast-high into one of the huge pools, and examine the worm-eaten surface of the rock above and below the brim. In such remote places – spots where I could never venture, being left, a slightly timorous Andromeda, chained to a safer level of the cliff – in these extreme basins, there used often to lurk a marvellous profusion of animal and vegetable forms. My Father would search for the roughest and most corroded points of rock, those offering the best refuge for a variety of creatures, and would then chisel off fragments as low down in the water as he could. These pieces of rock were instantly plunged in the salt water of jars which we had brought with us for the purpose. When as much had been collected as we could carry away – my Father always dragged about an immense square basket, the creak of whose handles I can still fancy that I hear – we turned to trudge up the long climb home. Then all our prizes were spread out, face upward, in shallow pans of clean sea-water.

In a few hours, when all dirt had subsided, and what living creatures we had brought seemed to have recovered their composure, my work began. My eyes were extremely keen and powerful, though they were vexatiously near-sighted. Of no use in examining objects at any distance, in investigating a minute surface my vision was trained to be invaluable. The shallow pan, with our spoils, would rest on a table near the window, and I, kneeling on a chair opposite the light, would lean over the surface till everything was within an inch or two of my eyes. Often I bent, in my zeal, so far forward that the water touched the tip of my nose and gave me a little icy shock. In this attitude – an idle spectator might have formed the impression that I was trying to wash my head and could not quite summon up resolution enough to plunge – in this odd pose I would remain for a long time, holding my breath and examining with extreme care every atom of rock, every swirl of detritus. This was a task which my Father could only perform by the help of a lens, with which, of course, he took care to supplement my examination. But that my survey was of use, he has himself most handsomely testified in his *Actinologia Britannica*, where he expresses his debt to the 'keen and well-practised eye of my little son'. Nor, if boasting is not to be excluded, is it every eminent biologist, every proud and masterful F.R.S., who can lay his hand on his heart and swear that, before reaching the age of ten years, he had added, not merely a new species, but a new genus to the British fauna. That however, the author of these pages can do, who, on 29 June 1859, discovered a tiny atom, – and ran in the greatest agitation to announce the discovery of that object 'as a form with which he was unacquainted', – which figures since then on all lists of sea-anemones as *phellia murocincta*, or the walled corklet. Alas! that so fair a swallow should have made no biological summer in afterlife.

Source: Edmund Gosse, *Father and Son*, London, Heinemann, 1907.

In Praise of Rust

Besides being the greatest British art critic, a superb draughtsman and a pioneer social reformer, John Ruskin (1819–1900) was a keen amateur geologist. He believed that scientific knowledge (of rocks, botanical forms, clouds) was essential for artists. This excerpt is from a lecture he gave at Tunbridge Wells (hence the reference to 'your spring') in 1858.

You will probably know that the ochreous stain, which, perhaps, is often thought to spoil the basin of your spring, is iron in a state of rust: and when you see rusty iron in other places you generally think, not only that it spoils the places it stains, but that it is spoiled itself – that rusty iron is spoiled iron.

For most of our uses it generally is so; and because we cannot use a rusty knife or razor so well as a polished one, we suppose it to be a great defect in iron that it is subject to rust. But not at all. On the contrary, the most perfect and useful state of it is that ochreous stain; and therefore it is endowed with so ready a disposition to get itself into that state. It is not a fault in the iron, but a virtue, to be so fond of getting rusted, for in that condition it fulfils its most important functions in the universe, and most kindly duties to mankind. Nay, in a certain sense, and almost a literal one, we may say that iron rusted is Living; but when pure or polished, Dead. You all probably know that in the mixed air we breathe, the part of it essentially needful to us is called oxygen; and that this substance is to all animals, in the most accurate sense of the word, 'breath of life' [Genesis 2:7]. The nervous power of life is a different thing; but the supporting element of the breath, without which the blood, and therefore the life, cannot be nourished, is this oxygen. Now it is this very same air which the iron breathes when it gets rusty. It takes the oxygen from the atmosphere as eagerly as we do, though it uses it differently. The iron keeps all that it gets; we, and other animals, part with it again; but the metal absolutely keeps what it has once received of this aërial gift; and the ochreous dust which we so much despise is, in fact, just so much nobler than pure iron, in so far as it is *iron and the air.* Nobler, and more useful – for, indeed, as I shall be able to show you presently – the main service of this metal, and of all other metals, to us, is not in making knives, and scissors, and pokers, and pans, but in making the ground we feed from, and nearly all the substances first needful to our existence. For these are all nothing but metals and oxygen – metals with breath put into them. Sand, lime, clay, and the rest of the earths – potash and soda, and the rest of the alkalies – are all of them metals which have undergone this, so to speak, vital change, and have been rendered fit for the service of man by permanent unity with the purest air which he himself breathes. There is only one metal which does not rust readily; and that in its influence on Man hitherto, has caused Death rather than Life; it will not be put to its right use till it is made a pavement of, and so trodden under foot.

Is there not something striking in this fact, considered largely as one of the types, or lessons, furnished by the inanimate creation? Here you have your hard, bright, cold, lifeless metal – good enough for swords and scissors – but not for food. You think, perhaps, that your iron is wonderfully useful in a pure form, but how would you like the world, if all your meadows, instead of grass, grew nothing but iron wire – if all your arable ground, instead of being made of sand and clay, were suddenly turned into flat surfaces of steel – if the whole earth, instead of its green and glowing sphere, rich with forest and flower, showed nothing but the image of the vast furnace of a ghastly engine – a globe of black, lifeless, excoriated metal? It would be that, – probably it was once that; but assuredly it would be, were it not that all the substance of which it is made sucks and breathes the brilliancy of the atmosphere; and, as it breathes, softening from its merciless hardness, it falls into fruitful and beneficent dust; gathering itself again into the earths from which we feed, and the stones with which we build; – into the rocks that frame the mountains, and the sands that bind the sea.

Hence, it is impossible for you to take up the most insignificant pebble at your feet, without being able to read, if you like, this curious lesson in it. You look upon it at first as if it were earth only. Nay, it answers, 'I am not earth – I am earth and air in one; part of that blue heaven which you love, and long for, is already in me; it is all my life – without it I should be nothing, and able for nothing; I could not minister to you, nor nourish you – I should be a cruel and helpless thing; but, because there is, according to my need and place in creation, a kind of soul in me, I have become capable of good, and helpful in the circles of vitality.'

Thus far the same interest attaches to all the earths, and all the metals of which they are made; but a deeper interest and larger beneficence belong to that ochreous earth of iron which stains the marble of your springs. It stains much besides that marble. It stains the great earth wheresoever you can see it, far and wide – it is the colouring substance appointed to colour the globe for the sight, as well as subdue it to the service of man. You have just seen your hills covered with snow, and, perhaps, have enjoyed, at first, the contrast of their fair white with the dark blocks of pine woods; but have you ever considered how you would like them always white – not pure white, but dirty white – the white of thaw, with all the chill of snow in it, but none of its brightness? That is what the colour of the earth would be without its iron; that would be its colour, not here or there only, but in all places, and at all times. Follow out that idea till you get it in some detail. Think first of your pretty gravel walks in your gardens, and fine, like plots of sunshine between the yellow flower-beds; fancy them all suddenly turned to the colour of ashes. That is what they would be without iron ochre. Think of your winding walks over the common, as warm to the eye as they are dry to the foot, and imagine them all laid down suddenly with gray cinders. Then pass beyond the common into the country, and pause at the first ploughed field that you see sweeping up the hill sides in the sun, with its deep brown furrows, and wealth of ridges all a-glow, heaved aside by the ploughshare, like deep folds of a mantle of russet velvet – fancy it all changed suddenly into grisly furrows in a field of mud. That is what it would be without iron. Pass on, in fancy, over hill and dale, till you reach the bending line of the sea shore; go down upon its breezy beach – watch the white foam flashing among the amber of it, and all the blue sea embayed in belts of gold: then fancy those circlets of far sweeping shore suddenly put into mounds of mourning – all those golden sands turned into gray slime; the fairies no more able to call to each other, 'Come unto these yellow sands'; [*The Tempest*, I: ii: 376] but, 'Come unto these drab sands.' That is what they would be, without iron.

Iron is in some sort, therefore, the sunshine and light of landscape, so far as that light depends on the ground . . .

All those beautiful violet veinings and variegations of the marbles of

Sicily and Spain, the glowing orange and amber colours of those of Siena, the deep russet of the Rosso antico, and the blood-colour of all the precious jaspers that enrich the temples of Italy; and, finally, all the lovely transitions of tint in the pebbles of Scotland and the Rhine, which form, though not the most precious, by far the most interesting portion of our modern jewellers' work; – all these are painted by Nature with this one material only, variously proportioned and applied – the oxide of iron that stains your Tunbridge springs . . .

A nobler colour than all these – the noblest colour ever seen on this earth – one which belongs to a strength greater than that of the Egyptian granite, and to a beauty greater than that of the sunset or the rose – is still mysteriously connected with the presence of this dark iron. I believe it is not ascertained on what the crimson of blood actually depends; but the colour is connected, of course, with its vitality, and that vitality with the existence of iron as one of its substantial elements.

Is it not strange to find this stern and strong metal mingled so delicately in our human life that we cannot even blush without its help?

Source: John Ruskin, *The Two Paths*, London, Smith, Elder & Co., 1859.

The Devil's Chaplain

Charles Darwin (1809–82) was terrified by his own ideas. He was already convinced, by the late 1830s, that mankind and the other animal species had not been separately created by God, but had evolved from a common ancestor – probably, Darwin speculated, a bisexual mollusc with a vertebra but no head. But the blow to Christianity and to the dignity of man inherent in such a theory would, he feared, encourage atheistic agitators and socialist revolutionaries. 'What a book a Devil's Chaplain might write on the clumsy, wasteful, blundering low & horridly cruel works of nature!' he confided to his private notebook. Wedded to respectability and social order, he refrained for twenty years from publishing his epoch-making work *On the Origin of Species by Means of Natural Selection* (1859). Only when he discovered that a younger biologist, Alfred Russel Wallace (1823–1913), had reached similar conclusions independently did he steel himself and go public.

The crucial experience of Darwin's life had been his five-year voyage (1831–6), as resident naturalist aboard the survey ship HMS *Beagle*, to South America and the islands of the Pacific and South Atlantic. His *Journal* (published in 1839) records his elation on encountering the New World:

Bahia, or San Salvador. Brazil, Feb. 29th 1832 – The day has passed delightfully. Delight itself, however, is a weak term to express the feelings of a naturalist who, for the first time, has wandered by himself in a Brazilian forest. The elegance of the grasses, the novelty of the parasitical plants, the beauty of the flowers, the glossy green of the foliage, but above all the general luxuriance of the vegetation, filled me with admiration. A most paradoxical mixture of sound and silence pervades the shady parts of the wood. The noise from the insects is so loud, that it may be heard even in a vessel anchored several hundred yards from the shore; yet within the recesses of the forest a universal silence appears to reign. To a person fond of natural history, such a day as this brings with it a deeper pleasure than he can ever hope to experience again. After wandering about for some hours, I returned to the landing-place; but, before reaching it, I was overtaken by a tropical storm. I tried to find shelter under a tree, which was so thick that it would never have been penetrated by common English rain; but here, in a couple of minutes, a little torrent flowed down the trunk. It is to this violence of the rain that we must attribute the verdure at the bottom of the thickest woods: if the showers were like those of a colder clime, the greater part would be absorbed or evaporated before it reached the ground.

In February 1834 the *Beagle* reached Tierra del Fuego, where the sight of the savage inhabitants shocked Darwin and permanently affected his thinking, making him more receptive to the notion of man's animal ancestry.

While going one day on shore near Wollaston Island, we pulled alongside a canoe with six Fuegians. These were the most abject and miserable creatures I anywhere beheld. On the east coast the natives . . . have guanaco cloaks, and on the west they possess seal-skins. Amongst these central tribes the men generally have an otter-skin, or some small scrap about as large as a pocket-handkerchief, which is barely sufficient to cover their backs as low down as their loins. It is laced across the breast by strings, and according as the wind blows, it is shifted from side to side. But these Fuegians in the canoe were quite naked, and even one full-grown woman was absolutely so. It was raining heavily, and the fresh water, together with the spray, trickled down her body. In another harbour not far distant, a woman, who was suckling a recently-born child, came one day alongside the vessel, and remained there out of mere curiosity, whilst the sleet fell and thawed on her naked bosom, and on the skin of her naked baby! These poor wretches were stunted in their growth, their hideous faces bedaubed with white paint, their skins filthy and greasy, their hair entangled, their voices discordant, and their gestures violent. Viewing such men, one can hardly make oneself believe that they are fellow-creatures, and inhabitants of the same world. It is a common subject of conjecture what pleasure in life some of the lower animals can enjoy: how much more reasonably the same question may be asked with respect to these barbarians! At night, five or six human beings, naked and scarcely protected from the wind and rain of this tempestuous climate, sleep on the wet ground coiled up like animals. Whenever it is low water, winter or summer, night or day, they must rise to pick shell-fish from the rocks; and the women either dive to collect sea-eggs, or sit patiently in their canoes, and with a baited hair-line without any hook, jerk out little fish. If a seal is killed, or the floating carcass of a putrid whale discovered, it is a feast; and such miserable food is assisted by a few tasteless berries and fungi. Their country is a broken mass of wild rocks, lofty hills, and useless forests: and these are viewed through mists and endless storms. The habitable land is reduced to the stones on the beach; in search of food they are compelled unceasingly to wander from spot to spot, and so steep is the coast, that they can only move about in their wretched canoes. They cannot know the feeling of having a home, and still less that of domestic affection; for the husband is to the wife a brutal master to a laborious slave. Was a more horrid deed ever perpetrated, than that witnessed on the west coast by Byron [John Byron (1723–86), navigator], who saw a wretched mother pick up her bleeding dying infant-boy, whom her husband had mercilessly dashed on the stones for dropping a basket of sea-eggs! How little can the higher powers of the mind be brought into play: what is there for imagination to picture, for reason to compare, for judgment to decide upon? To knock a limpet from the rock does not require even cunning, that lowest power of the mind. Their skill in some respects may be compared to the instinct of animals; for it is not improved by experience: the canoe, their most ingenious work, poor as it is, has remained the same, as we know from Drake, for the last two hundred and fifty years.

For man and the other species to have evolved from a primitive life-form would require, Darwin realized, an enormous lapse of time, far greater than was traditionally thought of as the age of the earth. This stimulated his interest in geology, the science that was re-dating the earth in the nineteenth century. He took a copy of Lyell's *Principles of Geology* (see p. 71) on the voyage, and made his own geological observations. Crossing the Andes by the Uspallata Pass in March 1835, he was excited to find a grove of fossilized trees – 'snow white columns like Lot's wife' – at 7,000 feet. Since the range on which they stood was composed of alternate layers of submarine lava and volcanic sandstone, he was able to reconstruct the cataclysmic upheavals of the earth's crust that this petrified forest bore witness to.

In the central part of the range, at an elevation of about seven thousand feet, I observed on a bare slope some snow-white projecting columns. These were petrified trees, eleven being silicified, and from thirty to forty converted into coarsely-crystallized white calcareous spar. They were abruptly broken off, the upright stumps projecting a few feet above the ground. The trunks measured from three to five feet each in circumference. They stood a little way apart from each other, but the whole formed one group. Mr Robert Brown has been kind enough to examine the wood: he says it belongs to the fir tribe, partaking of the character of the Araucarian family, but with some curious points of affinity with the yew. The volcanic sandstone in which the trees were embedded, and from the lower part of which they must have sprung, had accumulated in successive thin layers around their trunks; and the stone yet retained the impression of the bark.

It required little geological practice to interpret the marvellous story which this scene at once unfolded; though I confess I was at first so much astonished, that I could scarcely believe the plainest evidence. I saw the spot where a cluster of fine trees once waved their branches on the shores of the Atlantic, when that ocean (now driven back 700 miles) came to the foot of the Andes. I saw that they had sprung from a volcanic soil which had been raised above the level of the sea, and that subsequently this dry land, with its upright trees, had been let down into the depths of the ocean. In these depths, the formerly dry land was covered by sedimentary beds, and these again by enormous streams of submarine lava – one such mass attaining the thickness of a thousand feet; and these deluges of molten stone and aqueous deposits five times alternately had been spread out. The ocean which received such thick masses, must have been profoundly deep; but again the subterranean forces exerted themselves, and I now beheld the bed of that ocean, forming a chain of mountains more than seven thousand feet in height. Nor had those antagonist forces been dormant, which are always at work wearing down the surface of the land: the great piles of strata had been intersected by many wide valleys, and the trees, now changed into silex, were exposed projecting from the volcanic soil, now changed into rock whence formerly, in a green and budding state, they had raised their lofty heads. Now, all is utterly irreclaimable and desert; even the lichen cannot adhere to the stony casts of former trees. Vast, and scarcely comprehensible as such changes must ever appear, yet they have all occurred within a period, recent when compared with the history of the Cordillera; and the Cordillera itself is absolutely modern as compared with many of the fossiliferous strata of Europe and America.

The rampant abundance of life, as illustrated by the Brazilian forests, the kinship between men and animals disclosed by the Fuegians, and the vast time-span suggested by the petrified trees of the Andes, were all necessary conditions for the theory of evolution Darwin proposed in the *Origin of Species*. According to this theory, the evolution of the different plant and animal species can be traced to the chance variations that occur in the offspring of living creatures. Only some of these variations can survive, since life-forms reproduce in such insupportable profusion. Which will survive, and develop into new species, is determined by the process Darwin called natural selection.

It may be said that natural selection is daily and hourly scrutinizing, throughout the world, every variation, even the slightest; rejecting that which is bad, preserving and adding up all that is good; silently and insensibly working, whenever and wherever opportunity offers, at the improvement of each organic being in relation to its organic and inorganic conditions of life.

Viewed in this way, even the seemingly peaceful life of a quiet woodland reveals itself as the result of a ceaseless battle for survival.

What a struggle between the several kinds of trees must here have gone on during long centuries, each annually scattering its seeds by the thousand; what war between insect and insect – between insects, snails, and other animals with birds and beasts of prey – all striving to increase, and all feeding on each other or on the trees or their seeds and seedlings, or on the other plants which first clothed the ground and thus checked the growth of the trees!

Darwin regards the 'victorious' forms of life as 'higher' than the 'beaten' forms, and at times describes the battle for survival in nationalistic or imperialistic terms.

From the extraordinary manner in which European productions have recently spread over New Zealand, and have seized on places which must have been previously occupied, we may believe, if all the animals and plants of Great Britain were set free in New Zealand, that in the course of time a multitude of British forms would become thoroughly naturalized there, and would exterminate many of the natives. On the other hand, from what we see now occurring in New Zealand, and from hardly a single inhabitant of the southern hemisphere having become wild in any part of Europe, we may doubt, if all the productions of New Zealand were set free in Great Britain, whether any considerable number would be enabled to seize on places now occupied by our native plants and animals. Under this point of view, the productions of Great Britain, may be said to be higher than those of New Zealand. Yet the most skilful naturalist from an examination of the species of the two countries could not have foreseen this result.

The simple but ingenious experiments Darwin carried out to support his theory are among the most attractive parts of the *Origin*. He was anxious, for example, to explain how species of fresh-water plants could spread over different continents, and to remote oceanic islands, without being separately created in each location by God.

I think favourable means of dispersal explain this fact. I have before mentioned that earth occasionally, though rarely, adheres in some quantity to the feet and beaks of birds. Wading birds, which frequent the muddy edges of ponds, if suddenly flushed, would be the most likely to have muddy feet. Birds of this order I can show are the greatest wanderers, and are occasionally found on the most remote and barren islands in the open ocean; they would not be likely to alight on the surface of the sea, so that the dirt would not be washed off their feet; when making land, they would be sure to fly to their natural fresh-water haunts. I do not believe that botanists are aware how charged the mud of ponds is with seeds: I have tried several little experiments, but will here give only the most striking case: I took in February three table-spoonfuls of mud from three different points, beneath water, on the edge of a little pond; this mud when dry weighed only $6\frac{3}{4}$ ounces; I kept it covered up in my study for six months, pulling up and counting each plant as it grew; the plants were of many kinds, and were altogether 537 in number; and yet the viscid mud was all contained in a breakfast cup! Considering these facts, I think it would be an inexplicable circumstance if water-birds did not transport the seeds of fresh-water plants to vast distances, and if consequently the range of these plants was not very great. The same agency may have come into play with the eggs of some of the smaller fresh-water animals.

A less successful experiment, which did not, however, shake Darwin's faith in his general principle, was designed to show that in the evolutionary battle for life no species can be unselfish.

The instinct of each species is good for itself, but has never, as far as we can judge, been produced for the exclusive good of others. One of the strongest instances of an animal apparently performing an action for the sole good of another, with which I am acquainted, is that of aphides voluntarily yielding their sweet excretion to ants: that they do so voluntarily, the following facts show. I removed all the ants from a group of about a dozen aphides on a dock-plant, and prevented their attendance during several hours. After this interval, I felt sure that the aphides would want to excrete. I watched them for some time through a lens, but not one excreted; I then tickled and stroked them with a hair in the same manner, as well as I could, as the ants do with their antennae; but not one excreted. Afterwards I allowed an ant to visit them, and it immediately seemed, by its eager way of running about, to be well aware what a rich flock it had discovered; it then began to play with its antennae on the abdomen first of one aphis and then of another; and each aphis, as soon as it felt the antennae, immediately lifted up its abdomen and excreted a limpid drop of sweet juice, which was eagerly devoured by the ant. Even the quite young aphides behaved in this manner, showing that the action was instinctive, and not the result of experience. But as the execretion is extremely viscid, it is probably a convenience to the aphides to have it removed; and therefore probably the aphides do not instinctively excrete for the sole good of the ants. Although I do not believe that any animal in the world performs an action for the exclusive good of another of a distinct species, yet each species tries to take advantage of the instincts of others, as each takes advantage of the weaker bodily structure of others.

To avoid controversy, Darwin said almost nothing in the *Origin* about the evolution of mankind. However, the human implications of his theory were apparent, and the fact that we are descended from monkeys quickly became the popular version of evolution theory. Darwin had, in fact, been observing the similarities between apes and human children for many years, as a letter to his grandmother in 1838 reveals:

Two days since, when it was very warm, I rode to the Zoological Society, & by the greatest piece of good fortune it was the first time this year, that the Rhinoceros was turned out. – Such a sight has seldom been seen, as to behold the rhinoceros kicking & rearing, (though neither end reached any great height) out of joy . . . I saw also the Ourang-outang in great perfection: the keeper showed her an apple, but would not give it her, whereupon she threw herself on her back, kicked & cried, precisely like a naughty child. – She then looked very sulky & after two or three fits of pashion, the keeper said, 'Jenny if you will stop bawling & be a good girl, I will give you the apple.' – She certainly understood every word of this, &, though like a child, she had great work to stop whining, she at last succeeded, & then got the apple, with which she jumped into an arm chair & began eating it, with the most contented countenance imaginable. –

In *The Descent of Man* (1871), which applies the theories of the *Origin* to mankind, Darwin frankly expresses a preference for ape-ancestry.

The main conclusion arrived at in this work, namely, that man is descended from some lowly organized form, will, I regret to think, be highly distasteful to many. But there can hardly be a doubt that we are descended from barbarians. The astonishment which I felt on first seeing a party of Fuegians on a wild and broken shore will never be forgotten by me, for the reflection at once rushed into my mind – such were our ancestors. These men were absolutely naked and bedaubed with paint, their long hair was tangled, their mouths frothed with excitement, and their expression was wild, startled, and distrustful. They possessed hardly any arts, and like wild animals lived on what they could catch; they had no government, and were merciless to every one not of their own small tribe. He who has seen a savage in his native land will not feel much shame, if forced to acknowledge that the blood of some more humble creature flows in his veins. For my own part I would as soon be descended from that heroic little monkey, who braved his dreaded enemy in order to save the life of his keeper, or from that old baboon, who descending from the mountains, carried away in triumph his young comrade from a crowd of astonished dogs – as from a savage who delights to torture his enemies, offers up bloody sacrifices, practises infanticide without remorse, treats his wives like slaves, knows no decency, and is haunted by the grossest superstitions.

Opponents of Darwin's theory pointed to man's moral sense and conscience as distinctive human attributes that must have been implanted by God. Darwin, however, argued that they could quite conceivably have evolved as a consequence of man's social nature and intelligence.

The following proposition seems to me in a high degree probable – namely, that any animal whatever, endowed with well-marked social instincts, the parental and filial affections being here included, would inevitably acquire a moral sense or conscience, as soon as its intellectual powers had become as well, or nearly as well developed, as in man. For, *firstly*, the social instincts lead an animal to take pleasure in the society of its fellows, to feel a certain amount of sympathy with them, and to perform various services for them. The services may be of a definite and evidently instinctive nature; or there may be only a wish and readiness, as with most of the higher social animals, to aid their fellows in certain general ways. But these feelings and services are by no means extended to all the individuals of the same species, only to those of the same association. *Secondly*, as soon as the mental faculties had become highly developed, images of all past actions and motives would be incessantly passing through the brain of each individual: and that feeling of dissatisfaction, or even misery, which invariably results . . . from any unsatisfied instinct, would arise, as often as it was perceived that the enduring and always present social instinct had yielded to some other instinct, at the time stronger, but neither enduring in its nature, nor leaving behind it a very vivid impression. It is clear that many instinctive desires, such as that of hunger, are in their nature of short duration; and after being satisfied, are not readily or vividly recalled. *Thirdly*, after the power of language had been acquired, and the wishes of the community could be expressed, the common opinion how each member ought to act for the public good, would naturally become in a paramount degree the guide to action. But it should be borne in mind that however great weight we may attribute to public opinion, our regard for the approbation and disapprobation of our fellows depends on sympathy, which, as we shall see, forms an essential part of the social instinct, and is indeed its foundation-stone. *Lastly*, habit in the individual would ultimately play a very important part in guiding the conduct of each member; for the social instinct, together with sympathy, is, like any other instinct, greatly strengthened by habit, and so consequently would be obedience to the wishes and judgment of the community . . .

It may be well first to premise that I do not wish to maintain that any strictly social animal, if its intellectual faculties were to become as active and as highly developed as in man, would acquire exactly the same moral sense as ours. In the same manner as various animals have some sense of beauty, though they admire widely different objects, so they might have a sense of right and wrong, though led by it to follow widely different lines of conduct. If, for instance, to take an extreme case, men were reared under precisely the same conditions as hive-bees, there can hardly be a doubt that our unmarried females would, like the worker-bees, think it a sacred duty to kill their brothers, and mothers would strive to kill their fertile daughters; and no one would think of interfering. Nevertheless, the bee, or any other social animal, would gain in our supposed case, as it appears to me, some feeling of right or wrong, or a conscience. For each individual would have an inward sense of possessing certain stronger or more enduring instincts, and others less strong or enduring; so that there would often be a struggle as to which impulse should be followed; and satisfaction, dissatisfaction, or even misery would be felt, as past impressions were compared during their incessant passage through the mind. In this case an inward monitor would tell the animal that it would have been better to have followed the one impulse rather than the other. The one course ought to have been followed, and the other ought not; the one would have been right and the other wrong.

Accepting the usual Victorian assumptions about the different mental powers of men and women, Darwin argued that this, too, could be traced to the evolutionary process, and to that subdivision of natural selection that he called sexual selection.

With respect to differences of this nature between man and woman, it is probable that sexual selection has played a highly important part. I am aware that some writers doubt whether there is any such inherent difference; but this is at least probable from the analogy of the lower animals which present other secondary sexual characters. No one disputes that the bull differs in disposition from the cow, the wild-boar from the sow, the stallion from the mare, and, as is well known to the keepers of menageries, the males of the larger apes from the females. Woman seems to differ from man in mental disposition, chiefly in her greater tenderness and less selfishness; and this holds good even with savages, as shewn by a well-known passage in Mungo Park's *Travels*, and by statements made by many other travellers. Woman, owing to her maternal instincts, displays these qualities towards her infants in an eminent degree; therefore it is likely that she would often extend them towards her fellow-creatures. Man is the rival of other men; he delights in competition, and this leads to ambition which passes too easily into selfishness. These latter qualities seem to be his natural and unfortunate birthright. It is generally admitted that with woman the powers of intuition, of rapid perception, and perhaps of imitation, are more strongly marked than in man; but some, at least, of these faculties are characteristic of the lower races, and therefore of a past and lower state of civilization.

The chief distinction in the intellectual powers of the two sexes is shewn by man's attaining to a higher eminence, in whatever he takes up, than can woman – whether requiring deep thought, reason, or imagination, or merely the use of the senses and hands. If two lists were made of the most eminent men and women in poetry, painting, sculpture, music (inclusive both of composition and performance), history, science, and philosophy, with half-a dozen names under each subject, the two lists would not bear comparison . . .

Amongst the half-human progenitors of man, and amongst savages, there have been struggles between the males during many generations for the possession of the females. But mere bodily strength and size would do little for victory, unless associated with courage, perseverance, and determined energy. With social animals, the young males have to pass through many a contest before they win a female, and the older males have to retain their females by renewed battles. They have, also, in the case of mankind, to defend their females, as well as their young, from enemies of all kinds, and to hunt for their joint subsistence. But to avoid enemies or to attack them with success, to capture wild animals, and to fashion weapons, requires the aid of the higher mental faculties, namely, observation, reason, invention, or imagination. These various faculties will thus have been continually put to the test and selected during manhood; they will, moreover, have been strengthened by use during this same period of life. Consequently in accordance with the principle often alluded to, we might expect that they would at least tend to be transmitted chiefly to the male offspring at the corresponding period of manhood.

Now, when two men are put into competition, or a man with a woman, both possessed of every mental quality in equal perfection, save that one has higher energy, perseverance, and courage, the latter will generally become more eminent in every pursuit, and will gain the ascendancy. He may be said to possess genius – for genius has been declared by a great authority to be patience; and patience, in this sense, means unflinching, undaunted perseverance. But this view of genius is perhaps deficient; for without the higher powers of the imagination and reason, no eminent success can be gained in many subjects. These latter faculties, as well as the former, will have been developed in man, partly through sexual selection, – that is, through the contest of rival males, and partly through natural selection, – that is, from success in the general struggle for life; and as in both cases the struggle will have been during maturity, the characters gained will have been transmitted more fully to the male than to the female offspring. It accords in a striking manner with this view of the modification and re-inforcement of many of our mental faculties by sexual selection, that, firstly, they notoriously undergo a considerable change at puberty, and, secondly, that eunuchs remain throughout life inferior in these same qualities. Thus man has ultimately become superior to woman. It is, indeed, fortunate that the law of the equal transmission of characters to both sexes prevails with mammals; otherwise it is probable that man would have become as superior in mental endowment to woman, as the peacock is in ornamental plumage to the peahen.

The year after *The Descent of Man*, Darwin continued his investigation of the physical and mental resemblances between humans and non-humans in *The Expression of the Emotions in Man and Animals* (1872). One emotional expression not yet developed by any animal except the human was, he found, the blush, and this led him to probe its nature and cause.

I was desirous to learn how far down the body blushes extend; and Sir J. Paget, who necessarily has frequent opportunities for observation, has kindly attended to this point for me during two or three years. He finds that with women who blush intensely on the face, ears, and nape of neck, the blush does not commonly extend any lower down the body. It is rare to see it as low down as the collar-bones and shoulder-blades; and he has never himself seen a single instance in which it extended below the upper part of the chest. He has also noticed that blushes sometimes die away downwards, not gradually and insensibly, but by irregular ruddy blotches. Dr Langstaff has likewise observed for me several women whose bodies did not in the least redden while their faces were crimsoned with blushes. With the insane, some of whom appear to be particularly liable to blushing, Dr J. Crichton Browne has several times seen the blush extend as far down as the collar-bones, and in two instances to the breasts. He gives me the case of a married woman, aged twenty-seven, who suffered from epilepsy. On the morning after her arrival in the Asylum, Dr Browne, together with his assistants, visited her whilst she was in bed. The moment that he approached, she blushed deeply over her cheeks and temples; and the blush spread quickly to her ears. She was much agitated and tremulous. He unfastened the collar of her chemise in order to examine the state of her lungs; and then a brilliant blush rushed over her chest, in an arched line over the upper third of each breast, and extended downwards between the breasts nearly to the ensiform cartilage of the sternum. This case is interesting, as the blush did not thus extend downwards until it became intense by her attention being drawn to this part of her person. As the examination proceeded she became composed, and the blush disappeared; but on several subsequent occasions the same phenomena were observed.

The foregoing facts show that, as a general rule, with English women, blushing does not extend beneath the neck and upper part of the chest. Nevertheless Sir J. Paget informs me that he has lately heard of a case, on which he can fully rely, in which a little girl, shocked by what she imagined to be an act of indelicacy, blushed all over her abdomen and the upper parts of her legs. Moreau also relates, on the authority of a celebrated painter, that the chest, shoulders, arms, and whole body of a girl, who unwillingly consented to serve as a model, reddened when she was first divested of her clothes.

It is a rather curious question why, in most cases the face, ears, and neck alone redden, inasmuch as the whole surface of the body often tingles and grows hot. This seems to depend, chiefly, on the face and adjoining parts of the skin having been habitually exposed to the air, light, and alternations of temperature, by which the small arteries not only have acquired the habit of readily dilating and contracting, but appear to have become unusually developed in comparison with other parts of the surface . . . Nevertheless it may be doubted whether the habitual exposure of the skin of the face and neck, and its consequent power of reaction under stimulants of all kinds, is by itself sufficient to account for the much greater tendency in English women of these parts than of others to blush; for the hands are well supplied with nerves and small vessels, and have been as much exposed to the air as the face or neck, and yet the hands rarely blush . . . Of all parts of the body, the face is most considered and regarded, as is natural from its being the chief seat of expression and the source of the voice. It is also the chief seat of beauty and of ugliness, and throughout the world is the most ornamented. The face, therefore, will have been subjected during many generations to much closer and more earnest self-attention than any other part of the body; and in accordance with the principle here advanced we can understand why it should be the most liable to blush.

The effect of Darwin's theory of evolution on man's self-image has been momentous. Sigmund Freud, in his essay 'A Difficulty in Psycho-Analysis', compares it to the Copernican revolution which dealt an irreparable blow to human narcissism by removing the earth from the centre of the universe.

In the course of the development of civilization man acquired a dominating position over his fellow-creatures in the animal kingdom. Not content with this supremacy, however, he began to place a gulf between his nature and theirs. He denied the possession of reason to them, and to himself he attributed an immortal soul, and made claims to a divine descent which permitted him to break the bond of community between him and the animal kingdom. Curiously enough, this piece of arrogance is still foreign to children, just as it is to primitive and primaeval man. It is the result of a later, more pretentious stage of development. At the level of totemism primitive man had no repugnance to tracing his descent from an animal ancestor. In myths, which contain the precipitate of this ancient attitude of mind, the gods take animal shapes, and in the art of earliest times they are portrayed with animals' heads. A child can see no difference between his own nature and that of animals. He is not astonished at animals thinking and talking in fairy-tales; he will transfer an emotion of fear which he feels for his human father onto a dog or a horse, without intending any derogation of his father by it. Not until he is grown up does he become so far estranged from animals as to use their names in vilification of human beings.

We all know that little more than half a century ago the researches of Charles Darwin and his collaborators and forerunners put an end to this presumption on the part of man. Man is not a being different from animals or superior to them; he himself is of animal descent, being more closely related to some species and more distantly to others. The acquisitions he has subsequently made have not succeeded in effacing the evidences, both in his physical structure and in his mental dispositions, of his parity with them. This was the second, the *biological* blow to human narcissism.

Sources: Charles Darwin, *Journal of Researches into the Geology and Natural History of the Various Countries Visited by H.M.S. Beagle*, London, Henry Colburn, 1839, *On the Origin of Species by Means of Natural Selection*, London, John Murray, 1859, *The Descent of Man*, London, John Murray, 1871, *The Expression of the Emotions in Man and Animals*, London, John Murray, 1872. *The Standard Edition of the Complete Psychological Works of Sigmund Freud. Translated from the German under the General Editorship of James Strachey*, volume XVII, London, The Hogarth Press and the Institute of Psychoanalysis, 1955.

The Discovery of Prehistory

Prehistory was discovered in the mid-nineteenth century in Denmark. This account of the process is from Daniel J. Boorstin's *The Discoverers* (1983). Dr Boorstin was for twenty-five years a Professor of American History at Chicago, and became Librarian of Congress in 1975.

Surviving *objects* had a special power to help people grasp the past. But the buried relics in Rome and Greece simply documented a past familiar from sacred or classical literature. The discovery of prehistory through objects would reach back far beyond the written word and vastly extended the dimensions of human history.

A strange series of coincidences gave the leading role in this discovery to a Danish businessman, Christian Jürgensen Thomsen (1788–1865). Without the erudition of a Scaliger or the mathematical genius of a Newton, he was a man of superlative common sense, richly endowed with the virtues of the dedicated amateur. His passion for curious objects was matched by his talent for awakening the curiosity of the new museum public. Born in Copenhagen, the eldest of six sons of a prosperous shipowner, he was trained for business. He came to know the family of a Danish consul who had served in Paris during the French Revolution, and who had brought back collections purchased from the panicked aristocracy. When young Christian, still only fifteen, helped his friends unpack their treasures, they gave him a few old coins to begin his own collection, and by the time he was nineteen he was a respected numismatist. In 1807, when the British fleet bombarded Copenhagen harbor to keep the Danish fleet from Napoleon, buildings went up in flames, and Christian joined the emergency fire brigade. Working through the night, he rescued the coins of a leading numismatist whose house was hit, and carried them to safety with the Keeper of the Royal Cabinet of Antiquities.

Copenhagen's newly established Royal Commission for the Preservation of Danish Antiquities was being flooded by miscellaneous old objects sent in by public-spirited citizens. The aged secretary of the commission could not face the accumulating pile. It was time for a younger man – and an opportunity made to order for Thomsen, then twenty-seven and known for his own beautifully organized collection of coins. 'Mr Thomsen is admittedly only a dilettante,' the bishop on the commission conceded, 'but a dilettante with a wide range of knowledge. He has no university degree, but in the present state of scientific knowledge I hardly consider that fact as being a disqualification.' Accordingly, young Thomsen was honored with the post of unpaid nonvoting secretary. As it turned out, Thomsen's lack of academic learning equipped him with the naïveté that archaeology needed at that moment.

The dusty shelves of the commission's storerooms overflowed with unlabeled odd bundles. How could Thomsen put them in order? 'I had no previous example on which to base the ordering of such a collection,' Thomsen confessed, nor had he money to hire a professor to classify objects by academic categories. So he applied the commonsense procedures learned in his father's shipping warehouse. Opening the parcels, first he separated them into objects of stone, of metal, and of pottery. Then he subdivided these according to their apparent use as weapons, tools, food containers, or religious objects. With no texts to guide him, he simply looked at the objects, then asked himself what questions would be asked by museum visitors who saw them for the first time.

When Thomsen opened his museum to the public in 1819, visitors saw the objects sorted into three cabinets. The first contained objects of stone; the second, objects of bronze; the third, objects of iron. This exercise in museum housekeeping led Thomsen to suspect that objects made of similar materials might be relics of the same era. To his amateur eye it seemed that the objects of stone might be older than similar metal objects, and that the bronze objects might be older than those of iron. He shared this elementary suggestion with learned antiquarians, to whom he later modestly gave credit for the idea.

His notion was not entirely novel, but the similar notions found in classical authors were fanciful and misleading. In the Beginning, according to Hesiod, Cronos created men of the Golden Age who never grew old. Labor, war, and injuries were unknown. They eventually became guardian spirits on earth. Then in the Silver Age, when men lost their reverence for the gods, Zeus punished them and buried them among the dead. The Bronze Age, which followed (when even houses were made of bronze), was a time of endless strife. After the brief interlude of a Heroic Age of godlike leaders in their Isles of the Blessed, came Hesiod's own unfortunate Iron Age. Yet worse was still in store for mankind, a future of men born senile, and of universal decay.

Thomsen was not well enough educated to try to fit his museum objects into this appealing literary scheme. He was more interested in objects than in words. There were already 'too many books,' he complained, and he was not eager to add his own. But finally, in 1836, he produced his practical *Guide to Scandinavian Antiquities*, which outlined his famous Three-Age System. This, his only book, translated into English, French, and German, and spread across Europe, was an invitation to 'Pre-History.'

It was hard for European scholars at the time to imagine that human experience before writing could have been divided into the epochs that Thomsen suggested. It seemed more logical to assume that stone tools were always used by the poor, while their betters always used bronze or iron. Thomsen's commonsense scheme did not please the pedants. If there was a Stone Age, they scoffed, then why not also an Age of Crockery, a Glass Age, and a Bone Age? Thomsen's scheme, refined but not abandoned by scholars in the next century, proved to be more than an exercise in museum management. It carried the plain message that human history had somehow developed in homogeneous stages that reached across the world. And he arranged the objects in his museum according to his 'principle of progressive culture.'

Thomsen showed how much was to be learned, not only from those ancient sculptures that embodied Winckelmann's ideal of beauty but even from the simple tools and crude weapons of anonymous prehistoric man. Opening his collections free to everybody, Thomsen offered lively talks about the everyday experience of people in the remote past. A deft lecturer, he would hide some interesting little object behind his coattails, then suddenly produce it at the point in his story when that kind of object – a bronze utensil or an iron weapon – first appeared in history.

Following Thomsen's hints, archaeologists discovered and explored the trash heaps of the past. Their paths into history no longer ran only through the gold-laden tombs of ancient kings, but also through the buried kitchen middens ('middens,' from an Old Scandinavian word for muck or dung-hill). The first excavation of these unlikely sources was the work mainly of Thomsen's disciple Jens Jacob Worsaae (1821–85). At the age of fifteen he had become Thomsen's museum assistant and during the next four years spent his holidays digging into the ancient barrows of Jutland with the aid of two laborers paid by his parents. With his athletic temperament and his outdoor enthusiasms he was the ideal complement to the museum-oriented Thomsen. In 1840, when he was only nineteen, using stratigraphy and the field evidence from Danish barrows and peat bogs, he published an article confirming Thomsen's Three-Age theory and assigning prehistoric objects to a Stone Age, a Bronze Age, or an Iron Age. He, too, was suggesting latitudes of time, throughout Denmark and beyond. A dozen years later, in 1853, the Swiss archaeologist Ferdinand Keller (1800–81),when exploring the lake dwellings of Lake Zurich, concluded that 'in Switzerland the three ages of stone, bronze, and iron, are quite as well represented as in Scandinavia.'

Some obvious difficulties plagued these prophets of prehistory. How could you stretch human experience to fill the thousands of years of the past opened by Buffon and the geologists? How much neater to fit all pre-Christian history into the comfortable 4004 years BC defined by Archbishop Ussher! And then there were new problems created by the geologists, who now revealed that northern Europe had been covered by ice when Stone Age men were living in caves in southern France. To correlate all these facts required a still more sophisticated approach to the early human past. If the Stone Age people of southern Europe advanced northward only after the retreat of the glaciers, then the three universal stages were reached at different times in different places.

To make the Three-Age scheme fit the whole human past in Europe was not easy. The so-called Age of Stone in Thomsen's museum was represented by polished stone artifacts of the kind people would be tempted to send in as curios. Meanwhile, Worsaae, out in the field, was hinting that the Age of Stone was far more extensive and more ancient than was suggested by these skillfully polished stone implements. On the digging sites each object unearthed could be studied not as an isolated curio but among all the remains of a Stone Age community. And these too might provide clues to other Stone Age communities across the world.

Worsaae's opportunity came in 1849, when a wealthy Dane named

Olsen was trying to improve his large estate called Meilgaard on the northeast coast of Jutland. Building a road, he sent his workmen in search of gravel for surfacing material. When they dug into a bank a half-mile from the shore, they found no gravel but luckily hit an eight-foot layer of oyster-shells, which was even better for their purpose. Mixed with the shells they found pieces of flint and animal bones. One small bone object two and a half inches long caught their attention. Shaped like a four-fingered hand, it was plainly the work of human craft. Perhaps it had been made for a comb.

Olsen, the proprietor, sharing the popular interest in antiquities which had been stimulated by Thomsen, sent the object to the museum in Copenhagen, where Worsaae's curiosity was aroused. Shell heaps recently turned up elsewhere in Denmark had brought to light flaked flint, odd pottery fragments, and crude stone objects similar to the Meilgaard comb. Perhaps this mound of oyster shells 'had been a sort of eating-place for the people of the neighborhood in the earliest prehistoric times. This would account for the ashes, the bones, the flints and the potsherds.' Perhaps here, at long last, modern man might visit an authentic Stone Age community. And actually imagine Stone Age men and women at their everyday meals. Worsaae observed that the shells had all been opened, which would not have been the case if they were merely washed up from the shore.

When other scholars disagreed, each with his own theory, the Danish Academy of Sciences appointed a commission. Worsaae, with a zoologist and a geologist, was assigned to interpret these shell heaps found along the ancient Danish shore. These 'shell middens,' the commission concluded, were really kitchen middens, which meant that now for the first time the historian could enter into the daily life of ancient peoples. Trash heaps might be gateways to prehistory. Such a discovery could not have been made indoors in a museum, but only on the spot in the field. Since the crudely crafted artifacts of the kitchen middens were never polished, unlike the polished stone artifacts of a later Stone Age, they were not likely to be noticed by laymen or sent to a museum. The kitchen middens opened another vast epoch of human prehistory – an early Stone Age, which extended behind the later Stone Age of polished stonework.

Thomsen and his museum collaborators had done their work of publicizing archaeology so well that the question now raised – whether the Stone Age really should be divided into two clearly defined stages –

was no longer an arcane conundrum for university professors. The issue was hotly debated in the public proceedings of the Danish Academy. Worsaae's opponents insisted that the shell heaps were only the picnic sites of the Stone Age visitors who had left their best implements elsewhere. The king of Denmark, Frederick VII, who shared the growing interest in antiquities, had excavated middens on his own estate and even wrote a monograph with his interpretation. In 1861, to 'settle' the issue, he summoned the leading scholars to a full-dress public meeting at Meilgaard, where he would preside. This royal conclave, no routine academic conference, would be celebrated with the panoply of a coronation. Besides hearing a debate, all those invited would witness the ritual excavation of a new portion of the mound. In the mid-June heat, archaeologists dug into the celebrated mound from eight in the morning till six in the evening, wearing their official 'archaeologist's' uniform out of respect for the King. When King Frederick had appointed Worsaae curator of his private collection of antiquities in 1858, he had playfully designated this archaeologist's uniform (high collar and tight-fitting jacket, topped off by a pillbox hat), which was now de rigueur at the diggings.

The lords of surrounding estates entertained the King and his party with banquets and dancing to band music every night. In honor of their royal visitor the neighbors created triumphal arches, and the King was accompanied everywhere by his mounted guard in full livery. A royal welcome to the Old Stone Age!

Early in the meeting it was agreed that Worsaae had won his scholarly point, which now would be proclaimed in royal company and for the whole nation. 'I had the especial satisfaction,' Worsaae wrote, 'of seeing that, among the many hundred stone implements discovered among the oysters, not a single specimen was found with any traces of polishing or of superior culture.' And he reported with relish how a human fillip was added to the formal splendor. 'Only at the last minute, after we had frequently remarked on this fact, did two polished axes turn up, of a completely different type, which some practical joker had inserted in the heap to cheat us.' The practical joker, it was widely assumed, was King Frederick himself.

Seldom has so drab an epoch of history been so splendidly inaugurated. But now, to the royal Danish imprimatur was added the near-unanimous agreement of scholars across Europe. What came to be called the Culture of Kitchen Middens (*c.* 4000–*c.* 2000 BC) was discovered in due course across the northern European coasts, and in Spain, Portugal, Italy, and North Africa. In southern Africa, northern Japan, in the islands of the Pacific, and in the coastal regions of both Americas, Kitchen Middens cultures seemed to have persisted into a later era. Once identified and placed in the chronicle of human development, the middens provided revealing latitudes of time – and a new vividness for the prehistoric past.

Worsaae, who became professor of archaeology at Copenhagen, and then succeeded Thomsen as director of the museum, is often called 'the first professional archaeologist.' His mentor Thomsen called him a 'heaven stormer.' Worsaae accurately praised Thomsen's Three-Age System as 'the first clear ray . . . shed across the Universal prehistoric gloom of the North and the World in general.' (Not in the heavily documented realms of recent history but in the dark recesses of earliest times would mankind first discover the 'universality' of history. The first discovery of the community of all human experience in eras and epochs, the worldwide phenomena of human history, was made when 'prehistory' was parsed into the three ages: Stone, Bronze, and Iron. And as Worsaae explored the boundaries between the three ages, he began to raise some profound questions that were explosive for fundamentalist Christians. One of these was the problem, still agitated by anthropologists: independent invention or cultural diffusion?

The disturbing notion, suggested by bold thinkers from Buffon to Darwin – that man had existed long before the Biblical date of Creation in 4004 BC – was beginning to be accepted by the scientific community. But the remote antiquity of man was popularized not so much by a theory as by the discovery of a vast and undeniable subject matter, a new dark continent of time, prehistory. More persuasively than a theory, the artifacts themselves seemed to bear witness to a chronology of prehistory that argued the evolution of man's culture.

Gradually, as the word 'prehistory' came into use in the European languages, the idea entered popular consciousness. The exhibition in Hyde Park in 1851, which purported to survey all the works of humankind, still gave no glimpse of prehistory. Then, at the Universal Exhibition in Paris in 1867, the Hall of the History of Labor showed an extensive collection of artifacts from all over Europe and from Egypt. The official guide to Prehistoric Walks at the Universal Exhibition offered three lessons from the new science: the law of the progress of humanity; the law of similar development; and the high antiquity of man. In that same year the announcement of the first Congrès International Préhistorique de Paris brought the first official use of the word 'prehistoric.'

Source: Daniel J. Boorstin, *The Discoverers*, New York, Random House, 1983.

Chains and Rings: Kekule's Dreams

Nineteenth-century chemists were puzzled to find that many organic substances with the same chemical formula had widely different properties. Though they seemed to be chemically identical, they were in fact different substances. It gradually became evident that this was because the arrangement of the atoms within their molecular structure was different. The scientist who made this breakthrough was Friedrich August Kekule (1829–96). He said that the fundamental theory of organic molecular structure came to him in a dream.

During my stay in London I resided for a considerable time in Clapham Road in the neighborhood of Clapham Common. I frequently, however, spent my evenings with my friend Hugo Müller at Islington at the opposite end of the metropolis. We talked of many things but most often of our beloved chemistry. One fine summer evening I was returning by the last bus, 'outside,' as usual, through the deserted streets of the city, which are at other times so full of life. I fell into a reverie, and lo, the atoms were gamboling before my eyes! Whenever, hitherto, these diminutive beings had appeared to me, they had always been in motion; but up to that time I had never been able to discern the nature of their motion. Now, however, I saw how, frequently, two smaller atoms united to form a pair; how a larger one embraced the two smaller ones; how still larger ones kept hold of three or even four of the smaller; whilst the whole kept whirling in a giddy dance. I saw how the larger ones formed a chain, dragging the smaller ones after them but only at the ends of the chain . . . The cry of the conductor: 'Clapham Road,' awakened me from my dreaming: but I spent a part of the night in putting on paper at least sketches of these dream forms. This was the origin of the 'Structural Theory.'

Kekule published his *Theory of Molecular Structure* in 1858, explaining how carbon atoms link together to form chains, just as his dream had told him.

However, his theory failed to cover the whole field of organic chemistry. One important group of substances, related to the coal-tar hydrocarbon benzene, failed to fit his theory. These were known as 'aromatic' compounds, because many of them occur in fragrant oils and aromatic spices. Kekule brooded over the problem of the aromatic compounds for a further seven years, trying to devise a structural formula that would account for their peculiar chemical characteristics. Then he had another dream.

Something similar happened with the benzene theory. During my stay in Ghent I resided in elegant bachelor quarters in the main thoroughfare. My study, however, faced a narrow side-alley and no daylight penetrated it. For the chemist who spends his day in the laboratory this mattered little. I was sitting writing at my textbook but the work did not progress; my thoughts were elsewhere. I turned my chair to the fire and dozed. Again the atoms were gamboling before my eyes. This time the smaller groups kept modestly in the background. My mental eye, rendered more acute by repeated visions of the kind, could now distinguish larger structures of manifold conformation: long rows, sometimes more closely fitted together all twining and twisting in snake-like motion. But look! What was that? One of the snakes had seized hold of its own tail, and the form whirled mockingly before my eyes. As if by a flash of lightning I awoke; and this time also I spent the rest of the night in working out the consequences of the hypothesis.

Kekule's realization that carbon atoms form rings as well as chains was made public in 1866. His sketch of the benzene ring shows that it consists of six carbon atoms each carrying a hydrogen atom. Later research has confirmed that all organic Nature is based on the carbon chain and the carbon ring, and that life itself depends on the capacity of carbon atoms to form molecular chains and rings as they did in Kekule's dreams.

Source: Kekule's reminiscence reprinted in: O. T. Benfey, *Journal of Chemical Education*, vol. 35, 1958, p. 211.

On a Piece of Chalk

Thomas Henry Huxley (1825–95), an ardent Darwinian, was the greatest Victorian scientific popularizer. He coined the word 'agnostic' for disbelievers like himself, and his book *Man's Place in Nature* (1863) impressed humanity's ape-origins on the public imagination. Its frontispiece showed a queue of skeletons, with man at the front, and progressively more stooping apes behind. His lectures drew huge crowds – 2,000 were turned away in January 1866 when he inaugurated the 'Sunday Evenings for the People' at St Martin's Hall (Jenny Marx, Karl's daughter, squeezed in and found it 'packed to suffocation'). His most famous moment came at a British Association meeting in June 1860, when he clashed with Bishop 'Soapy Sam' Wilberforce. The bishop inquired whether it was on his grandfather's or his grandmother's side that he was descended from an ape. Huxley retorted that if he were asked whether he would rather have an ape as ancestor, or a man who, possessed of great means and faculties, employed them for the purpose of introducing ridicule into scientific debate, he would unhesitatingly choose the ape. Thirteen years later, when Wilberforce was pitched on his head while riding and killed, Huxley commented, 'For once reality and his brain came into contact, and the result was fatal.'

When the young H. G. Wells won a scholarship to the Royal College of Science, it was the teaching of Huxley ('a yellow-faced, square-jawed old man, with bright little brown eyes') that inspired him. So without Huxley's scientific imagination we might never have had science fiction, the genre Wells virtually invented. 'A Piece of Chalk' was originally a lecture given to working men at a meeting of the British Association in Norwich in 1868.

If a well were sunk at our feet in the midst of the city of Norwich, the diggers would very soon find themselves at work in that white substance almost too soft to be called rock, with which we are all familiar as 'chalk'.

Not only here, but over the whole county of Norfolk, the well-sinker might carry his shaft down many hundred feet without coming to the end of the chalk; and, on the sea coast, where the waves have pared away the face of the land which breasts them, the scarped faces of the high cliffs are often wholly formed of the same material. Northward, the chalk may be followed as far as Yorkshire; on the south coast it appears abruptly in the picturesque western bays of Dorset, and breaks into the Needles of the Isle of Wight; while on the shores of Kent it supplies that long line of white cliffs to which England owes her name of Albion.

Were the thin soil which covers it all washed away, a curved band of white chalk, here broader, and there narrower, might be followed diagonally across England from Lulworth in Dorset, to Flamborough Head in Yorkshire – a distance of over 280 miles as the crow flies . . .

Attaining as it does in some places a thickness of more than a thousand feet, the English chalk must be admitted to be a mass of considerable magnitude. Nevertheless, it covers but an insignificant portion of the whole area occupied by the chalk formation of the globe, much of which has the same general characters as ours, and is found in detached patches, some less, and others more extensive, than the English. Chalk occurs in north-west Ireland; it stretches over a large part of France – the chalk which underlies Paris being in fact a continuation of that of the London basin; it runs through Denmark and central Europe, and extends southward to North Africa; while eastward it appears in the Crimea and in Syria, and may be traced as far as the Sea of Aral, in Central Asia. If all the points at which true chalk occurs were circumscribed, they would lie within an irregular oval about 3,000 miles in long diameter – the area of which would be as great as that of Europe, and would many times exceed that of the largest existing inland sea – the Mediterranean . . .

Thus the chalk is no unimportant element in the masonry of the earth's crust . . . What is this widespread component of the surface of the earth? and whence did it come?

You may think this no very hopeful inquiry. You may not unnaturally suppose that the attempt to solve such problems as these can lead to no result, save that of entangling the enquirer in vague speculations, incapable of refutation and of verification. If such were really the case, I should have selected some other subject than a 'piece of chalk' for my discourse. But, in truth, after much deliberation I have been unable to think of any topic which would so well enable me to lead you to see how solid is the foundation upon which some of the most startling conclusions of physical science rest.

A great chapter of the history of the world is written in the chalk . . . To the unassisted eye chalk looks like a very loose and open kind of stone. But it is possible to grind a slice of chalk down so thin that you can see through it – until it is thin enough, in fact, to be examined with any magnifying power that may be thought desirable . . . When placed under the microscope, the general mass of it is made up of very minute granules; but, imbedded in this matrix are innumerable bodies, some smaller and some larger, but on a rough average not more than a hundredth of an inch in diameter, having a well-defined shape and structure. A cubic inch of some specimens of chalk may contain hundreds of thousands of these bodies, compacted together with incalculable millions of granules.

The examination of a transparent slice gives a good notion of the manner in which the components of the chalk are arranged, and of their relative proportions. But, by rubbing up some chalk with a brush in water and then pouring off the milky fluid, so as to obtain sediments of different degrees of fineness, the granules and the minute rounded bodies may be pretty well separated from one another, and submitted to microscopic examination, either as opaque or as transparent objects. By combining the views obtained in these various methods, each of the rounded bodies may be proved to be a beautifully-constructed calcareous fabric, made up of a number of chambers, communicating freely with one another. The chambered bodies are of various forms. One of the commonest is something like a badly-grown raspberry, being formed of a number of nearly globular chambers of different sizes congregated together. It is called *Globigerina*, and some specimens of chalk consist of little else than *Globigerinae* and granules. Let us fix our attention upon the *Globigerina*. It is the spoor of the game we are tracking. If we can learn what it is and what are the conditions of its existence, we shall see our way to the origins and the past history of the chalk . . .

It so happens that calcareous skeletons, exactly similar to the *Globigerinae* of the chalk, are being formed, at the present moment, by minute living creatures, which flourish in multitudes, literally more numerous than the sands of the sea-shore, over a large extent of that part of the earth's surface which is covered by the ocean . . . *Globigerinae* of every size, from the smallest to the largest, are associated together in the Atlantic mud, and the chambers of many are filled by a soft animal matter. This soft substance is, in fact, the remains of the creature to which the *Globigerinae* shell, or rather skeleton, owes its existence – and which is an animal of the simplest imaginable description. It is, in fact, a mere particle of living jelly without defined parts of any kind – without a mouth, nerves, muscles, or distinct organs, and only manifesting its vitality to ordinary observation by thrusting out and retracting from all parts of its surface long filamentous processes, which serve for arms and legs. Yet this amorphous particle, devoid of everything which, in the higher animals, we call organs, is capable of feeding, growing, and multiplying; of separating from the ocean the small proportion of carbonate of lime which is dissolved in sea water; and of building up that substance into a skeleton for itself, according to a pattern which can be imitated by no other known agency . . .

The important points for us are that the living *Globigerinae* are exclusively marine animals, the skeletons of which abound at the bottoms of deep seas; and that there is not a shadow of reason for believing that the habits of the *Globigerinae* of the chalk differed from those of the existing species. But if this be true, there is no escaping the conclusion that the chalk itself is the dried mud of an ancient deep sea.

In working over the soundings [samples of mud from the floor of the Atlantic, collected for Huxley by HMS *Cyclops* in 1857], I was surprised to find that many of what I have called the 'granules' of that mud were not, as one might have been tempted to think at first, the mere powder and waste of *Globigerinae*, but that they had a definite form and size. I termed these bodies *coccoliths*, and doubted their organic nature. Dr Wallich verified my observation, and added the interesting discovery that, not infrequently, bodies similar to these coccoliths were aggregated together into spheroids, which he termed *coccospheres*. So far as we knew, these bodies, the nature of which is extremely puzzling and problematical, were peculiar to the Atlantic soundings. But, a few years ago, Mr Sorby, in making a careful examination of the chalk by means of thin sections, observed that much of its granular basis possesses a definite form. Comparing these formed particles with those in the Atlantic soundings, he found the two to be identical . . . Here was a further and most interesting confirmation, from internal evidence, of the essential identity of the chalk with modern deep-sea mud. *Globigerinae*, coccoliths and coccospheres are found as the chief constituents of both . . .

When we consider that the remains of more than three thousand distinct species of aquatic animals have been discovered among the fossils of the chalk, that the great majority of them are of such forms as are now met with only in the sea, and that there is no reason to believe that any one of them inhabited fresh water – the collateral evidence that the chalk represents an ancient sea-bottom acquires as great force as the proof derived from the nature of the chalk itself. I think you will allow that I did not overstate my case when I asserted that we have as strong grounds for believing that all the vast area of dry land at present occupied by the chalk was once at the bottom of the sea, as we have for any matter of history whatever; while there is no justification for any other belief.

No less certain is it that the time during which the countries which we now call south-east England, France, Germany, Poland, Russia, Egypt, Arabia, Syria, were more or less completely covered by a deep sea, was of considerable duration. We have already seen that the chalk is, in places, more than a thousand feet thick. I think you will agree with me that it must have taken some time for the skeletons of animalcules of a hundredth of an inch in diameter to heap up such a mass as that. I have said that throughout the thickness of the chalk the remains of other animals are scattered. These remains are often in the most exquisite state of preservation. The valves of the shell-fishes are commonly adherent; the long spines of some of the sea-urchins, which would be detached by the smallest jar, often remain in their places. In a word, it is certain that these animals have lived and died when the place which they now occupy was the surface of as much of the chalk as had then been deposited; and that each has been covered up by the layer of *Globigerinae* mud upon which the creatures embedded a little higher up have, in like manner, lived and died . . .

Huxley now turns his attention to the strata above the chalk layer, among them the glacial deposits known as boulder clay and drift.

At one of the most charming spots on the coast of Norfolk, Cromer, you will see the boulder clay forming a vast mass, which lies upon the chalk, and must consequently have come into existence after it . . . The chalk, then, is certainly older than the boulder clay. If you ask how much, I will again take you no further than the same spot upon your own coasts for evidence. I have spoken of the boulder clay and drift as resting upon the chalk. That is not strictly true. Interposed between the chalk and the drift is a comparatively insignificant layer, containing vegetable matter. But that layer tells a wonderful history. It is full of stumps of trees standing as they grew. Fir-trees are there with their cones, and hazel-bushes with their nuts; there stand the stools of oak and yew trees, beeches and alders. Hence this stratum is appropriately called the 'forest-bed'.

It is obvious that the chalk must have been upheaved and converted into dry land before the timber trees could grow upon it. As the bolls of some of these trees are from two to three feet in diameter, it is no less clear that the dry land thus formed remained in the same condition for long ages. And not only do the remains of stately oaks and well-grown firs testify to the duration of this condition of things, but additional evidence to the same effect is afforded by the abundant remains of elephants, rhinoceroses, hippopotamuses, and other great wild beasts, which it has yielded to the zealous search of such men as the Rev. Mr Gunn. When you look at such a collection as he has formed, and bethink you that these elephantine bones did veritably carry their owners about, and these great grinders crunch, in the dark woods of which the forest-bed is now the only trace, it is impossible not to feel that they are as good evidence of the lapse of time as the annual rings of the tree stumps.

Thus there is writing upon the wall of cliffs at Cromer, and whoso runs may read it. It tells us, with an authority that cannot be impeached, that the ancient sea-bed of the chalk sea was raised up, and remained dry land until it was covered with forests, stocked with the great game the spoils of which have rejoiced your geologists. How long it remained in that condition cannot be said; but 'the whirligig of time brought in its revenges' in those days as in these. That dry land, with the bones and teeth of generations of long-lived elephants hidden away among the gnarled roots and dry leaves of its ancient trees, sank gradually to the bottom of the icy sea, which covered it with huge masses of drift and boulder clay. Sea-beasts, such as the walrus, now restricted to the extreme north, paddled about where birds had twittered among the topmost twigs of the fir trees. How long this state of things endured we know not, but at length it came to an end. The upheaved glacial mud hardened into the soil of modern Norfolk. Forests grew once more, the wolf and the beaver replaced the reindeer and the elephant; and at length what we call the history of England dawned.

Thus you have, within the limits of your own county, proof that the chalk can justly claim a very much greater antiquity than even the oldest physical traces of mankind . . . Evidence which cannot be rebutted, and which need not be strengthened, though if time permitted I might infinitely increase its quantity, compels you to believe that the earth, from the time of the chalk to the present day, has been the theatre of a series of changes as vast in their amount, as they were slow in their progress. The area on which we stand has been first sea and then land, for at least four alterations; and has remained in each of these conditions for a period of great length.

Nor have these wonderful metamorphoses of sea into land, and of land into sea, been confined to one corner of England. During the chalk period, or 'cretaceous epoch', not one of the present great physical features of the globe was in existence. Our great mountain ranges, Pyrenees, Alps, Himalayas, Andes, have all been upheaved since the chalk was deposited, and the cretaceous sea flowed over the sites of Sinai and Ararat . . .

I must ask you to believe that there is no less conclusive proof that a still more prolonged succession of similar changes occurred, before the chalk was deposited. Nor have we any reason to think that the first term in the series of these changes is known. The oldest sea-beds preserved to us are sands, and mud, and pebbles, the wear and tear of rocks which were formed in still older oceans.

But great as is the magnitude of these physical changes of the world, they have been accompanied by a no less striking series of modifications in its living inhabitants. All the great classes of animals, beasts of the field, fowls of the air, creeping things, and things which dwell in the waters, flourished upon the globe long ages before the chalk was deposited. Very few, however, if any, of these ancient forms of animal life were identified with those which now live. Certainly not one of the higher animals was of the same species as any of those now in existence. The beasts of the field, in the days before the chalk, were not our beasts of the field, nor the fowls of the air such as those which the eye of man has seen flying, unless his antiquity dates infinitely further back than we at present surmise. If we could be carried back into those times, we should be as one suddenly set down in Australia before it was colonized. We should see mammals, birds, reptiles, fishes, insects, snails, and the like, clearly recognized as such, and yet not one of them would be just the same as those with which we are familiar, and many would be extremely different.

From that time to the present the population of the world has undergone slow and gradual, but incessant, changes. There has been no grand catastrophe – no destroyer has swept away the forms of life of one period, and replaced them by a totally new creation: but one species has vanished and another has taken its place; creatures of one type of structure have diminished, those of another have increased as time has passed on. And thus, while the differences between the living creatures of the time before the chalk and those of the present day appear startling, if placed side by side, we are led from one to the other by the most gradual progress, if we follow the course of Nature through the whole series of those relics of her operations which she has left behind. It is by the population of the chalk sea that the ancient and modern inhabitants of the world are most completely connected. The groups which are dying out flourish side by side with the groups which are now the dominant forms of life. Thus the chalk contains remains of those strange flying and swimming reptiles, the pterodactyl, the ichthyosaurus, and the plesiosaurus, which are found in no later deposits, but abounded in preceding ages. The chambered cells called ammonites and belemnites, which are so characteristic of the period preceding the cretaceous, in like manner die with it.

But amongst these fading remainders of a previous state of things, are some very modern forms of life, looking like Yankee pedlars among a tribe of Red Indians. Crocodiles of modern type appear; bony fishes, many of them very similar to existing species, almost supplant the forms of fish which predominate in more ancient seas; and many kinds of living shell-fish first become known to us in the chalk . . .

There is not a shadow of a reason for believing that the physical changes of the globe, in past times, have been effected by other than natural causes. Is there any more reason for believing that the concomitant modifications in the forms of the living inhabitants of the globe have been brought about in other ways? . . . Science gives no countenance to such a wild fancy; nor can even the perverse ingenuity of a commentator pretend to discover this sense in the simple words in which the writer of Genesis records the proceedings of the fifth and sixth days of the Creation.

A small beginning has led us to a great ending. If I were to put the bit of chalk with which we started into the hot but obscure flame of burning hydrogen, it would presently shine like the sun. It seems to me that this physical metamorphosis is no false image of what has been the result of our subjecting it to a jet of fervent, though no-wise brilliant, thought tonight. It has become luminous, and its clear rays, penetrating the abyss of the remote past, have brought within our ken some stages of the evolution of the earth. And in the shifting 'without haste, but without rest' of the land and sea, as in the endless variation of the forms assumed by living beings, we have observed nothing but the natural product of the forces originally possessed by the substance of the universe.

Source: T. H. Huxley, *Collected Essays*, London, Macmillan & Co., 1894–1908.

Siberia Breeds a Prophet

In 1869 the Russian chemist Dmitri Ivanovitch Mendeléeff – or Mendeleyev – (1834–1907) began a new chapter in the history of chemistry by devising a periodic table of the elements according to their relative atomic weights. Making the periodic table dull for non-specialist readers is very easy, and many would-be popular science writers have done it. Making it as absorbing as Bernard Jaffe does takes a great deal of knowledge and imagination. Jaffe's book *Crucibles*, from which this extract comes, was originally published in America in 1930, and appropriately won the Francis Bacon Award for the Humanizing of Knowledge. Since then it has been through many reprintings and four updated editions.

Out of Russia came the patriarchal voice of a prophet of chemistry. 'There is an element as yet undiscovered. I have named it eka-aluminium. By properties similar to those of the metal aluminium you shall identify it. Seek it, and it will be found.' Startling as was this prophecy, the sage of Russia was not through. He predicted another element resembling the element boron. He was even bold enough to state its atomic weight. And before that voice was stilled, it foretold the discovery of a third element whose physical and chemical properties were thoroughly described. No man, not even the Russian himself, had beheld these unknown substances.

This was the year 1869. The age of miracles was long past. Yet here was a distinguished scientist, holding a chair of chemistry at a famous university, covering himself with the mantle of the prophets of old. Had he gathered this information from inside the crystal glass of some sorcerer? Perhaps, like the seer of ancient times, he had gone to the top of a mountain to bring down the tablets of these new elements. But this oracle disdained the robes of a priest. Rather did he announce his predictions from the stillness of his chemical laboratory, where midst the smoke, not of a burning bush, but of the fire of his furnace, he had seen visions of a great generalization in chemistry.

Chemistry had already been the object of prophecy. When Lavoisier heated some tin in a sealed flask and found it to change in appearance and weight, he saw clearly a new truth, and foretold other changes. Lockyer a year before had looked through a new instrument – the spectroscope devised by Bunsen and Kirchhof. Through this spectroscope he had gazed at the bright colored lines of a new element ninety-three million miles away. Since it was present in the photosphere of the sun he called it *helium* and predicted its existence on our earth. Twenty-one years later, William Hillebrand of the United States Geological Survey, came across this gas in the rare mineral cleveite.

But the predictions of the Russian were more astounding. He had made no direct experiments. He had come to his conclusions seemingly out of thin air. There had gradually been born in the fertile mind of this man the germ of a great truth. It was a fantastic seed but it germinated with surprising rapidity. When the flower was mature, he ventured to startle the world with its beauty.

In 1884 Sir William Ramsay had come to London to attend a dinner given in honor of William Perkin, the discoverer of the dye mauve.

> I was very early at the dinner [Ramsay recalled] and was putting off time looking at the names of people to be present, when a peculiar foreigner, every hair of whose head acted in independence of every other, came up bowing. I said, 'We are to have a good attendance, I think?' He said, 'I do not spik English.' I said, 'Vielleicht sprechen Sie Deutsch?' He replied, 'Ja ein wenig. Ich bin Mendeléeff.' Well, we had twenty minutes or so before anyone else turned up and we talked our mutual subject fairly out. He is a nice sort of fellow but his German is not perfect. He said he was raised in East Siberia and knew no Russian until he was seventeen years old. I suppose he is a Kalmuck or one of those outlandish creatures.

This 'outlandish creature' was Mendeléeff, the Russian prophet to whom the world listened. Men went in search of the missing elements he described. In the bowels of the earth, in the flue dust of factories, in the waters of the oceans, and in every conceivable corner they hunted. Summers and winters rolled by while Mendeléeff kept preaching the truth of his visions. Then, in 1875, the first of the new elements he foretold was discovered. In a zinc ore mined in the Pyrenees, Lecoq de Boisbaudran came upon the hidden eka-aluminium. This Frenchman analyzed and reanalyzed the mineral and studied the new element in every possible way to make sure there was no error. Mendeléeff must indeed be a prophet! For here was a metal exactly similar to his eka-aluminium. It yielded its secret of two new lines to the spectroscope, it was easily fusible, it could form alums, its chloride was volatile. Every one of these characteristics had been accurately foretold by the Russian. Lecoq named it *gallium* after the ancient name of his native country.

But there were many who disbelieved. 'This is one of those strange guesses which by the law of averages must come true,' they argued. Silly to believe that new elements could be predicted with such accuracy! One might as well predict the birth of a new star in the heavens. Has not Lavoisier, the father of chemistry, declared that 'all that can be said upon the nature and number of the elements is confined to discussions entirely of a metaphysical nature? The subject only furnishes us with indefinite problems.'

But then came the news that Winkler, in Germany, had stumbled over another new element, which matched the eka-silicon of Mendeléeff. The German had followed the clue of the Russian. He was looking for a dirty gray element with an atomic weight of about 72, a density of 5.5, an element which was slightly acted upon by acids. From the silver ore, argyrodite, he isolated a grayish white substance with atomic weight of 72.3 and a density of 5.5. He heated it in air and found its oxide to be exactly as heavy as had been predicted. He synthesized its ethide and found it to boil at exactly the temperature that Mendeléeff had prefigured. There was not a scintilla of doubt about the fulfilment of Mendeléeff's second prophecy. The spectroscope added unequivocal testimony. Winkler announced the new element under the name of *germanium* in honor of his fatherland. The sceptics were dumbfounded. Perhaps after all the Russian was no charlatan!

Two years later the world was completely convinced. Out of Scandinavia came the report that Nilson had isolated eka-boron. Picking up the scent of the missing element in the ore of euxenite, Nilson had tracked it down until the naked element, exhibiting every property foreshadowed for it, lay before him in his evaporating dish. The data were conclusive. The whole world of science came knocking at the door of the Russian in St Petersburg.

Dmitri Ivanovitch Mendeléeff came of a family of heroic pioneers.

More than a century before his birth, Peter the Great had started to westernize Russia. Upon a marsh of pestilence he reared a mighty city which was to be Russia's window to the West. For three-quarters of a century Russia's intellectual march eastward continued, until in 1787 in Tobolsk, Siberia, the grandfather of Dmitri opened up the first printing press, and with the spirit of a pioneer published the first newspaper in Siberia, the *Irtysch*. In this desolate spot, settled two centuries before by the Cossacks, Dmitri was born on February 7, 1834. He was the last of a family of seventeen children.

Misfortune overtook his family. His father, director of the local high school, became blind, and soon after died of consumption. His mother, Maria Korniloff, a Tartar beauty, unable to support her large family on a pension of five hundred dollars a year, reopened a glass factory which her family was the first to establish in Siberia. Tobolsk at this time was an administrative center to which Russian political exiles were taken. From one of these prisoners of the revolt of 1825, a 'Decembrist' who married his sister, Dimitri learned the rudiments of natural science. When fire destroyed the glass factory, little Dmitri, pet of his ageing mother – she was already fifty-seven – was taken to Moscow in the hope that he might be admitted to the University. Official red tape prevented this. Determined that her son should receive a good scientific education, his mother undertook to move to St Petersburg, where he finally gained admittance to the Science Department of the Pedagogical Institute, a school for the training of high school teachers. Here he specialized in mathematics, physics and chemistry. The classics were distasteful to this blue-eyed boy. Years later, when he took a hand in the solution of Russia's educational problems, he wrote, 'We could live at the present time without a Plato, but a double number of Newtons is required to discover the secrets of nature, and to bring life into harmony with its laws.'

Mendeléeff worked diligently at his studies and graduated at the head of his class. Never very robust during these early years, his health gradually weakened, and the news of his mother's death completely unnerved him. He had come to her as she lay on her death bed. She spoke to him of his future: 'Refrain from illusions, insist on work and not on words. Patiently search divine and scientific truth.' Mendeléeff never forgot those words. Even as he dreamed, he always felt the solid earth beneath his feet.

His physician gave him six months to live. To regain his health, he was ordered to seek a warmer climate. He went to the south of Russia and obtained a position as science master at Simferopol in the Crimea. When the Crimean War broke out he left for Odessa, and at the age of twenty-two he was back in St Petersburg as a privat-docent. An appointment as privat-docent meant nothing more than permission to teach, and brought no stipend save a part of the fees paid by the students who attended the lectures. Within a few years he asked and was granted permission from the Minister of Public Instruction to study in France and Germany. There was no opportunity in Russia for advanced work in science. At Paris he worked in the laboratory of Henri Regnault and, for another year, at Heidelberg in a small private laboratory built out of his meager means. Here he met Bunsen and Kirchhof from whom he learned the use of the spectroscope, and together with Kopp attended the Congress of Karlsruhe, listening to the great battle over the molecules of Avogadro. Cannizarro's atomic weights were to do valiant service for him in the years to come. Mendeléeff's attendance at this historic meeting ended his *Wanderjahre*.

The next few years were very busy ones. He married, completed in sixty days a five-hundred-page textbook on organic chemistry which earned him the Domidoff Prize, and gained his doctorate in chemistry for a thesis on *The Union of Alcohol with Water*. The versatility of this gifted teacher, chemical philosopher and accurate experimenter was soon recognized by the University of St Petersburg, which appointed him full professor before he was thirty-two.

Then came the epoch-making year of 1869. Mendeléeff had spent twenty years reading, studying and experimenting with the chemical elements. All these years he had been busy collecting a mass of data from every conceivable source. He had arranged and rearranged this data in the hope of unfolding a secret. It was a painstaking task. Thousands of scientists had worked on the elements in hundreds of laboratories scattered over the civilized world. Sometimes he had to spend days searching for missing data to complete his tables. The number of the elements had increased since the ancient artisans fashioned instruments from their gold, silver, copper, iron, mercury, lead, tin, sulfur and carbon. The alchemists had added six new elements in their futile search for the seed of gold and the elixir of life. Basil Valentine, a German physician, in the year when Columbus was discovering America had rather fancifully described antimony. In 1530

Georgius Agricola, another German, talked about bismuth in his *De Re Metallica*, a book on mining which was translated into English for the first time by a (later) President of the United States, Herbert Hoover, and his wife in 1912. Paracelsus was the first to mention the metal zinc to the Western World. Brandt discovered glowing phosphorus in urine, and arsenic and cobalt were soon added to the list of the elements.

Before the end of the eighteenth century, fourteen more elements were discovered. In faraway Choco, Colombia, a Spanish naval officer, Don Antonio de Ulloa, had picked up a heavy nugget while on an astronomical mission, and had almost discarded it as worthless before the valuable properties of the metal platinum were recognized. This was in 1735. Then came lustrous nickel, inflammable hydrogen, inactive nitrogen, life-giving oxygen, death-dealing chlorine, manganese, used among other things for burglar-proof safes, tungsten, for incandescent lamps, chromium, for stainless steel, molybdenum and titanium, so useful in steel alloys, tellurium, zirconium, and uranium, heaviest of all the elements. The nineteenth century had hardly opened when Hatchett, an Englishman, discovered columbium (niobium) in a black mineral that had found its way from the Connecticut Valley to the British Museum. And thus the search went on, until in 1869 sixty-three different elements had been isolated and described in the chemical journals of England, France, Germany and Sweden.

Mendeléeff gathered together all the data on these sixty-three chemical elements. He did not miss a single one. He even included fluorine whose presence was known, but which had not yet been isolated because of its tremendous activity. Here was a list of all the chemical elements, every one of them consisting of different Daltonian atoms. Their atomic weights, ranging from 1 (hydrogen) to 238 (uranium), were all dissimilar. Some, like oxygen, hydrogen, chlorine and nitrogen, were gases. Others, like mercury and bromine, were liquids under normal conditions. The rest were solids. There were some very hard metals like platinum and iridium, and soft metals like sodium and potassium. Lithium was a metal so light that it could float on water. Osmium, on the other hand, was twenty-two and a half times as heavy as water. Here was mercury, a metal which was not a solid at all, but a liquid. Copper was red, gold yellow, iodine steel gray, phosphorus white, and bromine red. Some metals, like nickel and chromium, could take a very high polish; others like lead and aluminium, were duller. Gold, on exposure to the air, never tarnished, iron rusted very easily, iodine sublimed and changed into a vapor. Some elements united with one atom of oxygen, others with two, three or four atoms. A few, like potassium and fluorine, were so active that it was dangerous to handle them with unprotected fingers. Others could remain unchanged for ages. What a maze of varying, dissimilar, physical characteristics and chemical properties!

Could some order be found in this body of diverse atoms? Was there any connection between these elements? Could some system of evolution or development be traced among them, such as Darwin, ten years before, had found among the multiform varieties of organic life? Mendeléeff wondered. The problem haunted his dreams. Constantly his mind reverted to this puzzling question.

Mendeléeff was a dreamer and a philosopher. He was going to find the key to this heterogeneous collection of data. Perhaps nature had a simple secret to unfold. And while he believed it to be 'the glory of God to conceal a thing,' he was firmly convinced that it was 'the honor of kings to search it out.' And what a boon it would prove to his students!

He arranged all the elements in the order of increasing atomic weights, starting with the lightest, hydrogen, and completing his table with uranium, the heaviest. He saw no particular value in arranging the elements in this way; it had been done previously. Unknown to Mendeléeff, an Englishman, John Newlands, had three years previously read, before the English Chemical Society at Burlington House, a paper on the arrangement of the elements. Newlands had noticed that each succeeding eighth element in his list showed properties similar to the first element. This seemed strange. He compared the table of the elements to the keyboard of a piano with its eighty-eight notes divided into periods or octaves of eight. 'The members of the same group of elements,' he said, 'stand to each other in the same relation as the extremities of one or more *octaves* in music.' The members of the learned society of London laughed at his Law of Octaves. Professor Foster ironically inquired if he had ever examined the elements according to their initial letters. No wonder – think of comparing the chemical elements to the keyboard of a piano! One might as well compare the sizzling of sodium as it skims over water to the music of the heavenly spheres. 'Too fantastic,' they agreed, and J. A. R. Newlands almost went down to oblivion.

Mendeléeff was clear-visioned enough not to fall into such a pit. He took sixty-three cards and placed on them the names and properties of the elements. These cards he pinned on the walls of his laboratory. Then he carefully re-examined the data. He sorted out the similar elements and pinned their cards together again on the walls. A striking relationship was thus made clear.

Mendeléeff now arranged the elements into seven groups, starting with lithium (at. wt. 7), and followed by beryllium (at. wt. 9), boron (11), carbon (12), nitrogen (14), oxygen (16) and fluorine (19). The next element in the order of increasing atomic weight was sodium (23). This element resembled lithium very closely in both physical and chemical properties. He therefore placed it below lithium in his table. After placing five more elements he came to chlorine, which had properties very similiar to fluorine, under which it miraculously fell in his list. In this way he continued to arrange the remainder of the elements. When his list was completed he noticed a most remarkable order. How beautifully the elements fitted into their places! The very active metals lithium, sodium, potassium, rubidium and caesium fell into one group (No. 1). The extremely active non-metals, fluorine, chlorine, bromine and iodine, all appeared in the seventh group.

Mendeléeff had discovered that the properties of the elements 'were periodic functions of their atomic weights,' that is, their properties repeated themselves periodically after each seven elements. What a simple law he had discovered! But here was another astonishing fact. All the elements in Group I united with oxygen two atoms to one. All the atoms of the second group united with oxygen atom for atom. The elements in Group III joined with oxygen two atoms to three. Similar uniformities prevailed in the remaining groups of elements. What in the realm of nature could be more simple? To know the properties of one element of a certain group was to know, in a general way, the properties of all the elements in that group. What a saving of time and effort for his chemistry students!

Could his table be nothing but a strange coincidence? Mendeléeff wondered. He studied the properties of even the rarest of the elements. He re-searched the chemical literature lest he had, in the ardor of his work, misplaced an element to fit in with his beautiful edifice. Yes, here was a mistake! He had misplaced iodine, whose atomic weight was recorded as 127, and tellurium, 128, to agree with his scheme of things. Mendeléeff looked at his Periodic Table of the Elements and saw that it was good. With the courage of a prophet he made bold to say that the atomic weight of tellurium was wrong; that it must be between 123 and 126 and not 128, as its discoverer had determined. Here was downright heresy, but Dmitri was not afraid to buck the established order of things. For the present, he placed the element tellurium in its proper position, but with its false atomic weight. Years later his action was upheld, for further chemical discoveries proved his position of tellurium to be correct. This was one of the most magnificent prognostications in chemical history.

Perhaps Mendeléeff's table was now free from flaws. Again he examined it, and once more he detected an apparent contradiction. Here was gold with the accepted atomic weight of 196.2 placed in a space which rightfully belonged to platinum, whose established atomic weight was 196.7. The fault-finders got busy. They pointed out this discrepancy with scorn. Mendeléeff made brave enough to claim that the figures of the analysts, and not his table, were inaccurate. He told them to wait. He would be vindicated. And again the balance of the chemist came to the aid of the philosopher, for the then-accepted weights were wrong and Mendeléeff was again right. Gold had an atomic weight greater than platinum. This table of the queer Russian was almost uncanny in its accuracy!

Mendeléeff was still to strike his greatest bolt. Here were places in his table which were vacant. Were they always to remain empty or had the efforts of man failed as yet to uncover some missing elements which belonged in these spaces? A less intrepid person would have shrunk from the conclusion that this Russian drew. Not this Tartar, who would not cut his hair even to please his Majesty, Czar Alexander III. He was convinced of the truth of his great generalization, and did not fear the blind, chemical sceptics.

Here in Group III was a gap between calcium and titanium. Since it occurred under boron, the missing element must resemble boron. This was his eka-boron which he predicted. There was another gap in the same group under aluminium. This element must resemble aluminium, so he called it eka-aluminium. And finally he found another vacant space between arsenic and eka-aluminium, which appeared in the fourth group. Since its position was below the element silicon, he called it eka-silicon. Thus he predicted three undiscovered elements and left it to his chemical contemporaries to verify his prophecies. Not such remarkable guesses after all – at least not to the genius Mendeléeff!

In 1869 Mendeléeff, before the Russian Chemical Society, presented his paper *On the Relation of the Properties to the Atomic Weights of the Elements*. In a vivid style he told them of his epoch-making conclusions. The whole scientific world was overwhelmed. His great discovery, however, had not sprung forth overnight full grown. The germ of this important law had begun to develop years before. Mendeléeff admitted that 'the law was the direct outcome of the stock of generalizations of established facts which had accumulated by the end of the decade 1860–1870.' De Chancourtois in France, Strecher in Germany, Newlands in England, and Cooke in America had noticed similarities among the properties of certain elements. But no better example could be cited of how two men, working independently in different countries, can arrive at the same generalization, than the case of Lothar Meyer, who conceived the Periodic Law at almost the same time as Mendeléeff. In 1870 there appeared in *Liebig's Annalen* a table of the elements by Lothar Meyer which was almost identical with that of the Russian. The time was ripe for this great law. Some wanted the boldness or the genius necessary 'to place the whole question at such a height that its reflection on the facts could be clearly seen.' This was the statement of Mendeléeff himself. Enough elements had been discovered and studied to make possible the arrangement of a table such as Mendeléeff had prepared. Had Dmitri been born a generation before, he could never, in 1840, have enunicated the Periodic Law.

'The Periodic Law has given to chemistry that prophetic power long regarded as the peculiar dignity of the sister science, astronomy.' So wrote the American scientist Bolton. Mendeléeff had made places for more than sixty-three elements in his Table. Three more he had predicted. What of the other missing building blocks of the universe? Twenty-five years after the publication of Mendeléeff's Table, two Englishmen, following a clue of Cavendish, came upon a new group of elements of which even the Russian had never dreamed. These elements constituted a queer company – the Zero Group as it was later named. Its members, seven in number, are the most unsociable of all the elements. Even with that ideal mixer, potassium, they will normally not unite. Fluorine, most violent of all the non-metals, cannot shake these hermit elements out of their inertness. Moissan tried sparking them with fluorine but failed to make them combine. (Xenon tetrafluoride and several other 'noble' compounds were prepared in 1962. They are no longer regarded as non-reactive.)

Besides, they are all gases, invisible and odorless. Small wonder they had remained so long hidden.

True, the first of these noble gases, as they were called, had been observed in the sun's chromosphere during a solar eclipse in August, 1868, but as nothing was known about it except its orange yellow spectral line, Mendeléeff did not even include it in his table. Later, Hillebrand described a gas expelled from cleveite. He knew enough about it to state that it differed from nitrogen but he failed to detect its real nature. Then Ramsay, obtaining a sample of the same mineral, bottled the gas expelled from it in a vacuum tube, sparked it and detected the spectral line of helium. The following year Kayser announced the presence of this gas in very minute amounts, one part in 185,000, in the earth's atmosphere.

The story of the discovery and isolation of these gases from the air is one of the most amazing examples of precise and painstaking researches in the whole history of science. Ramsay had been casually introduced to chemistry while convalescing from an injury received in a football game. He had picked up a textbook in chemistry and turned to the description of the manufacture of gunpowder. This was his first lesson in chemistry. Rayleigh, his co-worker, had been urged to enter either the ministry or politics, and when he claimed that he owed a duty to science, was told his action was a lapse from the straight and narrow path. Such were the initiations of these two Englishmen into the science which brought them undying fame. They worked with gases so small in volume that it is difficult to understand how they could have studied them in their time. Rayleigh, in 1894, wrote to Lady Frances Balfour: 'The new gas has been leading me a life. I had only about a quarter of a thimbleful. I now have a more decent quantity but it has cost about a thousand times its weight in gold. It has not yet been christened. One pundit suggested "aeron," but when I have tried the effect privately, the answer has usually been, "When may we expect Moses?"' It was finally christened argon, and if not Moses, there came other close relatives: neon, krypton, xenon and finally radon. These gases were isolated by Ramsay and Travers from one hundred and twenty tons of air which had been liquefied. Sir William Ramsay used a micro-balance which could detect a difference in weight of one fourteen-trillionth of an ounce. He worked with a millionth of a gram of invisible, gaseous radon – the size of a tenth of a pin's head.

Besides these six Zero Group elements, some of which are doing effective work in argon and neon incandescent lamps, in helium-filled dirigibles, in electric signs, and in replacing the nitrogen in compressed air to prevent the 'bends' among caisson workers, seventeen other elements were unearthed. So that, a year after Mendeléeff died in 1907, eighty-six elements were listed in the Periodic Table, a fourfold increase since the days of Lavoisier . . .

To the end, Mendeléeff clung to scientific speculations. He published an attempt towards a chemical conception of the ether. He tried to solve the mystery of this intangible something which was believed to pervade the whole universe. To him ether was material, belonged to the zero Group of Elements, and consisted of particles a million times smaller than the atoms of hydrogen.

Two years after he was laid beside the grave of his mother and son, the American Pattison Muir declared that 'the future will decide whether the Periodic Law is the long looked for goal, or only a stage in the journey: a resting place while material is gathered for the next advance.' Had Mendeléeff lived a few more years, he would have witnessed the beginnings of the final development of his Periodic Table by a young Englishman at Manchester [Henry Mosely, who discovered the Law of Atomic Numbers, and was killed at Gallipoli in 1915 aged twenty-six].

The Russian peasant of his day never heard of the Periodic Law, but he remembered Dmitri Mendeléeff for another reason. One day, to photograph a solar eclipse, he shot into the air in a balloon, 'flew on a bubble and pierced the sky.' But to every boy and girl of the Soviet Union today Mendeléeff is a national hero. A special Mendeléeff stamp in his honor was issued in 1957 on the fiftieth anniversary of his death, and a new transuranium element, Number 101, created in 1955, was named *mendelevium* to commemorate his classic contribution to the science of chemistry.

Source: Bernard Jaffe, *Crucibles: The Story of Chemistry from Ancient Alchemy to Nuclear Fission*, new and revised updated fourth edition, New York, Dover Publications, 1976.

Socialism and Bacteria

The great French chemist and microbiologist Louis Pasteur (1822–95) established that putrefaction and fermentation are caused by micro-organisms. He introduced vaccination when he showed, in 1881, that sheep and cows vaccinated with the baccilli of anthrax became immune to the disease. 'Pasteurization ', the heat-treatment of milk to destroy bacteria, such as those of tuberculosis, typhoid and brucellosis, was his invention. This account is from David Bodanis's *Web of Words* (1988). The 'Maxwell' to whom he refers is James Clerk Maxwell (see p. 167), whose kinetic theory explained that the pressure of a gas is due to the incessant impacts of the gas molecules on the walls of the container.

It was dinner time in the Pasteur house, and Louis was at it again. With his wife, daughters and sole son sitting in mortified silence around the table; with the usual dinner guest, Monsieur Loir, at the table with them; with the best tablecloth laid, the right plates out, the first course on, and the long-suffering maids in position at the side; with everyone set to begin the meal, the Professor began his hunt.

'He minutely inspected the bread that was served to him', Monsieur Loir wrote much later, in old age, 'and placed on the tablecloth everything he found in it: small fragments of wool, of cockroaches, of flour worms . . . I tried to find in my own piece of bread from the same load the objects found by Pasteur, but could not discover anything. All the others ate the same bread without finding anything in it.'

Then Pasteur went to work on the glasses. He lifted them up, peered at them closely, and wiped down each one he was going to use, hoping to remove all the contaminating dirt, which again no one else could see. He kept his fingers clean for the wiping, by refusing to shake hands with strangers or even friends during the day. The family waited, the maids and guest waited too, for all were used to the great man's obsession. 'This search took place at almost every meal', Loir continued, 'and is perhaps the most extraordinary memory that I have of Pasteur.'

What ever was going on? Had Pasteur gone bonkers, nuts, off his rocker? At first it's tempting to think so. If Mme Pasteur came home from the Galeries Lafayette and started tearing apart the family's food in search of non-existent wool and cockroaches, so that when her children returned they found her on the floor, legs out, hat askew and surrounded by great mounds of food in the kitchen, we could imagine that they would consider seeking professional help. But when it was their father who embarrassed them with his hunt through the food they took it as normal. To some extent this was because he was the greatest scientist in France, and so had the prerogatives of the gifted. But I suspect even more important was that this pre-dinner hunting ritual matched almost exactly what Pasteur talked about when it was over and he finally looked up.

There are many accounts surviving of what personal conversation with Pasteur was like. In his loud voice, and with his sombre expression (there is only one known drawing, photo, engraving, or sculpture of Pasteur smiling), Pasteur would continually harp on two themes. The first of course was his laboratory work. During dinner at home he would recount with great satisfaction details of the mice he had eviscerated that day, or the purées of vaccinated spinal cord he had prepared, or whatever else he had done in his continuing, remorseless battle against the bacteria. Those bacteria were tiny infecting creatures that most people couldn't see, but which were always there, ready to pounce, to enter us and take over and grow. The hunt inside the dinner bread was no aberration with them around.

After the account of the day's laboratory work had run dry, Pasteur's monotone would turn to his second topic: politics. It was the only interest he held as strongly as bacteria. Some of his views were shared with all Frenchmen of his time, such as his great hatred of Germany, especially after the invasion of 1870–1. It was so strong that he devoted months of free work to the perfecting of French beer, so loyal patriots wouldn't have to drink that Boche muck again. Yet his main political view was not quite so universally shared. Pasteur was an extreme reactionary in politics. He ran (unsuccessfully) for the Senate on an extreme right-wing ticket, and in his letters recorded that the social high point of his life was a one-week visit with Louis Napoleon, at the Emperor's Palace in Compiègne.

The reason was simple. Pasteur had a horror of democracy. There was ordered society, which was good, especially if led by a strong man, and there was also a curious anti-society, a disordered thing of raw uncultivated bodies: the mob. That was a collection of small infecting creatures that decent people didn't ordinarily see, but which was always there, ready to pounce, to enter our society and take over and grow. It was what Pasteur and most right-wing Frenchmen thought had created the French Revolution, surging into existence on the streets of Paris; it was what had produced the Terror against the aristocracy, and the uprisings of 1830, 1848, and then – what Pasteur called a *saturnale* – the brief workers' takeover of the Commune in 1871.

Would someone coming late to the table know which of his two enemies Pasteur was going on about? The language of Pasteur and conservatives generally against the masses of the people was almost exactly like the language Pasteur had developed to use against bacteria. Both were everywhere, small swarming things ready to strike, to grow and propagate. They would destroy us in doing so, subvert our inner structure, have us collapse in disorder, and turn us into – the worst of all possible fates – a thing no different from the seething mass that had attacked. Let the mob take Paris and without the King or Emperor to shore us up we would dissolve into aimless bodies no different from the mob; let the bacterial mob take our physical body and we would decay into a putrefying bacterial mass no different from the attackers here either. If unpleasant entities such as the people or bacteria had to exist, then they must be kept firmly in their place. The people, and especially the workers, were safe only if kept in passive Catholic trade unions, or state-run clubs, or other trustworthy bureaucratic bounds. The bacteria, in all their unpleasant and quick-to-grow varieties, were safe only if restricted to one slot in the Great Chain of Being, that of the decomposer of dead bodies, destroying order only after all life in it had naturally gone, and returning its atoms to the soil for rebirth. Outside of that, though, and they were terrible.

Which came first? There is some evidence that for Pasteur it was fear of the mob. His ideas about bacteria appeared pretty much fully formed in his first writings (1857) on the process of fermentation. In trying to explain how grapes turned into wine, and similar processes, he predicted the existence of living microbes, all apparently identical, yet autonomous, and which competed among each other in an attempt to grow on their target medium until they had fully taken it over. It turned out to be a good guess, but when he made it there was little evidence to back this or indeed any other detailed idea. Pasteur's other descriptions of bacteria, again generally before there was clear evidence to demonstrate it, also matched the view that extreme conservatives took of society. One was that the infection had to be stopped early (think of putting people who even might be revolutionaries in prison); another was that apparently weak individual organisms could cause the demise of large, complex bodies, i.e., that outside bugs could cause inside infection.

To us such views are standard, but at the time medical tradition thought otherwise. We have to imagine the scientific world before the germ theory of disease. When bacteria were found in wounds or sick people it was really thought of as an unimportant by-product of the true disease, which came somehow mysteriously from within and had to run its course. This is why doctors were so upset when Pasteur and others suggested that by not washing their hands between touching diseased corpses and touching healthy or somewhat healthy patients, they might be spreading disease. To the doctors this was preposterous. How could minute organisms cause disease in creatures so much larger? All authorities brought up in the old tradition concurred. Queen Victoria's medical advisors saw no need to clean up the no doubt typhoid-full cesspools near the water sources at Balmoral, from which she and the unfortunate Albert were encouraged to drink. Even Florence Nightingale never believed in 'infection', and was always against what in later life she called the 'germ-fetish'.

It was mere common sense – but for Pasteur it was a common sense which he saw, he *felt*, must be mistaken. An investigator with the standard medical view in mind, let alone one with a brain swept clean of all pre-hypotheses, could never have developed the whole concept of infecting microbes from the small evidence with which Pasteur began. But someone disposed to push forth this idea of small swarming things always ready to destroy order and take over; someone primed to find it anywhere he looked: he would be the one more likely to come up with the germ theory of disease.

Such similarities between social and scientific views have long been common – what better place to get fresh ideas than to just look around you? – and were especially so in the nineteenth century, when so many fields were being set up for the first time. When German professors discovered the approach of several million sperm to the human egg, which only one successfully penetrated, they described it as following the morally sound marriage patterns of the time. On one side there was a passive, waiting egg; on the other a crowd of rushing, eager sperm suitors, of which only the luckiest and strongest one would make it all the way into her affection – just as the professors might hope would happen to their own no doubt properly brought-up daughters.

From the pure evidence they had to work with this is almost all unjustified interpolation. The microscopes of the time could barely get any detail on the egg and its fine movements, and only produced a series of isolated, blurry images. From those static images one could just as well imagine the female with her egg being not passive but taking a more Boadicean approach to her men. This indeed is the standard view today: video microscope images and better *in vivo* techniques show that the sperm don't head towards the eggs, but rush around randomly in all directions; it's the woman's body that directs them in, sometimes helped by an actively slurping cervix. Once drawn closer the sperm are dragged over the final approach by chemical trails the egg sprays out to energize and pull in a particular one. But this for the proper professors, if not their eager-to-boogie daughters, is not what they would have liked to see.

Maxwell's development of the kinetic theory of gases also seems to have come from his sharing in a standard view of society at the time. It was hard to tell what each individual in the great nation of England under Victoria was going to do, but somehow you could be sure that the end result of all those millions fussing, scurrying, slipping and interacting would be to man the navy, rule over the colonies, maintain a large coal industry, and do all those other things England was known for. This strangely cohesive power of the multitude, even though you could never tell what all the individuals in it were doing, was being described in detail by the new science of Social Statistics, and it was by explicit acknowledgement to it that Maxwell worked out his theory of gases where the scurrying molecules also were described only by overall statistics, and not individual biographies.

This sort of explanation sounds good, but it could become too deterministic. Should not every French conservative of the time who was aware of the problems of fermentation and disease have struck his head and said, 'Quelle bêtise! Of course the problem must be due to multitudes of blindly swarming bacteria! What else would make sense of my political phenomenology and analogical thinking?' In the Hollywood version, some of the big words judiciously dropped, that's no doubt how it would be. But as we know, such mass discovery did not occur: most French brows remained unslapped. Why Pasteur happened to be especially sensitive to this aspect of political society and worked it into his answers to the problem of disease, is a matter for the psychologist or biographer to answer. Our question now rather is: why did so many people at the time – so many of our own great-grandparents – go along with him? For the bacterial concept was not one of those scientific ideas, such as quantum mechanics, which ordinary people have a difficult time taking up. Rather it was like momentum, or computers: quickly accepted by all.

The first thing to note in an explanation is that, for humans, thinking by analogy is almost inescapable. Everything that works at one level we're keen to try to see in another one. I remember as a kid, when first learning about the solar system model of the atom, immediately wondering if our solar system was an atom in a larger being. Perhaps the gentle reader remembers the same.

Even easier is to compare what we see with our actual physical body. That, after all, is what we have to spend our lives immersed in. Children who draw the windows of houses to look like eyes so that the whole family home becomes like a larger body are doing just this. It is a very old technique, and was given wide spread in our culture through the notion of the Body of Christ. For long centuries that body was not just an analogy to society, but in the *corpus mysticum* was actually identified with the whole body of Christian society.

When we do compare the world to a body, we end up having to take into account that our own physical body is limited, both in prowess and, especially, in the fact that it will in time come to an end. Religion provides one consolation for this, but whenever men have strayed from religion there has been a need to find another consolation. Frequently this has meant finding something in the outside world to identify with that would provide that missing but so desired escape from mortality. In the late sixteenth century, legal and administrative documents began to note that the king had a natural body, which was certain to decay, but that the political body he was identified with was oh so very much better than that material one. Even in that early period those identifications with the Body Politic seem to be phrased wistfully, as if realizing it was only a second best.

In Pasteur's era the problem was becoming especially severe. Life was increasingly under rational control, so each loss of life seemed more objectionable, wrong. There also seems to have been a decrease in genuine popular belief in religion. The conjunction meant that there was an especially strong interest in altered forms of the body that had any sort of immortality to offer. One of these was patriotism, a continuation of that Kingly identification with the whole mass of living creatures in a political unit. But another, not a consolation but still a terrible fascination, was that mass of small creatures, that whole distorted society in miniature, which yet also happened to be immortal: the bacteria. Organisms known to science before that – cows, humans, daffodils – were not immortal. These were. The first journalists and royalty who peered through the microscopes in Pasteur's or Koch's laboratory to see the bacteria consistently reported this fascination.

Along with these factors of individual psychology, there were changes in the whole society to make Pasteur's concept so readily picked up in this particular era. The increased life expectancy meant that population was growing, a lot. Also there was a great amount of internal migration, from one country to another, and from the land to the city. There were perhaps 100 million more people in Europe in 1900 than in 1870. Strange things happened. In 1830 a swampy settlement by one of the American Great Lakes had a population of under 100. By 1890 it was the city of Chicago, with a population of one million.

There were not enough accepted institutions to handle all these new bodies. Guilds were gone, upper and middle society seemed closed, and so enormous numbers were left in between: working, or joining trade unions, or just being – always in those great numbers, always milling and jumbling and getting in the way of the established citizens and of each other. One would not need to have been M. Pasteur to be attuned to swarming masses with that going on.

Source: David Bodanis, *Web of Words: The Ideas Behind Politics*, London, Macmillan, 1988.

God and Molecules

The Scot James Clerk Maxwell (1831–79) has been ranked with Newton and Einstein as a scientific innovator. He was the first to produce a unified theory of electricity and magnetism, showing that these two phenomena always co-exist, and he formulated the concept of electromagnetic waves (of which heat, light, radio waves and X-rays are all examples). Following Maxwell's lead, the German physicist Heinrich Hertz (1857–94) produced electromagnetic waves in the laboratory, and was the first to broadcast and receive radio waves.

Cultured, widely-read and humorous (he once wrote an analysis of George Eliot's *Middlemarch* claiming that it was in fact a solar myth) Maxwell was also a Christian, as this excerpt from his *Discourse on Molecules* (1873) indicates.

In the heavens we discover by their light, and by their light alone, stars so distant from each other that no material thing can ever have passed from one to another; and yet this light, which is to us the sole evidence of the existence of these distant worlds, tells us also that each of them is built up of molecules of the same kinds as those which we find on earth. A molecule of hydrogen, for example, whether in Sirius or in Arcturus, executes its vibrations in precisely the same time.

Each molecule therefore throughout the universe bears impressed upon it the stamp of a metric system as distinctly as does the metre of the Archives at Paris, or the double royal cubit of the temple of Karnac.

No theory of evolution can be formed to account for the similarity of molecules, for evolution necessarily implies continuous change, and the molecule is incapable of growth or decay, of generation or destruction.

None of the processes of Nature, since the time when Nature began, have produced the slightest difference in the properties of any molecule. We are therefore unable to ascribe either the existence of the molecules or the identity of their properties to any of the causes which we call natural.

On the other hand, the exact equality of each molecule to all others of the same kind gives it, as Sir John Herschel [English astronomer 1792–1871] has well said, the essential character of a manufactured article, and precludes the idea of its being eternal and self-existent.

Thus we have been led, along a strictly scientific path, very near to the point at which Science must stop, – not that Science is debarred from studying the internal mechanism of a molecule which she cannot take to pieces, any more than from investigating an organism which she cannot put together. But in tracing back the history of matter, Science is arrested when she assures herself, on the one hand, that the molecule has been made, and, on the other, that it has not been made by any of the processes we call natural . . .

Natural causes, as we know, are at work, which tend to modify, if they do not at length destroy, all the arrangements and dimensions of the earth and the whole solar system. But though in the course of ages catastrophes have occurred and may yet occur in the heavens, though ancient systems may be dissolved and new systems evolved out of their ruins, the molecules out of which these systems are built – the foundation-stones of the material universe – remain unbroken and unworn. They continue this day as they were created – perfect in number and measure and weight; and from the ineffaceable characters impressed on them we may learn that those aspirations after accuracy in measurement, and justice in action, which we reckon among our noblest attributes as men, are ours because they are essential constituents of the image of Him who in the beginning created, not only the heaven and the earth, but the materials of which heaven and earth consist.

Source: Lewis Campbell and William Garnett, *The Life of James Clerk Maxwell. With a Selection from his Correspondence and Occasional Writings and a Sketch of his Contributions to Science*, London, Macmillan, 1882.

Inventing Electric Light

In 1876 the American technological genius and rags-to-riches folk hero Thomas Alva Edison (1847–1931) set up the world's first industrial research laboratory in the remote hamlet of Menlo Park, New Jersey. During the six years he and his team worked there he secured patents for scores of inventions, including the phonograph, the telephone (an improvement on Alexander Graham Bell's invention), the electric pen (a stencil duplicator), and the electric light bulb. Incandescent electric light had been the despair of inventors for fifty years, and, as one of the Menlo Park assistants Francis Jehl recalls, Edison spent fourteen months searching for a suitable filament.

The hunt was a long, tedious one. Many materials which at first seemed promising fell down under later tests and had to be laid aside. Every experiment was recorded methodically in the notebooks. In many there was simply the name of the fiber and after it the initials 'T. A.,' meaning 'Try Again.'

Literally hundreds of experiments were made on different sorts of fiber; for the master seemed determined to exhaust them all. Threads of cotton, flax, jute silks, cords, manila hemp and even hard woods were tried.

Some of the fibers being worked at the moment were piled conveniently on top of the chest; and today you may see them still in the same spot. Others were stored in jars along the shelves. An examination of the labels on the jars as they stand today on the shelves along the east wall of the restored laboratory will give an idea of what an infinite variety were examined.

Chinese and Italian raw silk both boiled out and otherwise treated were among those used. Others included horsehair, fish line, teak, spruce, boxwood, vulcanized rubber, cork, celluloid, grass fibres from everywhere, linen twine, tar paper, wrapping paper, cardboard, tissue paper, parchment, holly wood, absorbent cotton, rattan, California redwood, raw jute fiber, corn silk, and New Zealand flax.

The most interesting material of all that we used in our researches after a successful filament was the hair from the luxurious beards of some of the men about the laboratory. There was the great 'derby,' in which we had a contest between filaments made from the beards of [John] Kruesi and J. U. Mackenzie, to see which would last the longer in a lamp. Bets were placed with much gusto by the supporters of the two men, and many arguments held over the rival merits of their beards.

Kruesi, you know, was a cool mountaineer from Switzerland possessed of a bushy black beard. Mackenzie was the station master at Mt. Clemens, Michigan, who had taught telegraphy to the chief in the early days after the young Edison had saved the life of Mackenzie's small son Jimmy. His beard, or rather, his burnsides, were stiff and bristling.

As I now recall, he won the contest, though some claimed that an unfair advantage was given him; that less current was used on the filament made from his beard than on that from Kruesi's. Be that as it may, both burned out with considerable rapidity.

At last, on 21 October, 1879, Edison made a bulb that did not burn out. Its filament was of carbonized cotton sewing thread, and Edison and Jehl sat up all night watching it shine. The first commercial bulb, which followed swiftly, had a horseshoe filament of carbonized paper. The *New York Herald* reporter Marshall Fox, who visited the laboratory, explained how the filament was prepared in an article published on 21 December, 1879:

Edison's electric light, incredible as it may appear, is produced from a little piece of paper – a tiny strip of paper that a breath would blow away. Through this little strip of paper is passed an electric current, and the result is a bright, beautiful light, like the mellow sunset of an Italian autumn.

'But paper instantly burns, even under the trifling heat of a tallow candle!' exclaims the sceptic, 'and how, then, can it withstand the fierce heat of an electric current.' Very true, but Edison makes the little piece of paper more infusible than platinum, more durable than granite. And this involves no complicated process. The paper is merely baked in an oven until all its elements have passed away except its carbon framework. The latter is then placed in a glass globe connected with the wires leading to the electricity producing machine, and the air exhausted from the globe. Then the apparatus is ready to give out a light that produces no deleterious gases, no smoke, no offensive odors – a light without flame, without danger, requiring no matches to ignite, giving out but little heat, vitiating no air, and free from all flickering; a light that is a little globe of sunshine, a veritable Aladdin's lamp. And this light, the inventor claims, can be produced cheaper than that from the cheapest oil.

The first public demonstration of electric light took place soon afterwards, on 31 December, 1879. As the *New York Herald* reported:

Edison's laboratory was tonight thrown open to the general public for the inspection of his electric light. Extra trains were run from east and west, and notwithstanding the stormy weather, hundreds of persons availed themselves of the privilege. The laboratory was brilliantly illuminated with twenty-five electric lamps, the office and counting room with eight, and twenty others were distributed in the street leading to the depot and in some of the adjoining houses. The entire system was explained in detail by Edison and his assistants, and the light was subjected to a variety of tests. Among others the inventor placed one of the electric lamps in a large glass jar filled with water and turned on the current, the little horseshoe filament when this submerged burned with the same bright steady illumination as it did in the air, the water not having the slightest effect upon it. The lamp was kept thus under water for four hours. Another test was turning the electric current on and off on one of the lamps with great rapidity and as many times as it was calculated the light would be turned on and off in actual household illuminations in a period of thirty years, and no perceptible variation either in the brilliancy, steadiness or durability of the lamp occurred.

Three years later the Pearl Street Central Power Station was completed in New York – the first of the world's great cities to be electrically lit. The coming of electric light was widely seen as banishing the fear and superstition that darkness had bred. The German historian Emil Ludwig proclaimed:

> When Edison, the father of the American Nation, the greatest living benefactor of mankind, snatched up the spark of Prometheus in his little pear-shaped glass bulb, it meant that fire had been discovered for the second time, that mankind had been delivered again from the curse of night.

The same point is made, though less poetically, in Conan Doyle's *The Hound of the Baskervilles* (1902) when the new heir to Baskerville Hall arrives from North America and remarks, on viewing his spooky ancestral home:

> It's enough to scare any man. I'll have a row of electric lamps up here inside of six months, and you won't know it again with a thousand-candlepower Swan and Edison right here in front of the hall door.

Source: Francis Jehl (1860–1941), *Menlo Park Reminiscences, Volume One*, published by the Edison Institute, Dearborn, Michigan, 1937.

Bird's Custard: The True Story

This touching piece of social history is by Nicholas Kurti, FRS, Emeritus Professor of Physics at Oxford.

It is widely believed that Bird's custard is one of the earliest examples of 'convenience foods' or of regrettable substitutes designed purely to reduce the cost and the time of preparation of a dish. Nothing could be further from the truth. Indeed, the invention of Bird's custard is a shining example of alleviating a deprivation caused by cruel nature.

Alfred Bird, whose father taught astronomy at Eton, was born in 1811 in Birmingham and in 1837 established himself as an analytical and retail pharmaceutical chemist there. When he married Elizabeth Lavinia Ragg he faced a challenge which was to influence his career. His young wife suffered from a digestive disorder which prevented her from eating anything prepared with eggs or with yeast. But Elizabeth Lavinia was apparently yearning for custard to go with her favourite fruit pies so Alfred Bird started experimenting in his shop. The result was the custard powder bearing his name and based on cornflour, which when mixed with milk produced, after heating, a sauce reminiscent in appearance, taste and consistency of a genuine egg-and-milk custard sauce.

The young wife was overjoyed and this substitute custard became the normal accompaniment to puddings at the Birds' dinner table, though, when they entertained, genuine custard sauce was offered to their guests. Then came an occasion when, whether by accident or by design, 'Bird's custard' was served and Alfred must have been gratified to hear his guests declare that it was the best custard they had ever tasted!

This then was the beginning of the firm Alfred Bird and Sons Ltd of Birmingham which for 120 years remained a family business, first under the chairmanship of the founder, then of his son, Sir Alfred Bird Bt and then of his grandson Sir Robert Bird Bt. While the firm's main product remained custard powder Alfred Bird's other invention to circumvent his wife's digestive troubles, namely baking powder, was also manufactured and was used during the Crimean war so that British troops could be given fresh, palatable bread.

Alfred Bird was a Fellow of the Chemical Society and, a few months after his death on 2 December 1878, a brief obituary was published in the *Journal of the Chemical Society*, Vol. 35, p. 206, 1879. It described at some length Bird's interest in physics and meteorology, thus: 'He constructed a beautiful set of harmonized glass bowls extending over 5 octaves which he used to play with much skill'; and 'in 1859 he constructed a water barometer with which he was fond of observing and showing to others the minute oscillations of the atmospheric pressure'. But of Bird's Custard Powder – not a word!

Source: *But the Crackling Was Superb: An Anthology on Food and Drink by Fellows and Foreign Members of the Royal Society.* Nicholas and Giana Kurti, Adam Hilger, Bristol, IOP Publishing, 1988.

Birth Control: The Diaphragm

Birth control is not new. The methods used by the ancient Greeks – abstinence, abortion, withdrawal and extended breastfeeding – remained the commonest forms of fertility-limitation until the arrival of the oral contraceptive and the IUD in the 1960s. Condoms (made of linen, animal bladders, or fine skins) were in use from the late sixteenth century. But since they were regarded mainly as a safeguard against venereal disease (and were available in brothels in several European capitals by the end of the eighteenth century), they remained disreputable.

Public defence of birth control is first found in late-eighteenth century France. In the early nineteenth century it drew strongly on the theories of Malthus (see p. 54). It was not, at first, linked with women's rights but with restriction of the irresponsible fertility of the poor. The introduction of a relatively reliable 'scientific' female contraceptive may have helped to change this emphasis during the later nineteenth century. By 1918 Marie Stopes was arguing, in her bestseller *Married Love*, that the wife had as much right to sexual pleasure as her husband, and she opened the first English birth control clinic in London's Holloway Road in March 1921.

This extract is from Angus McLaren's *History of Contraception* (1990).

The invention of the diaphragm did represent a significant innovation in fertility control. Nineteenth-century doctors popularized the use of pessaries to correct prolapsed uteruses; it was a short step to employ them as a barrier method of birth control. Such a device was presumably what Dr Edward B. Foote meant when referring to an Indian-rubber 'womb veil'. The German physician W. P. J. Mensinga provided a clearer account of his diaphragm in 1882; a soft rubber shield which the woman inserted into the vagina to block entry to the uterus. Mensinga's explicit intent was to protect unhealthy women from undesired pregnancies. The diaphragm was, when accompanied by douching, an effective female contraceptive; unfortunately its expense and the fact that it had to be fitted by a physician long restricted its use to a middle-class clientele.

Commercial houses began at the turn of the century to develop acidic powders and jellies to block and kill sperm. Easier to use was the soluble quinine pessary or suppository developed by the Rendell company in England in the 1880s and popularized by Dr Henry Arthur Allbutt. Similar home-made products which countered conception with both a barrier and a spermicide were soon being made from cocoa butter or glycerine by innovative housewives across Europe and North America.

Diaphragms and pessaries to be fully effective had to be followed by douching. Douching after intercourse with a vaginal syringe to destroy 'the fecundating property of the sperm by chemical agents' was recommended by the Massachusetts doctor Charles Knowlton in his *Fruits of Philosophy*, published in 1832. Knowlton was prosecuted for obscenity, but his douching advice was repeated by others, such as Frederick Hollick in 1850 in his *Marriage Guide*. Simple cold water was suggested by some; Knowlton stressed the need to add a restringent or acidic agent such as alum, various sulphates or vinegar. Douches, like diaphragms, were regarded as providing a woman with contraceptive independence. By mid-century they were readily available in pharmacies and drug stores and sold via respectable mail-order catalogues, purportedly for purposes of hygiene. They were, for this reason, promoted by the German Health Insurance Programme and provided free to members of local Funds. Douching did entail expenses and required both a privacy and a water supply that was not available to many working-class couples. Perhaps this was just as well, since simply douching after intercourse was in fact less effective than coitus interruptus.

In the latter decades of the nineteenth century, contraceptives and abortifacients were advertised in newspapers and magazines, sold in barber shops, rubber good stores and pharmacies, and brought to villages by itinerant pedlars and to working-class neighbourhoods by door-to-door hucksters. Irish doctors were astonished at the display by London chemists of 'antigestatory appliances' and 'orchitological literature'.

Source: Angus McLaren, *A History of Contraception: From Antiquity to the Present Day*, Oxford, Basil Blackwell, 1990.

Headless Sex: The Praying Mantis

Several early naturalists observed instances of a female praying mantis eating the male during copulation. L. O. Howard sent the following account to the American magazine *Science* in 1886.

A few days since, I brought a male of *Mantis carolina* to a friend who had been keeping a solitary female as a pet. Placing them in the same jar, the male, in alarm, endeavored to escape. In a few minutes the female succeeded in grasping him. She first bit off his left front tarsus, and consumed the tibia and femur. Next she gnawed out his left eye. At this the male seemed to realize his proximity to one of the opposite sex, and began to make vain endeavors to mate. The female next ate up his right front leg, and then entirely decapitated him, devouring his head and gnawing into his thorax. Not until she had eaten all of his thorax except about three millimetres, did she stop to rest. All this while the male had continued his vain attempts to obtain entrance at the valvules, and he now succeeded, as she voluntarily spread the parts open, and union took place. She remained quiet for four hours, and the remnant of the male gave occasional signs of life by a movement of one of his remaining tarsi for three hours. The next morning she had entirely rid herself of her spouse, and nothing but his wings remained.

The female was apparently full-fed when the male was placed with her, and had always been plentifully supplied with food.

The extraordinary vitality of the species which permits a fragment of the male to perform the act of impregnation is necessary on account of the rapacity of the female, and it seems to be only by accident that a male ever escapes alive from the embraces of his partner.

In the *Biological Bulletin* for October 1935, a later researcher, K. D. Roeder, demonstrated that removal of the male praying mantis's head actually improved its sexual performance. This, he showed, was because the subesophageal ganglion (near the head) normally inhibits the copulatory movement of the abdomen. Once the subesophageal ganglion has been removed, by decapitating the insect, it will copulate with almost anything.

In his experiment, Roeder beheaded eighteen male mantises. Decapitation, he reported, is followed by a preliminary stage of shock, lasting for about ten minutes, after which the insects begin vigorous copulatory movements:

> If they encounter any rounded object, such as a pencil or the observer's finger, it is immediately grasped by the forelegs, while the other legs steady the body. Violent attempts are made to copulate with the object.

Decapitating females, Roeder found, does not have such dramatic results, though it does cause some muscular activity in the abdomen:

> A decapitated female will readily accept a male, decapitated or otherwise, and actual copula results sooner than when both insects are intact. The pair remain together for about four hours.

Roeder illustrated his article with a photograph of a headless male mantis copulating with a headless female.

Sources: *Science*, New York: The Science Co., 1886. *The Biological Bulletin*, October 1935, published by the Marine Biological Laboratory, printed and issued by Lancaster Press Inc., Prince and Lemon Sts, Lancaster, Pa.

The World as Sculpture

William James (1842–1910), elder brother of the novelist Henry James, was a pioneer psychologist, and founder of the first US psychological laboratory. In his *Principles of Psychology* (1890) he coined the term 'stream of consciousness' (later used to describe the technique of writers like James Joyce and Virginia Woolf) to characterize the chaotic flow of human mental activity. As a philosopher, he was one of the founders of Pragmatism, inventing the term 'neutral monism' for the theory that the ultimate constituents of reality are individual momentary experiences. In *A Pluralistic Universe* (1909) he suggests that the substance of reality (or 'all-form') may never get totally collected, and that 'a distributive form of reality, the each-form, is as acceptable as the all-form'. This extract from the *Principles*, positing the innumerable possible worlds consequent on such a view, anticipates the multiple worlds of quantum theory (see p. 278).

The mind, in short, works on the data it receives very much as a sculptor works on his block of stone. In a sense the statue stood there from eternity. But there were a thousand different ones beside it, and the sculptor alone is to thank for having extricated this one from the rest. Just so the world of each of us, howsoever different our several views of it may be, all lay embedded in the primordial chaos of sensations, which gave the mere *matter* to the thought of all of us indifferently. We may, if we like, by our reasonings unwind things back to that black and jointless continuity of space and moving clouds of swarming atoms which science calls the only real world. But all the while the world *we* feel and live in will be that which our ancestors and we, by slowly cumulative strokes of choice, have extricated out of this, like sculptors, by simply rejecting certain portions of the given stuff. Other sculptors, other statues from the same stone! Other minds, other worlds from the same monotonous and inexpressive chaos! My world is but one in a million alike embedded, alike real to those who may abstract them. How different must be the worlds in the consciousness of ant, cuttle-fish or crab!

Source: William James, *The Principles of Psychology,* London, Macmillan, 1890.

The Discovery of X-Rays

On the evening of Friday 8 November 1895, the Professor of Physics at the University of Würzburg, Wilhelm Conrad Roentgen (1845–1923) was working late in his laboratory, after everyone else had gone home. He was preparing to carry out some experiments with an induction coil connected to the electrodes of a partially evacuated glass tube.

It had been known since 1858 that when electricity was discharged through the air or other gases in such tubes, the glass became phosphorescent. The 'cathode rays' causing this had been fancifully described by Sir William Crookes in 1878 as 'a stream of molecules in flight'. It was also known that if the tube had a thin metal-foil 'window' in it, the cathode rays would penetrate this and cause fluorescence a few centimetres beyond the tube. Roentgen thought cathode rays might be detectable outside the tube even when there was no metal-foil window. As a first step in investigating this he covered the entire tube in black cardboard, and drew the curtains to darken the room. To test that his cardboard shield would not let light through, he then turned on the high-voltage coil and passed a current through the tube.

What happened next is described by Roentgen's student Charles Nootnangle of Minneapolis, who had it from Roentgen himself a few days later:

> By chance he happened to note that a little piece of paper lying on his work table was sparkling as though a single ray of bright sunshine had fallen upon it lying in the darkness. At first he thought it was merely the reflection from the electric spark, but the reflection was too bright to allow that explanation. Finally he picked up the piece of paper and, examining it, found that the reflected light was given by a letter A which had been written on the paper with a platinocyanide [fluorescent] solution.

It was at once clear to Roentgen that the piece of chemically-treated paper could not have been made to fluoresce by cathode rays, since it was several feet away from the tube. Some other rays must be responsible – rays that were able to pass through the cardboard shield round the tube, and travel invisibly through the air. Since he had no idea what these rays were, Roentgen called them X-rays, and he began experimenting to see what other substances they would pass through. On 28 December 1895 he presented his 'Preliminary Communication', entitled *On a new Kind of Rays*, to the President of the Würzburg Physical and Medical Society.

The most striking feature of this phenomenon is the fact that an active agent here passes through a black cardboard envelope, which is opaque to the visible and the ultra-violet rays of the sun or of the electric arc; an agent, too, which has the power of producing active fluorescence. Hence we may first investigate the question whether other bodies also possess this property.

We soon discover that all bodies are transparent to this agent, though in very different degrees. I proceed to give a few examples: Paper is very transparent; behind a bound book of about one thousand pages I saw the fluorescent screen light up brightly, the printers' ink offering scarcely a notable hindrance. In the same way the fluorescence appeared behind a double pack of cards; a single card held between the apparatus and the screen being almost unnoticeable to the eye. A single sheet of tin-foil is also scarcely perceptible; it is only after several layers have been placed over one another that their shadow is distinctly seen on the screen. Thick blocks of wood are also transparent, pine boards two or three centimetres thick absorbing only slightly. A plate of aluminium about fifteen millimetres thick, though it enfeebled the action seriously, did not cause the fluorescence to disappear entirely. Sheets of hard rubber several centimetres thick still permit the rays to pass through them. (For brevity's sake I shall use the expression 'rays'; and to distinguish them from others of this name I shall call them 'X-rays'.) Glass plates of equal thickness behave quite differently, according as they contain lead (flint-glass) or not; the former are much less transparent than the latter. If the hand be held between the discharge-tube and the screen, the darker shadow of the bones is seen within the slightly dark shadow-image of the hand itself . . . Lead of a thickness of 1.5 millimetres is practically opaque . . .

I have observed, and in part photographed, many shadow-pictures of this kind, the production of which has a particular charm. I possess, for instance, photographs of the shadow of the profile of a door which separates the rooms in which, on one side, the discharge-apparatus was placed, on the other the photographic plate; the shadow of the bones of the hand; the shadow of a covered wire wrapped on a wooden spool; of a set of weights enclosed in a box; of a galvanometer in which the magnetic needle is entirely enclosed by metal; of a piece of metal whose lack of homogeneity becomes noticeable by means of the X-rays, etc. I have obtained a most beautiful photographic shadow-picture of the double barrels of a hunting-rifle with cartridges in place, in which all the details of the cartridges, the internal faults of the damask barrels, etc., could be seen most distinctly and sharply.

Roentgen's discovery was publicized in the world's press, and caused great excitement. A reporter, H. J. W. Dam, interviewed Roentgen in his laboratory and described what ensued for readers of *McClure's Magazine* (New York and London) in April 1896.

In addition to his own language he speaks French well and English scientifically, which is different from speaking it popularly. These three tongues being more or less within the equipment of his visitor, the conversation proceeded on an international or polyglot basis, so to speak, varying at necessity's demand.

'Now then,' he said smiling and with some impatience, when some personal questions at which he chafed were over, 'you have come to see the invisible rays.'

'Is the invisible visible?'

'Not to the eye, but its results are. Come in here.'

He led the way to a square room and indicated the induction coil with which his researches were made, an ordinary Ruhmkorff coil with a spark of from 4 to 6 in., charged by a current of twenty amperes. Two wires led from the coil through an open door into a smaller room on the right. In this room was a small table carrying a Crookes' tube connected with the coil. The most striking object in the room, however, was a huge and mysterious tin [actually zinc and lead] box about 7 ft. high and 4 ft. square. It stood on end like a huge packing case, its side being perhaps 5 in. from the Crookes' tube.

The professor explained the mystery of the tin box, to the effect that it was a device of his own for obtaining a portable dark room. When he began his investigations he used the whole room as was shown by the heavy blinds and curtains so arranged as to exclude the entrance of all interfering light from the windows. In the side of the tin box at the point immediately against the tube was a circular sheet of aluminium

1 mm. in thickness, and perhaps 18 in. diameter, soldered to the surrounding tin. To study his rays the professor had only to turn on the current, enter the box, close the door, and in perfect darkness inspect only such light or light effects as he had a right to consider his own, hiding his light, in fact, not under the Biblical bushel but in a more commodious box.

'Step inside,' said he, opening the door which was on the side of the box farthest from the tube. I immediately did so, not altogether certain whether my skeleton was to be photographed for general inspection or my secret thoughts held up to light on a glass plate. 'You will find a sheet of barium paper on the shelf,' he added, and then went away to the coil. The door was closed and the interior of the box became black darkness. The first thing I found was a wooden stool on which I resolved to sit. Then I found the shelf on the side next the tube, and then the sheet of paper prepared with barium platinocyanide. I was thus being shown the first phenomenon which attracted the discoverer's attention and led to the discovery, namely, the passage of rays, themselves wholly invisible, whose presence was only indicated by the effect they produced on a piece of sensitized photographic paper.

A moment later, the black darkness was penetrated by the rapid snapping sound of the high-pressure current in action, and I knew that the tube outside was glowing. I held the sheet vertically on the shelf, perhaps 4 in. from the plate. There was no change, however, and nothing was visible.

'Do you see anything?'

'No.'

'The tension is not high enough,' and he proceeded to increase the pressure by operating an apparatus of mercury in long vertical tubes acted upon automatically by a weight lever which stood near the coil. In a few moments the sound of the discharge again began, and then I made my first acquaintance with the roentgen rays.

The moment the current passed, the paper began to glow. A yellowish-green light spread all over its surface in clouds, waves, and flashes. The yellow-green luminescence, all the stranger and stronger in the darkness, trembled, wavered, and floated over the paper, in rhythm with the snapping of the discharge. Through the metal plate, the paper, myself, and the tin box, the visible rays were flying, with an effect strange, interesting, and uncanny. The metal plate seemed to offer no appreciable resistance to the flying force, and the light was as rich and full as if nothing lay between the paper and the tube.

'Put the book up,' said the professor.

I felt upon the shelf, in the darkness, a heavy book, 2 in. in thickness, and placed this against the plate. It made no difference. The rays flew through the metal and the book as if neither had been there, and the waves of light, rolling cloud-like over the paper, showed no change in brightness. It was a clear, material illustration of the ease with which paper and wood are penetrated. And then I laid the book and paper down, and put my eyes against the rays. All was blackness, and I neither saw nor felt anything. The discharge was in full force, and the rays were flying through my head, and, for all I knew, through the side of the box behind me. But they were invisible and impalpable. They gave no sensation whatever. Whatever the mysterious rays may be, they are not to be seen and are to be judged only by their works.

I was loath to leave this historical tin box, but the time pressed. I thanked the professor, who was happy in the reality of his discovery, and the music of his sparks. Then I said, 'Where did you first photograph living bones?'

'Here,' he said, leading the way into the room where the coil stood. He pointed to a table on which was another – the latter a small, short-legged wooden one, with more the shape and size of a wooden seat. It was 2 ft. square and painted coal black.

'How did you take the first hand photograph?'

The professor went over to a shelf by the window, where lay a number of prepared glass plates, closely wrapped in black paper. He put a Crookes' tube underneath the table, a few inches from the under side of its top. Then he laid his hand flat on the top of the table, and placed the glass plate loosely on his hand.

'You ought to have your portrait painted in that attitude,' I suggested.

'No, that is nonsense,' he said, smiling.

'Or be photographed.' This suggestion was made with a deeply hidden purpose.

The rays from the Röntgen eyes instantly penetrated the deeply hidden purpose. 'Oh, no,' said he, 'I can't let you make pictures of me. I am too busy.' Clearly the professor was entirely too modest to gratify the wishes of the curious world.

The reception of the discovery by the public was not entirely favourable. Photographing the skeleton of a living person was felt to be eerie. A Professor Czermak of Graz was so appalled to see an X-ray photograph of his skull that he could not sleep. 'He has not closed an eye since he saw his own death's head,' reported the *Grazer Tageblatt*. The possibility of seeing other people's internal organs was widely considered a threat to privacy. But enthusiasm outweighed disapproval, and many potential uses of the new technique were suggested. In Paris a Dr Baraduc claimed that he could photograph the human soul with X-rays, and presented 400 such plates at an exhibition in Munich. During 1896 the use of X-rays in medical diagnosis was rapidly explored worldwide, especially in the USA. Photographs of a human foetus, of a tubercular patient's lungs, and of the stomach, heart and other organs were published, and a Harvard professor, W. B. Cannon, watched pearl buttons pass down the oesophagus of a dog. The harmful effects of exposure to X-rays were soon noticed. Many cases of severe skin burns and loss of hair were reported, but no one appreciated the real danger. Noting their depilatory effect one enterprising Frenchman, M. Gaudoin of Dijon, offered to use X-rays to remove unwanted hair from women's faces, and had many clients.

Dramatic use is made of early responses to X-rays in Thomas Mann's novel *The Magic Mountain* (1924), which is set in a Swiss sanatorium in the years before the First World War. A student, Hans Castorp, has come to the sanatorium to visit his cousin Joachim, a patient there. The resident physician, Hofrat Behrens, takes him to the room containing the X-ray apparatus, where Joachim is to be examined.

They heard a switch go on. A motor started up, and sang furiously higher and higher, until another switch controlled and steadied it. The floor shook with an even vibration. The little red light, at right angles to the ceiling, looked threateningly across at them. Somewhere lightning flashed. And with a milky gleam a window of light emerged from the darkness: it was the square hanging screen, before which Hofrat Behrens bestrode his stool, his legs sprawled apart with his fists supported on them, his blunt nose close to the pane, which gave him a view of a man's interior organism.

'Do you see it, young man?' he asked. Hans Castorp leaned over his shoulder, but then raised his head again to look toward the spot where Joachim's eyes were presumably gazing in the darkness, with their gentle, sad expression. 'May I?' he asked.

'Of course,' Joachim replied magnanimously, out of the dark. And to the pulsation of the floor and the snapping and crackling of the forces at play, Hans Castorp peered through the lighted window, peered into Joachim's empty skeleton. The breastbone and spine fell together in a single dark column. The frontal structure of the ribs was cut across by the paler structure of the back. Above, the collar bones branched off on both sides, and the framework of the shoulder, with the joint and the beginning of Joachim's arm, showed sharp and bare through the soft envelope of flesh. The thoracic cavity was light, but blood vessels were to be seen, some dark spots, a blackish shadow.

'Clear picture,' said the Hofrat . . . 'Breathe deep,' he commanded. 'Deeper! Deep, I tell you!' And Joachim's diaphragm rose quivering, as high as it could; the upper parts of the lungs could be seen to clear up, but the Hofrat was not satisfied. 'Not good enough,' he said. 'Can you see the hilus glands? Can you see the adhesions? Look at the cavities here, that is where the toxins come from that fuddle him.' But Hans Castorp's attention was taken up by something like a bag, a strange, animal shape, darkly visible behind the middle column, or more on the right side of it – the spectator's right. It expanded and contracted regularly, a little after the fashion of a swimming jelly-fish.

'Look at his heart,' and the Hofrat lifted his huge hand again from his thigh and pointed with his forefinger at the pulsating shadow. Good God, it was the heart, it was Joachim's honour-loving heart, that Hans Castorp saw!

Sources: Charles Nootnangle, 'How Roentgen Discovered the X-Ray', *The Electrical Engineer*, New York, 22, 125, 5 August 1896; H. J. W. Dam, *McClure's Magazine*, New York and London, April 1896. Both quoted in Otto Glasser, *William Conrad Roentgen and the Early History of the Roentgen Rays*, London, John Ball, Sons and Danielsson Ltd, 1933. George F. Barker, tr. and ed., *Roentgen Rays, Memoirs by Roentgen, Stokes and J. J. Thomson*, Harper and Bros, New York and London, 1899. Thomas Mann, *The Magic Mountain*, tr. H. T. Lowe-Porter, London, Penguin Books in association with Secker and Warburg, 1960.

No Sun in Paris

Henri Becquerel (1852–1908), Professor of Physics at the Ecole Polytechnique in Paris, read about X-rays soon after their discovery. He thought that similar penetrating rays might be emitted by phosphorescent substances when exposed to sunlight. So he took some phosphorescent crystals of a uranium compound, in the form of a thin crust, and placed them on a photographic plate which he had previously wrapped in thick black paper to keep the light out. Then he exposed the whole thing to the sun for a few hours. When he developed the photographic plate he found a silhouette of the crystals in black on the negative – which seemed to confirm his idea that sunlight made the crystals emit radiation.

His discovery that they emitted radiation even in the dark was a matter of chance. The sun did not shine in Paris for several days, but, as he had set up his apparatus, he decided to develop the plate nevertheless – as he explains in his paper 'On the Radiation Emitted by Phosphorescence' (1896).

I particularly insist on the following fact, which appears to me exceedingly important and not in accord with the phenomena which one might expect to observe: the same encrusted crystals placed with respect to the photographic plates in the same conditions and acting through the same screens, but kept in the dark, still produce the same photographic effects. I may relate how I was led to make this observation: among the preceding experiments some had been made ready on Wednesday the 26th and Thursday the 27th of February and as on those days the sun only showed itself intermittently I kept my arrangements all prepared and put back the holders in the dark in the drawer of the case, and left in place the crusts of uranium salt. Since the sun did not show itself again for several days I developed the photographic plates on the 1st of March, expecting to find the images very feeble. The silhouettes appeared on the contrary with grcat intensity. I at oncc thought that the action might be able to go on in the dark, and I arranged the following experiment.

At the bottom of a box of opaque cardboard, I placed a

photographic plate, and then on the sensitive face I laid a crust of uranium salt which was convex, so that it only touched the emulsion at a few points; then alongside of it I placed on the same plate another crust of the same salt, separated from the emulsion by a thin plate of glass; this operation was carried out in the dark room, the box was shut, was then enclosed in another cardboard box, and put away in a drawer.

I did the same thing with a holder closed by an aluminium plate, in which I put a photographic plate and then laid on it a crust of uranium salt. The whole was enclosed in an opaque box and put in a drawer. After five hours I developed the plates, and the silhouettes of the encrusted crystals showed black, as in the former experiment, and as if they had been rendered phosphorescent by light. In the case of the crust which was placed directly on the emulsion, there was a slightly different action at the points of contact from that under the parts of the crust which were about a millimeter away from the emulsion; the difference may be attributed to the different distances of the sources of the active radiation. The action of the crust placed on the glass plate was very slightly enfeebled, but the form of the crust was very well reproduced. Finally, in passing through the plate of aluminium, the action was considerably enfeebled but nevertheless was very clear.

It is important to notice that this phenomenon seems not to be attributable to luminous radiation emitted by phosphorescence . . . The radiations of uranium salts are emitted not only when the substances are exposed to light but when they are kept in the dark, and for more than two months the same pieces of different salts, kept protected from all known exciting radiations, continued to emit, almost without perceptible enfeeblement, the new radiations. From the 3rd of March to the 3rd of May these substances were enclosed in a box of opaque cardboard. Since the 3rd of May they have been in a double box of lead, which has never left the dark room. A very simple arrangement makes it possible to slip a photographic plate under a black paper stretched parallel to the bottom of the box, on which rest the substances which are being tested, without exposing them to any radiation which does not pass through the lead.

In these conditions the substances studied continued to emit active radiation.

All the salts of uranium that I have studied, whether they become phosphorescent or not in the light, whether crystallized, cast or in solution, have given me similar results. I have thus been led to think that the effect is a consequence of the presence of the element uranium in these salts, and that the metal would give more intense effects than its compounds. An experiment made several weeks ago with the powdered uranium of commerce, which has been for a long time in my laboratory, confirmed this expectation; the photographic effect is notably greater than the impression produced by one of the uranium salts.

Becquerel thus became the first human being to observe the phenomenon later known as radioactivity, and to discover that uranium was a radioactive element.

Source: *A Source Book in Physics*, ed. W. F. Magie, New York, McGraw Hill, 1935.

The Colour of Radium

In 1897 Becquerel's paper on radiation (see p. 188) was read by a young scientist looking for a doctoral-thesis subject, Marie Curie. Born in Warsaw in 1867, she was the youngest of five children of two Polish intellectuals. She attended the 'Floating University' run by Polish teachers in defiance of their Russian rulers, and espoused forward-looking movements – socialism, positivism, science.

In 1891, after a six-year stint as a governess, she went to Paris and enrolled at the Sorbonne. As a student she led a life of monastic simplicity, surviving mainly on tea and bread-and-butter. She came top in the Master's degree in Physics in 1893, and, two years later, married a shy, unworldly young research chemist, Pierre Curie. They rented a tiny flat up four flights of stairs and devoted themselves to science.

It was at this point that Marie read Becquerel's paper, and decided to investigate radiation for her doctorate. She worked in a damp, glassed-in lumber room in the Rue Lhomond – the only space the School of Physics could find for her. Her first discovery was that uranium was not the only chemical element capable of radiation. Another element, thorium, also emitted spontaneous 'rays'. She gave this phenomenon the name 'radio-activity'. She then found that certain minerals containing uranium and thorium (pitch-blende, chalcolite, uranite) were much more radioactive than the amount of uranium and thorium in them could account for. There must, she concluded, be another highly radioactive substance present, and she formed the hypothesis that this was a previously undiscovered element.

She determined to isolate this, and in May 1898 her husband Pierre joined her in the search. They found that radioactivity was concentrated principally in two different chemical fractions of pitch-blende, indicating the presence of two new elements, not one. By July they were able to announce the discovery of the first, which they called 'polonium' ('from the name' as they explained, 'of the original country of one of us').

In her biography of her mother, assembled from letters, diaries and conversations, Marie's daughter Eve Curie has left a vivid impression of her parents' life during this momentous period.

Life was unchanged in the little flat in the Rue de la Glacière. Marie and Pierre worked even more than usual: that was all. When the heat of summer came, the young wife found time to buy some baskets of fruit in the markets and, as usual, she cooked and put away preserves for the winter, according to the recipes used in the Curie family. Then she locked the shutters on her windows, which gave on burnt leaves; she registered their two bicycles at the Orleans station, and, like thousands of other young women in Paris, went off on holiday with her husband and her child.

This year [1898] the couple had rented a peasant's house at Auroux, in Auvergne. Happy to breathe fresh air after the noxious atmosphere of the Rue Lhomond, the Curies made excursions to Mende, Puy, Clermont, Mont-Dore. They climbed hills, visited grottoes, bathed in rivers. Every day, alone in the country, they spoke of what they called their 'new metals', polonium and 'the other' – the one that remained to be found. In September they would go back to the damp workroom and the dull minerals; with freshened ardour they would take up their search again. . . .

In spite of their prosaic character – or perhaps because of it – some notes written by Mme Curie in that memorable year 1898 seem to us worth quoting. Some are to be found in the margins of a book called *Family Cooking*, with respect to a recipe for gooseberry jelly:

> I took eight pounds of fruit and the same weight in crystallised sugar. After boiling for ten minutes, I passed the mixture through a rather fine sieve. I obtained fourteen pots of very good jelly, not transparent, which 'took' perfectly.

In a school notebook covered with grey linen, in which the young mother had written little Irène's weight day by day, her diet and the appearance of her first teeth, we read under the date of July 20th, 1898, some days after the publication of the discovery of polonium:

> Irène says 'thanks' with her hand. She can walk very well now on all fours. She says 'Gogli, gogli, go.' She stays in the garden all day at Sceaux on a carpet. She can roll, pick herself up, and sit down.

On August 15th, at Auroux:

> Irène has cut her seventh tooth, on the lower left. She can stand for half a minute alone. For the past three days we have bathed

> her in the river. She cries, but today (fourth day) she stopped crying and played with her hands in the water.
>
> She plays with the cat and chases him with war cries. She is not afraid of strangers any more. She sings a great deal. She gets up on the table when she is in her chair.

Three months later, on October 17th, Marie noted with pride: 'Irène can walk very well, and no longer goes on all fours.'

On January 5th, 1899: 'Irène has fifteen teeth!'

Between these two notes – that of October 17th, 1898, in which Irène no longer goes on all fours, and that of January 5th, in which Irène has fifteen teeth – and a few months after the note on the gooseberry preserve, we find another note worthy of remark.

It was drawn up by Marie and Pierre Curie and a collaborator called G. Bémont. Intended for the Academy of Science, and published in the *Proceedings* of the session of December 26th, 1898, it announced the existence of a second new chemical element in pitch-blende.

Some lines of this communication read as follows:

> The various reasons we have just enumerated lead us to believe that the new radioactive substance contains a new element to which we propose to give the name of RADIUM.
>
> The new radioactive substance certainly contains a very strong proportion of barium; in spite of that its radioactivity is considerable. The radioactivity of radium, therefore, must be enormous.

Radium was present in pitch-blende in almost negligible quantities – one part to approximately ten million parts of the ore. To extract it, establish its atomic weight, and convince the many scientists who doubted the existence of the new element, was the huge task the Curies set themselves. Obtaining pitch-blende was an initial obstacle. It was a costly ore, treated at the St Joachimsthal mine in Bohemia for the extraction of uranium salts used in the manufacture of glass. The residue of this process was piled up in a no-man's-land, planted with pine trees, near the mine. The Curies worked out that polonium and radium would still be present in this slag-heap, and the Austrian government agreed to give the two French 'lunatics' a ton of it. Further supplies had to be paid for from their meagre savings. The dull brown ore arrived, still mixed with pine-needles, in sacks on a coal wagon, and the Curies processed it in an abandoned shed at the School of Physics that had replaced Marie's lumber room. Since the shed had no chimney to carry off noxious fumes, much of the work had to be done in the courtyard outside. 'I sometimes passed the whole day', Marie later wrote, 'stirring a boiling mass, with an iron rod nearly as big as myself. In the evening I was broken with fatigue.' The Curies worked for four years in these conditions, from 1898 to 1902. Determining the properties of radium was Pierre's allotted task; Marie's was extracting salts of pure radium from the ore. As her daughter explains:

In this division of labour Marie had chosen the 'man's job'. She accomplished the toil of a day labourer. Inside the shed her husband was absorbed by delicate experiments. In the courtyard, dressed in her old dust-covered and acid-stained smock, her hair blown by the wind, surrounded by smoke which stung her eyes and throat, Marie was a sort of factory all by herself.

> I came to treat as many as twenty kilogrammes of matter at a time [she writes], which had the effect of filling the shed with great jars of precipitates and liquids. It was killing work to carry the receivers, to pour off the liquids and to stir, for hours at a stretch, the boiling matter in a smelting basin.

Radium showed no intention of allowing itself to be known by human creatures. Where were the days when Marie naïvely expected the radium content of pitch-blende to be *one per cent*? The radiation of the new substance was so powerful that a tiny quantity of radium, disseminated through the ore, was the source of striking phenomena which could be easily observed and measured. The difficult, the impossible thing was to isolate this minute quantity, to separate it from the gangue in which it was so intimately mixed.

The days of work became months and years: Pierre and Marie were not discouraged. This material, which resisted them, which defended its secrets, fascinated them. United by their tenderness, united by their intellectual passions, they had, in a wooden shack, the 'anti-natural' existence for which they had both been made, she as well as he.

> At this period we were entirely absorbed by the new realm that was, thanks to an unhoped-for discovery, opening before us [Marie was to write]. In spite of the difficulties of our working conditions, we felt very happy. Our days were spent at the laboratory. In our humble shed there reigned a great tranquillity: sometimes, as we watched over some operation, we would walk

up and down, talking about work in the present and in the future; when we were cold a cup of hot tea taken near the stove comforted us. We lived in our single preoccupation as if in a dream.

Whenever Pierre and Marie, alone in this poor place, left their apparatus for a moment and quietly let their tongues run on, their talk about their beloved radium passed from the transcendent to the childish.

'I wonder what *It* will be like, what *It* will look like,' Marie said one day with the feverish curiosity of a child who has been promised a toy. 'Pierre, what form do you imagine *It* will take?'

'I don't know,' the physicist answered gently. 'I should like it to have a very beautiful colour . . .'

For the Congress of Physics of 1900, the Curies drew up a general report on radioactive substances that aroused great interest among European scientists. Other researchers and technicians joined them in their laboratory. The direction and execution of the project remained, however, their own.

Marie continued to treat, kilogramme by kilogramme, the tons of pitch-blende residue which were sent her on several occasions from St Joachimsthal. With her remarkable patience she was able to be, every day for four years, physicist, chemist, specialised worker, engineer and labouring man all at once. Thanks to her brain and muscle, the old tables in the shed held more and more concentrated products – products richer and richer in radium. Mme Curie was approaching the end: she no longer stood in the courtyard, enveloped in bitter smoke, to watch the heavy basins of material in fusion. She was now at the stage of purification and of the 'fractional crystallisation' of strongly radioactive solutions. But the poverty of her haphazard equipment hindered her work more than ever. It was now that she needed a spotlessly clean workroom and apparatus perfectly protected against cold, heat and dirt. In this shed, open to every wind, iron- and cold-dust was afloat which, to Marie's despair, became mixed with the products purified with so much care. Her heart sometimes constricted before these little daily accidents, which absorbed so much of her time and her strength.

Pierre was so tired of the interminable struggle that he would have been quite ready to abandon it. Of course, he did not dream of dropping the study of radium and of radioactivity. But he would willingly have renounced, for the time being, the special operation of preparing pure radium. The obstacles seemed insurmountable. Could they not resume this work later on, under better conditions? More attached to the meaning of natural phenomena than to their material reality, Pierre Curie was exasperated to see the paltry results to which Marie's exhausting effort had led. He advised an armistice.

He counted without his wife's character. Marie wanted to isolate radium and she *would* isolate it. She scorned fatigue and difficulties, and even the gaps in her own knowledge which complicated her task. After all, she was only a very young scientist: she still had not the certainty and great culture Pierre had acquired by twenty years' work, and sometimes she stumbled across phenomena or methods of calculation of which she knew very little and for which she had to make hasty studies.

So much the worse! With stubborn eyes under her great brow, she clung to her apparatus and her test-tubes.

In 1902, forty-five months after the day on which the Curies announced the probable existence of radium, Marie finally carried off the victory in this war of attrition: she succeeded in preparing a decigramme of pure radium, and made a first determination of the atomic weight of the new substance, which was 225.

The incredulous chemists – of whom there were still a few – could only bow before the facts, before the superhuman obstinacy of a woman.

Radium officially existed.

It was nine o'clock at night. Pierre and Marie Curie were in their little house at 108 Boulevard Kellermann, where they had been living since 1900. The house suited them well. From the boulevard, where three rows of trees half hid the fortifications, could be seen only a dull wall and a tiny door. But behind the one-storey house, hidden from all eyes, there was a narrow provincial garden, rather pretty and very quiet. And from the 'barrier' of Gentilly they could escape on their bicycles toward the suburbs and the woods. . . .

Old Dr Curie, who lived with the couple, had retired to his room. Marie had bathed her child and put her to bed, and had stayed for a long time beside the cot. This was a rite. When Irène did not feel her mother near her at night she would call out for her incessantly, with that 'Mé!' which was to be our substitute for 'Mamma' always. And Marie, yielding to the implacability of the four-year-old child, climbed the stairs, seated herself beside her and stayed there in the darkness until the young voice gave way to light, regular breathing. Only then would she go down again to Pierre, who was growing impatient. In spite of his kindness, he was the most possessive and jealous of husbands. He was so used to the constant presence of his wife that her least eclipse kept him from thinking freely. If Marie delayed too long near her daughter, he received her on her return with a reproach so unjust as to be comic:

'You never think of anything but that child!'

Pierre walked slowly about the room. Marie sat down and made some stitches on the hem of Irène's new apron. One of her principles was never to buy ready-made clothes for the child: she thought them too fancy and impractical. In the days when Bronya was in Paris the two sisters cut out their children's dresses together, according to patterns of their own invention. These patterns still served for Marie.

But this evening she could not fix her attention. Nervous, she got up; then, suddenly:

'Suppose we go down there for a moment?'

There was a note of supplication in her voice – altogether superfluous, for Pierre, like herself, longed to go back to the shed they had left two hours before. Radium, fanciful as a living creature, endearing as a love, called them back to its dwelling, to the wretched laboratory.

The day's work had been hard, and it would have been more reasonable for the couple to rest. But Pierre and Marie were not always reasonable. As soon as they had put on their coats and told Dr Curie of their flight, they were in the street. They went on foot, arm in arm, exchanging few words. After the crowded streets of this queer district, with its factory buildings, wastelands and poor tenements, they arrived in the Rue Lhomond and crossed the little courtyard. Pierre put the key in the lock. The door squeaked, as it had squeaked thousands of times, and admitted them to their realm, to their dream.

'Don't light the lamps!' Marie said in the darkness. Then she added with a little laugh:

'Do you remember the day when you said to me: "I should like radium to have a beautiful colour"?'

The reality was more entrancing than the simple wish of long ago.

Radium had something better than 'a beautiful colour'; it was spontaneously luminous. And in the sombre shed, where, in the absence of cupboards, the precious particles in their tiny glass receivers were placed on tables or on shelves nailed to the wall, their phosphorescent bluish outlines gleamed, suspended in the night.

'Look . . . Look!' the young woman murmured.

She went forward cautiously, looked for and found a straw-bottomed chair. She sat down in the darkness and silence. Their two faces turned toward the pale glimmering, the mysterous sources of radiation, toward radium – their radium. Her body leaning forward, her head eager, Marie took up again the attitude which had been hers an hour earlier at the bedside of her sleeping child.

Her companion's hand lightly touched her hair.

She was to remember for ever this evening of glow-worms, this magic.

From 1900 onwards the Curies had been in correspondence with scientists from all over the world, responding to requests for information. Research workers from other countries joined the search for unknown radioactive elements. In 1903 two English scientists, Ramsay and Soddy, demonstrated that radium continually disengaged a small quantity of gas, helium. This was the first known example of the transformation of atoms. A little later Rutherford and Soddy, taking up a hypothesis considered by Marie Curie as early as 1900, published their *Theory of Radioactive Transformation*, affirming that radioactive elements, even when they seemed unchangeable, were in a state of spontaneous evolution. Of this Pierre Curie wrote: 'Here we have a veritable theory of the transformation of elements, but not as the alchemists understood it. Inorganic matter must have evolved through the ages, following immutable laws.'

The excitement generated by the new element is caught in Eve Curie's account:

Prodigious radium! Purified as a chloride, it appeared to be a dull-white powder, which might easily be mistaken for common kitchen salt. But its properties, better and better known, seemed stupefying. Its radiation, by which it had become known to the Curies, passed all expectation in intensity; it proved to be two million times stronger than that of uranium. Science had already analysed and dissected it, subdividing the rays into three different kinds, which traversed the hardest and most opaque matter – undergoing modification, of course.

Only a thick screen of lead proved to be able to stop the insidious rays in their invisible flight.

Radium had its shadow, its ghost: it spontaneously produced a singular gaseous substance, the *emanation* of radium, which was also active and destroyed itself clearly even when enclosed in a glass tube, according to rigorous law. Its presence was to be proved in the waters of numerous thermal springs.

Another defiance of the theories which seemed the immovable basis of physics was that radium spontaneously gave off heat. In one hour it produced a quantity of heat capable of melting its own weight of ice. If it was protected against external cold it grew warmer, and its temperature would go up as much as ten degrees centigrade or more above that of the surrounding atmosphere.

What could it not do? It made an impression on photographic plates through black paper; it made the atmosphere a conductor of electricity and thus discharged electroscopes at a distance; it coloured the glass receivers which had the honour of containing it with mauve and violet; it corroded and, little by little, reduced to powder the paper or the cottonwool in which it was wrapped.

We have already seen that it was luminous.

> This luminosity cannot be seen by daylight [Marie wrote] but it can be easily seen in half-darkness. The light emitted can be strong enough to read by, using a little of the product for light in darkness. . .

Nor was this the end of the wonders of radium: it also gave phosphorescence to a large number of bodies incapable of emitting light by their own means.

Thus with the diamond:

> The diamond is made phosphorescent by the action of radium and can so be distinguished from imitations in paste, which have very weak luminosity.

And, finally, the radiation of radium was 'contagious' – contagious, like a persistent scent or a disease. It was impossible for an object, a plant, an animal or a person to be left near a tube of radium without immediately acquiring a notable 'activity' which a sensitive apparatus could detect. This contagion, which interfered with the results of precise experiments, was a daily enemy to Pierre and Marie Curie.

> When one studies strongly radioactive substances [Marie writes], special precautions must be taken if one wishes to be able to continue taking delicate measurements. The various objects used in a chemical laboratory, and those which serve for experiments in physics, all become radioactive in a short time and act upon photographic plates through black paper. Dust, the air of the room, and one's clothes all become radioactive. The air in the room is a conductor. In the laboratory where we work the evil has reached an acute stage, and we can no longer have any apparatus completely isolated.

Long after the death of the Curies, their working notebooks were to reveal this mysterious 'activity', so that after thirty or forty years the 'living activity' would still affect measuring apparatuses.

The property of radium that attracted most urgent interest was its medical potential. The Curies had noticed that it caused blisters and inflammation of the skin, and Henri Becquerel, carrying a glass tube of radium in his waistcoat pocket, had been badly burned. 'I love this radium, but I've got a grudge against it,' he complained to the Curies. Pierre studied the effects of radium on animals, and found that by destroying diseased cells it could offer a treatment for growths, tumours, and certain forms of cancer. This therapeutic method was called Curietherapy. Once it was made public, the industrial production of radium began, and clinics opened throughout the world. The Curies could have patented their production technique, and become rich. However they decided it would be 'contrary to the scientific spirit'. They were awarded the Nobel Prize in Physics in 1903, but they shunned publicity and continued their simple life-style. Einstein, who knew Mme Curie in later life, said that she was 'of all celebrated beings, the only one whom fame has not corrupted'. After Pierre's death in a road accident in 1906, she gave the gramme of radium they had prepared together – now worth more than a million gold francs – to her laboratory. One result of this generosity was that her later scientific work was hampered by lack of funds, as an American magazine editor, Mrs William Meloney, discovered when she visited Marie in 1920:

'America', she said, 'has about fifty grammes of radium. Four of them are in Baltimore, six in Denver, seven in New York.' She went on naming the location of every grain.

'And in France?' I asked.

'My laboratory has hardly more than a gramme.'

'*You* have only a gramme?'

'I? Oh, I have none. It belongs to my laboratory.'

... That week I learned that the market price of a gramme of radium was one hundred thousand dollars. I also learned that Mme Curie's laboratory, although practically a new building, was without sufficient equipment; that the radium held there was only for cancer treatment.

On her return to the States Mrs Meloney launched a public fund-raising campaign, and in 1921 Mme Curie travelled to New York to collect a gramme of radium, bought for her by the women of America.

Source: Eve Curie, *Madame Curie*, trans. Vincent Sheean, London, Heinemann, 1938.

The Innocence of Radium

During the First World War the dials of luminous clocks and watches were painted with a phosphorescent radioactive material. The women painters were in the habit of licking their brushes, and many died of a cancer called 'phossy jaw'. These facts lie behind Lavinia Greenlaw's poem 'The Innocence of Radium':

With a head full of Swiss clockmakers,
she took a job at a New Jersey factory
painting luminous numbers, copying the style
believed to be found in the candlelit backrooms
of snowbound alpine villages.

Holding each clockface to the light,
she would catch a glimpse of the chemist
as he measured and checked. He was old enough,
had a kind face and a foreign name
she never dared to pronounce: Sochocky.

For a joke she painted her teeth and nails,
jumped out on the other girls walking home.
In bed that night she laughed out loud
and stroked herself with ten green fingertips.
Unable to sleep, the chemist traced each number

on the face he had stolen from the factory floor.
He liked the curve of her eights;
the way she raised the wet brush to her lips
and, with a delicate purse of her mouth,
smoothed the bristle to a perfect tip.

Over the years he watched her grow dull.
The doctors gave up, removed half her jaw,
and blamed syphilis when her thighbone snapped

as she struggled up a flight of steps.
Diagnosing infidelity, the chemist pronounced

the innocence of radium, a kind of radiance
that could not be held by the body of a woman,
only caught between her teeth. He was proud
of his paint and made public speeches
on how it could be used by artists to convey

the quality of moonlight. Sochocky displayed
these shining landscapes on his walls;
his faith sustained alone in a room
full of warm skies that broke up the dark
and drained his blood of its colour.

His dangerous bones could not keep their secret.
Laid out for X-ray, before a single button was pressed,
they exposed the plate and pictured themselves
as a ghost, not a skeleton, a photograph
he was unable to stop being developed and fixed.

Source: Lavinia Greenlaw, *Night Photograph*, London, Faber and Faber, 1993.

The Secret of the Mosquito's Stomach

The solving of the malaria problem has been called the most dramatic episode in the history of medicine. Malaria had for centuries been the most common infective disease throughout the tropics, but its cause was unknown. Because it was common in swampy districts, it was thought to be due to poisonous emanations from marshes. The first step towards understanding came in 1880 when a French army doctor, Alphonse Laveran, observing a drop of blood from a malaria patient under the microscope, saw minute parasites living on the red corpuscles. How these entered the bloodstream remained a mystery until the Englishman Ronald Ross proved that they were transmitted by mosquitoes.

Ross (1857–1932) was the son of an Indian Army General. He wanted to be a writer, but was persuaded by his father to take up medicine. After qualifying, he entered the Indian Medical Service, but in his spare time he studied literature, taking up French, Italian and German, and writing poems, music, dramas and novels. Deeply moved by the misery and disease of rural India, which he described in his poetry, he determined to solve the malaria problem. His guru was Patrick Manson, the 'father of tropical medicine', whom he met in London in 1894. While working in China, Manson had shown that the parasite that causes elephantiasis in humans grew and developed inside the mosquito. One November afternoon, walking down Oxford Street with Ross, Manson remarked, 'Do you know, I have formed the theory that mosquitoes carry malaria.'

To test this hypothesis Ross began systematic research after his return to India, collecting and breeding mosquitoes, feeding them on the blood of malaria patients, dissecting them, and examining their organs microscopically. He received no help or encouragement from the authorities, British or Indian, who saw no value in his work, and he met with many failures since he was experimenting, at first, on the wrong (i.e. non-malaria carrying) species of mosquito.

In April 1897, while holidaying with his wife at Ootacamund, he visited a well-known malarial area, the Sigur Ghat, which was half an hour's cycle ride away. During his investigations there he was himself infected with malaria, and discovered a type of mosquito new to him (probably *Anopheles stephensi*), which he called the 'dapple-winged' or Type C (to distinguish it from Types A and B on which he had already worked).

As Ross relates below, the breakthrough came on 20 August 1897, after his return to work at Secunderabad, when he noticed the characteristic pigmented egg cells of the malaria parasite in the stomach of this dapple-winged species. The following year he found further stages of parasite development in the salivary glands of mosquitoes, and it became clear that malaria was transmitted by the mosquito bite.

Ross was awarded the Nobel Prize for medicine in 1902, and in 1926 the Ross Institute for Tropical Diseases at Putney Heath was founded in his honour by public subscription. He died in 1932. His last words were: 'I shall find out things, yes, yes!' On the sixtieth anniversary of his discovery his close friend John Masefield, by then Poet Laureate, published a commemorative poem in *The Times* (20 August 1957):

> Once on this August day, an exiled man
> Striving to read the hieroglyphics spelled
> By changing speckles upon glass, beheld
> A secret hidden since the world began.

I will try to reconstruct the events as exactly as I can out of my notebooks, letters and memories. On arrival at Secunderabad after the severe labour in Ootacamund I felt my first violent reaction against the microscope and could scarcely bring myself to look through mine for a month. The Great Monsoon seemed to have failed. The hot blast which, instead of it, struck us in June was followed by a suffocating stillness and the sky was filled with a haze of dust through which the sun glared like a foiled enchanter . . . Well do I remember those awful days – and nights. I spent the time doing almost nothing but (I believe) writing – or rather moulding in the mind – the stanzas of *In Exile*, Part VII:

> What ails the solitude?
>   Is this the Judgement Day?
> The sky is red as blood;
>   The very rocks decay
>
> And crack and crumble, and
>   There is a flame of wind
> Wherewith the burning sand
>   Is ever mass'd and thin'd . . .

The world is white with heat;
  The world is rent and riven;
The world and heavens meet;
  The lost stars cry in heav'n.

I do not boast my premonitions because they seldom come true! But at that time I was certainly much exalted in spirit and said to myself: 'One more effort and the thing will be done.' I remember especially a dreadful evening when I climbed one of the heaps of great boulders piled upon each other which dot the plain outside the station and saw the vulture and the dead jackal (mentioned in the poem) below. Then it was that the thought struck me: Why not see whether mosquitoes, fed on malaria blood as before, contain any of the mosquito parasites which I had found in the Sigur Ghat? I was at full work again on 21 July 1897 on the last lap . . . But the weather became very hot again in August. At first I toiled comfortably, but as failure followed failure I became exasperated and worked till I could hardly see my way home late in the afternoons. Well do I remember that dark, hot little office in the hospital at Begumpett with the necessary gleam of light coming in from under the eaves of the veranda. I did not allow the punka to be used because it blew about my dissected mosquitoes, which were partly examined without a cover glass, and the result was that swarms of flies and of 'eye flies' – minute little insects which try to get into one's ears and eyelids – tormented me at their pleasure, while an occasional stegomyia [mosquito] revenged herself on me for the death of her friends. The screws of my microscope were rusted with sweat from my forehead and hands and its last remaining eyepiece was cracked!

By 15 August 31 mosquitoes of types A and B, all bred from the larva and fed on malarial patients, had been scrupulously examined, not counting numerous unfed mosquitoes, bad dissections, partial dissections, and other studies . . . On the previous day I had written to my wife: 'I have failed in finding parasites in mosquitoes fed on malaria patients, but perhaps am not using the proper kind of mosquito.' Now, as if in answer, some Angel of Fate must have met one of my three 'mosquito men' [Indians hired by Ross to look for mosquitoes – adults and larvae] in his leisurely perambulations and must have put into his hand a bottle of mosquito larvae, some of which I saw at once were of a type different from the usual culex and stegomyia larvae. Next morning, the 16 August, when I went again to hospital after breakfast the hospital assistant (I regret I have forgotten his name) pointed out a small mosquito seated on the wall with its tail *sticking outwards*. I caught it by my method of placing the mouth of a bottle *slowly* over it – if one jabs the bottle quickly the insect always escapes sideways – and killed it with tobacco smoke. It had spots on the wings and was evidently like the insect which I had found in the rest house at Sigur and is described in my notebook as 'a brown mosquito, not brindled, with three black bars on wings caught in ward'. I dissected it at once and found nothing unusual; but while I was doing so – I remember the details well – the worthy hospital assistant ran in to say that there were a number of mosquitoes of the same class which had hatched out in the bottle that my men had brought me yesterday. Sure enough there they were: about a dozen big brown fellows, with fine tapered bodies and spotted wings, hungrily trying to escape through the gauze covering of the flask which the Angel of Fate had given to my humbler retainer – dappled-winged mosquitoes, type C, the first I had ever found in Secunderabad but larger than the one I had just caught on the wall. Immediately my patient, Husein Khan [a malaria sufferer, paid to feed mosquitoes with his blood] was stripped and put on the bed under the mosquito net. This was at 12.25 p.m. by my notebook; and in five minutes 10 of the new mosquitoes had gorged themselves on him and were caught by the hospital assistant, each in its separate test tube with a drop of water to drink and a loose lump of cotton wool to prevent escape – Husein Khan received one anna for each . . .

Next day, 17 August, two of my new beauties (*Anopheles stephensi*) were dead, but I dissected two of the survivors, mosquitoes 32 and 33. They are described in my notebook as 'large, legs, proboscis, and anterior border of wings spotted dark brown and white – brown spots on tail joint of body. Back of abdomen and thorax light brown, belly dark brown. Wings nearly white.' I was rather excited over the dissections, spoiled them, and found nothing . . .

The 20 August 1897 – the anniversary of which I always call Mosquito Day – was, I think, a cloudy, dull, hot day. I went to hospital at 7 a.m., examined my patients and attended to official correspondence but was much annoyed because my men had failed to bring any more larvae of the dappled-winged mosquitoes and still more because one of my three remaining anopheles had died during the night and had swelled up with decay. After a hurried breakfast at the mess I returned to dissect the cadaver (mosquito 36) but found nothing new in it. I then examined a small stegomyia which happened to have been fed on Husein Khan on the same day (the 16th) – mosquito 37 – which was also negative of course. At about 1 p.m. I determined to sacrifice the seventh anopheles (*Anopheles stephensi*) of the batch fed on the 16th, mosquito 37, although my eyesight was already fatigued. Only one more of the batch remained.

The dissection was excellent and I went carefully through the tissues, now so familiar to me, searching every micron with the same passion and care as one would search some vast ruined palace for a little hidden treasure. Nothing. No, these new mosquitoes also were going to be a failure: there was something wrong with the theory. But the stomach tissue still remained to be examined – lying there empty and flaccid before me on the glass slide, a great white expanse of cells like a large courtyard of flagstones, each one of which must be scrutinized – half an hour's labour at least. I was tired and what was the use? I must have examined the stomachs of a thousand mosquitoes by this time. But the Angel of Fate fortunately laid his hand on my head and I had scarcely commenced the search again when I saw a clear and almost perfectly circular outline before me of about 12 μm in diameter. The outline was much too sharp, the cell too small to be an ordinary stomach cell of a mosquito. I looked a little further. Here was another, and another exactly similar cell.

The afternoon was very hot and overcast and I remember opening the diaphragm of the substage condenser of the microscope to admit more light and then changing the focus. *In each of these cells there was a cluster of small granules, black as jet* and exactly like the black pigment granules of the plasmodium crescents [the crescent-shaped form of the malaria parasite]. As with that pigment, the granules numbered about 12 to 16 in each cell and became blacker and more visible when more light was admitted through the diaphragm. I laughed and shouted for the hospital assistant – he was away having his siesta . . .

Next day I went to hospital intensely excited. The last survivor of the batch fed on the 16th, mosquito 39, was alive. After looking through yesterday's specimen I slew and dissected it with a shaking hand. *There were the cells again*, 21 of them, just as before, *only now much larger*! Mosquito 38, the seventh of the batch fed on the 16th, was killed on the fourth day afterwards – that is, on the 20th. This one was killed on the 21st, the fifth day after feeding, and the cells had grown during the extra day. The cells were therefore parasites and, as they contained the characteristic malarial pigment, were almost certainly the malaria parasites growing in the mosquito's tissues.

The thing was really done. We had to discover two unknown quantities simultaneously – the kind of mosquito which carries the parasite and the form and position of the parasite within it. We could not find the first without knowing the second nor the third without knowing the first. By an extremely lucky observation I had now discovered both the unknown quantities at the same moment. The mosquito was the anopheles and the parasite lives in or on its gastric wall and can be recognized at once by the characteristic pigment. All the work on the subject which has been done since then by me and others during the last 25 years has been mere child's play which anyone could do after the clue was once obtained.

That evening I wrote to my wife: 'I have seen something very promising indeed in my new mosquitoes,' and I scribbled the following unfinished verses in one of my *In Exile* notebooks in pencil:

This day designing God
  Hath put into my hand
A wondrous thing. And God
  Be praised. At His command,

I have found thy secret deeds
  Oh million-murdering Death.

I know that this little thing
  A million men will save –
Oh death where is thy sting?
  Thy victory oh grave?

On the 22nd I wrote to my wife, after mentioning the poem again: 'I really think I have done the mosquito theory at last, having found something in mosquitoes fed on malaria patients exactly like the malaria parasite.' Then, or a few days later, I wrote the following amended verses on a separate slip of paper:

This day relenting God
  Hath placed within my hand

A wondrous thing; and God
Be praised. At His command,

Seeking His secret deeds
With tears and toiling breath,
I find thy cunning seeds,
O million-murdering Death.

I know this little thing
A myriad men will save.
O Death, where is thy sting?
Thy victory, O Grave?

About the same time the two subsequent sonnetelles of *In Exile* were added – also on separate slips for tentative arrangements; and I did not like to change them further when they were published 13 years later. The three final sonnetelles of *In Exile* had been written previously and the poem was now finished, though I did not know it then.

Ross's optimism was premature, though it seemed justified for a time. With the introduction of modern insecticides, notably DDT, at the end of the Second World War, malarial mosquitoes were almost wiped out in many parts of the world. But genes giving resistance to insecticides spread through the mosquito population, and malaria is a major killer disease once more.

Source: Ronald Ross, *Memoirs*, London, John Murray, 1923.

The Poet and the Scientist

Hugh MacDiarmid is the pseudonym of the poet Christopher Murray Grieve (1892–1978), founder of the National Party of Scotland. His poem 'Two Scottish Boys', with its four epigraphs, argues that poets need to be more like scientists. The two 'boys' he compares are the Celtic twilight poet William Sharp (1855–1905), who wrote under the pseudonym 'Fiona Macleod', and the physician and tropical medicine expert Sir Patrick Manson (1844–1922), nicknamed 'Mosquito Manson', who (see p. 204) first suggested to Ronald Ross that the mosquito was host to the malaria parasite. 'Bunyan's quag' (line 4) is the Slough of Despond in *The Pilgrim's Progress*; Sainte-Beuve (line 21) was a nineteenth-century French critic, and the French quotation is from an essay he published in 1857 about Flaubert's novel *Madame Bovary*.

Two Scottish Boys

Not only was Thebes built by the music of an Orpheus, but without the music of some inspired Orpheus was no city ever built, no work that man glories in ever done.

Thomas Carlyle

For the very essence of poetry is truth, and as soon as a word's not true it's not poetry, though it may wear the cast clothes of it.

George MacDonald

Poetry never goes back on you. Learn as many pieces as you can. Go over them again and again till the words come of themselves, and then you have a joy forever which cannot be stolen or broken or lost. This is much better than diamond rings on every finger . . . The thing you cannot get a pigeon-hole for is the finger-point showing the way to discovery.

Sir Patrick Manson

> Science is the Differential Calculus of the mind. Art the Integral Calculus; they may be beautiful when apart, but are greatest only when combined.
>
> Sir Ronald Ross

There were two Scottish boys, one roamed seashore and hill
Drunk with the beauty of many a lovely scene,
And finally lost in nature's glory as in a fog,
Tossing him into chaos, like Bunyan's quag in the Valley of the Shadow.
The other having shot a lean and ferocious cat
On his father's farm, was profoundly interested
In a tapeworm he found when he investigated
Its internal machinery in the seclusion of his attic room,
– A 'prologue to the omen coming on'!

For while the first yielded nothing but high-falutin nonsense,
Spiritual masturbation of the worst description,
From the second down the crowded years I saw
Heroism, power for and practice of illimitable good emerge,
Great practical imagination and God-like thoroughness,
And mighty works of knowledge, tireless labours,
Consummate skill, high magnanimity, and undying Fame,
A great campaign against unbroken servility,
Ceaseless mediocrity and traditional immobility,
To the end that European reason may sink back no more
Into the immemorial embraces of the supernatural . . .

Sainte-Beuve was right – the qualities we most need
(Most of all in sentimental Scotland) are indeed
'*Science, esprit d'observation, maturité, force,*
Un peu de dureté,' and poets who, like Gustave Flaubert,
(That son and brother of distinguished doctors) wield
Their pens as these their scalpels, and that their work
Should everywhere remind us of anatomists and physiologists.

Poet and therefore scientist the latter, while the former,
No scientist, was needs a worthless poet too.

Source: *The Complete Poems of Hugh MacDiarmid 1920–76*, London, Martin Brian & O'Keeffe, 1978.

Wasps, Moths and Fossils

The son of semi-literate peasant farmers in France's Massif Central, Jean-Henri Fabre (1823–1915) spent his early years on his grandparents' remote small-holding, since his parents could not afford to feed him. Starting his education at the village primary school, run by the local barber, he won a bursary to secondary school in Avignon, and became a schoolmaster in Ajaccio, Corsica, where he began to study plants and insects. He was almost entirely self-taught, receiving his only natural history lesson from a biologist who happened to be visiting Corsica and showed him how to dissect a snail.

Back in Avignon, teaching in the grammar school, Fabre made expeditions into the surrounding countryside and would sit motionless for hours watching insects, to the puzzlement of the yokels, who took him for a half-wit. When he was almost 50 years old he gave up schoolmastering and retired to the small village of Sérignan, near Orange. Here in his 'hermit's retreat', living on fruits, vegetables, and a little wine, he observed insects on a tract of stony ground in front of his house, and also in the surrounding plain, with its scrub of wild thyme and lavender, and on the slopes of Mont Ventoux.

His accounts of the creatures he studied – wasps, bees, dung beetles, gnats, spiders, scorpions – grew into the ten-volume *Souvenirs entomologiques*. Picturesque and informal, and enlivened by allusions to his eight children, the family dog, and other minor characters, these essays established Fabre's greatness as both poet and scientist. To Victor Hugo he was 'the insect's Homer'; to Charles Darwin, an 'incomparable observer'. A strain of callousness, even cruelty, in his writing, accentuated by his tendency to describe his insects as if they were people, and contrasting curiously with his humour and charm, enhances its dramatic quality.

A turning point in Fabre's life came when he read a monograph on parasitic wasps by Léon Dufour, which noted how a species of burrowing wasp (*Cerceris bupresticida*), common in the Landes, placed the bodies of a particular kind of beetle (*Buprestis bifasciata*) in its burrow for its grubs to feed on when they hatched out. Dufour could not make out why the dead beetles did not decay before the wasp-eggs hatched, and he assumed that the mother wasp must inject them with a preservative. The Sérignan region with its sandy soil was favourable for observing burrowing wasps, and Fabre first directed his attention to a species (*Cerceris major*) closely related to Dufour's, which preyed on large weevils. He found that the weevils left in the wasp's burrow as food were not dead, but paralysed by the mother wasp, which stung them with great accuracy in their thoracic ganglia, and was thus able to leave living food for her grubs. Pricking weevils in the same spot with a fine steel pen dipped in ammonia, Fabre found that he, like the wasp, could induce instant paralysis. Later experiments on other paralysed wasp-victims revealed that they were not only alive but conscious enough to eat, taking drops of sugar solution from the end of a straw.

In the first of the pieces that follow Fabre investigates the food-arrangements of a third species of burrowing wasp, the yellow-winged Sphex (*Sphex flavipennis*), which preys on crickets. The second piece shows him in less gruesome mood, surprised by moths. In the third, his imagination works on the least promising material, stone.

Wasps

There can be no doubt that the Sphex uses her greatest skill when immolating a cricket; it is therefore very important to explain the method by which the victim is sacrificed. Taught by my numerous attempts to observe the war tactics of the Cerceris, I immediately used on the Sphex the plan already successful with the former, *i.e.* taking away the prey and replacing it by a living specimen. This exchange is all the easier because the Sphex leaves her victim while she goes down her burrow, and the audacious tameness, which actually allows her to take from your fingertips, or even off your hand, the cricket stolen from her and now offered, conduces most happily to a successful result of the experiment by allowing the details of the drama to be closely observed.

It is easy enough to find living crickets; one has only to lift the first stone, and you find them, crouched and sheltering from the sun. These are the young ones of the current year, with only rudimentary wings, and which, not having the industry of the perfect insect, do not yet know how to dig deep retreats where they would be beyond the investigations of the Sphex. In a few moments I find as many crickets as I could wish, and all my preparations are made. I establish myself on the flat ground in the midst of the Sphex colony and wait.

A huntress comes, conveys her cricket to the mouth of her hole and goes down alone. The cricket is speedily replaced by one of mine, but placed at some distance from the hole. The Sphex returns, looks round, and hurries to seize her too distant prey. I am all attention. Nothing on earth would induce me to give up my part in the drama which I am about to witness. The frightened cricket springs away. The Sphex follows closely, reaches it, darts upon it. Then there is a struggle in the dust when sometimes conqueror, sometimes conquered is uppermost or undermost. Success, equal for a moment, finally crowns the aggressor. In spite of vigorous kicks, in spite of bites from its pincer-like jaws, the cricket is felled and stretched on its back.

The murderess soon makes her arrangements. She places herself body to body with her adversary, but in a reverse position, seizes one of the bands at the end of the cricket's abdomen and masters with her forefeet the convulsive efforts of its great hind-thighs. At the same moment her intermediate feet squeeze the panting sides of the vanquished cricket, and her hind ones press like two levers on its face, causing the articulation of the neck to gape open. The Sphex then curves her abdomen vertically, so as to offer a convex surface impossible for the mandibles of the cricket to seize, and one beholds, not without emotion, the poisoned lancet plunge once into the victim's neck, next into the jointing of the two front segments of the thorax, and then again towards the abdomen. In less time than it takes to tell, the murder is committed, and the Sphex, after setting her disordered toilette to rights, prepares to carry off her victim, its limbs still quivering in the death-throes. Let us reflect a moment on the admirable tactics of which I have given a faint sketch. The prey is armed with redoubtable mandibles, capable of disembowelling the aggressor if they can seize her, and a pair of strong feet, actual clubs, furnished with a double row of sharp spines, which can be used alternatively to enable the cricket to bound far away from an enemy or to overturn one by brutal kicks. Accordingly, note what precautions on the part of the Sphex before using her dart. The victim, lying on its back, cannot escape by using its hind levers, for want of anything to spring from, as of course it would were it attacked in its normal position. Its spiny legs, mastered by the fore-feet of the Sphex, cannot be used as offensive weapons, and its mandibles, held at a distance by the wasp's hind-feet, open threateningly but can seize nothing. But it is not enough for the Sphex to render it impossible for her victim to hurt her: she must hold it so firmly garrotted that no movement can turn the sting from the points where the drop of poison must be instilled, and probably it is in order to hinder any motion of the abdomen that one of the end segments is grasped. If a fertile imagination had had free play to invent a plan of attack it could not have devised anything better, and it is questionable whether the athletes of the classic palestra [wrestling-ground] when grappling an adversary would have assumed attitudes more scientifically calculated.

I have just said that the dart is plunged several times into the victim's body, once under the neck, then behind the prothorax, lastly near the top of the abdomen. It is in this triple blow that the infallibility, the infused science of instinct, appear in all their magnificence. First let us recall the chief conclusions to which the preceding study of the Cerceris have led us. The victims of Hymenoptera [the group of insects to which wasps belong] whose larva live on prey are not corpses, in spite of entire immobility. There is merely total or partial paralysis, and more or less annihilation of animal life, but vegetative life – that of the nutritive organs – lasts a long while yet, and preserves from decomposition the prey which the larvæ are not to devour for a considerable time. To produce this paralysis the predatory Hymenoptera use just those methods which the advanced science of our day might suggest to the experimental physiologist – namely, wounding, by means of a poisoned dart, those nervous centres which animate the organs of locomotion. We know too that the various centres or ganglia of the nervous chain in articulate animals act to a certain degree independently, so that injury to one only causes, at all events immediately, paralysis of the corresponding segment, and this in proportion as the ganglia are more widely separated and distant from each other. If, on the contrary, they are soldered together, injury to the common centre causes paralysis of all the segments where its ramifications spread. This is the case with Buprestids and Weevils, which the Cerceris paralyses by a single sting, directed at the common mass of the nerve centres in the thorax. But open a cricket, and what do we find to animate the three pairs of feet? We find what the Sphex knew long before the anatomist, three nerve centres far apart. Thence the fine logic of the three stabs. Proud science! humble thyself.

Crickets sacrificed by Sphex flavipennis are no more dead, in spite of all appearances, than are Weevils struck by a Cerceris. If one closely observes a cricket stretched on its back a week or even a fortnight or more after the murder, one sees the abdomen heave strongly at long intervals. Very often one can notice a quiver of the palpi and marked movements in the antennæ and the bands of the abdomen, which separate and then come suddenly together. By putting such crickets into glass tubes I have kept them perfectly fresh for six weeks. Consequently, the Sphex larvæ, which live less than a fortnight before enclosing themselves in their cocoons, are sure of fresh food as long as they care to feast.

The chase is over; the three or four crickets needed to store a cell are heaped methodically on their backs, their heads at the far end, their feet toward the entrance. An egg is laid on each. Then the burrow has to be closed. The sand from the excavation lying heaped before the cell door is promptly swept backward into the passage. From time to time fair-sized bits of gravel are chosen singly, the Sphex scratching in the fragments with her forefeet, and carrying them in her jaws to consolidate the pulverized mass. If none suitable are at hand, she goes to look for them in the neighbourhood, apparently choosing with such scrupulous care as a mason would show in selecting the best stones for a building. Vegetable remains and tiny bits of dead leaf are also employed. In a moment every outward sign of the subterranean dwelling is gone, and if one has not been careful to mark its position, it is impossible for the most attentive eye to find it again. This done, a new burrow is made, provisioned and walled up as soon as the Sphex has eggs to house. Having finished laying, she returns to a careless and vagabond life until the first cold weather ends her well-filled existence . . .

The egg of Sphex flavipennis is white, elongated, and cylindrical, slightly curved, and measuring three to four millimetres in length. Instead of being laid fortuitously on any part of the victim, it is invariably placed on one spot, across the cricket's breast – a little on one side, between the first and second pairs of feet. The eggs of the white bordered, and of the Languedocian Sphex occupy a like position . . . This chosen spot must possess some highly important peculiarity for the security of the young larva, as I have never known it vary.

Hatching takes place at the end of two or three days. A most delicate covering splits, and one sees a feeble maggot, transparent as crystal, somewhat attenuated and even compressed in front, slightly swelled out behind, and adorned on either side by a narrow white band formed by the chief trachea. The feeble creature occupies the same position as the egg; its head is, as it were, engrafted on the same spot where the front end of the egg was fixed, and the remainder of its body rests on the victim without adhering to it. Its transparency allows us readily to perceive rapid fluctuations within its body, undulations following one another with mathematical regularity, and which, beginning in the middle of the body, are impelled, some forward and some backward. These are due to the digestive canal, which imbibes long draughts of the juices drawn from the sides of the victim.

Let us pause a moment before a spectacle so calculated to arrest attention. The prey is laid on its back, motionless. The grub is a lost grub if torn from the spot whence it draws nourishment. Should it fall, all is over, for weak as it is, and without means of locomotion, how would it again find the spot where it should quench its thirst? The merest trifle would enable the victim to get rid of the animalcule gnawing at its entrails, yet the gigantic prey gives itself up without the least sign of protestation. I am well aware that it is paralysed, and has lost the use of its feet from the sting of its assassin, but at this early stage it preserves more or less power of movement and sensation in parts unaffected by the dart. The abdomen palpitates, the mandibles open and shut, the abdominal styles and the antennæ oscillate. What would happen if the grub fixed on one of the spots yet sensitive near the mandibles, or even on the stomach, which, being tenderer and more succulent, would naturally suggest itself as fittest for the first mouthful of the feeble grub? Bitten on the quick parts, the cricket would display at least some shuddering of the skin, which would detach and throw off the minute larva, for which probably all would be over, since it would risk falling into the formidable, pincer-like jaws.

But there is a part of the body where no such peril is to be feared – the thorax wounded by the sting. There and there only can the experimenter on a recent victim dig down the point of a needle – nay, pierce through and through without evoking any sign of pain. And there the egg is invariably laid – there the young larva always attacks its prey. Gnawed where pain is no longer felt, the cricket does not stir. Later, when the wound has reached a sensitive spot, it will move of course as much as it can; but then it is too late – its torpor will be too deep, and besides, its enemy will have gained strength. That is why the egg is always laid on the same spot, near the wounds caused by the sting on the thorax, not in the middle, where the skin might be too thick for the new-born grub, but on one side – toward the junction of the feet, where the skin is much thinner. What a judicious choice! what reasoning on the part of the mother when, underground, in complete darkness, she perceives and utilizes the one suitable spot for her egg!

I have brought up Sphex larvæ by giving them successively crickets taken from cells, and have thus been able, day by day, to follow the rapid progress of my nurslings. The first cricket – that on which the egg is laid – is attacked, as I have already said, toward the point where the dart first struck – between the first and second pairs of legs. At the end of a few days the young larva has hollowed a hole big enough for half its body in the victim's breast. One may then sometimes see the cricket, bitten to the quick, vainly move its antennæ and abdominal styles, open and close its empty jaws, and even move a foot, but the larva is safe and searches its vitals with impunity. What an awful nightmare for the paralysed cricket! This first ration is consumed in six or seven days; nothing is left but the outer integument, whose every portion remains in place. The larva, whose length is then twelve millimetres, comes out of the body of the cricket through the hole it had made in the thorax. During this operation it moults, and the skin remains caught in the opening. It rests, and then begins on a second ration. Being stronger it has nothing to fear from the feeble movements of the cricket, whose daily increasing torpor has extinguished the last shred of resistance, more than a week having passed since it was wounded; so it is attacked with no precautions, and usually at the stomach, where the juices are richest. Soon comes the turn of the third cricket, then that of the fourth, which is consumed in ten hours. Of these three victims there remains only the horny integument, whose various portions are dismembered one by one and carefully emptied. If a fifth ration be offered, the larva disdains or hardly touches it, not from moderation, but from an imperious necessity.

It should be observed that up to now the larva has ejected no excrement, and that its intestine, in which four crickets have been engulfed, is distended to bursting. Thus, a new ration cannot tempt its gluttony, and henceforward it only thinks about making a silken dwelling. Its repast has lasted from ten to twelve days without a pause. Its length now measures from twenty-five to thirty millimetres, and its greatest width from five to six. Its usual shape, somewhat enlarged behind and narrowed in front, agrees with that general in larvæ of Hymenoptera. It has fourteen segments, including the head, which is very small, with weak mandibles seemingly incapable of the part just played by them. Of these fourteen segments the intermediary ones are provided with stigmata. Its livery is yellowish-white, with countless chalky white dots.

We saw that the larva began on the stomach of the second cricket, this being the most juicy and fattest part. Like a child who first licks off the jam on his bread, and then bites the slice with contemptuous tooth, it goes straight to what is best, the abdominal intestines, leaving the flesh, which must be extracted from its horny sheath, until it can be digested deliberately. But when first hatched it is not thus dainty: it must take the bread first and the jam later, and it has no choice but to bite its first mouthful from the middle of the victim's chest, exactly where its mother placed the egg. It is rather tougher, but the spot is a secure one, on account of the deep inertia into which three stabs have thrown the thorax. Elsewhere, there would be, generally, if not always, spasmodic convulsions which would detach the feeble thing and expose it to terrible risks amid a heap of victims whose hind legs, toothed like a saw, might occasionally kick, and whose jaws could still grip. Thus it is motives of security, and not the habits of the grub, which determine the mother where to place its egg.

A suspicion suggests itself to me as to this. The first cricket, the ration on which the egg is laid, exposes the grub to more risks than do the others. First, the larva is still a weakly creature; next, the victim was only recently stung, and therefore in the likeliest state for displaying some remains of life. This first cricket has to be as thoroughly paralysed as possible, and therefore it is stabbed three times. But the others, whose torpor deepens as time passes, – the others which the larvæ only attack when grown strong, – have they to be treated as carefully? Might not a single stab, or two, suffice to bring on a gradual paralysis while the grub devours its first allowance? The poison is too precious to be squandered; it is powder and shot for the Sphex, only to be used economically. At all events, if at one time I have been able to see a victim stabbed thrice, at another I have only seen two wounds given. It is true that the quivering point of the Sphex's abdomen seemed seeking a favourable spot for a third wound; but if really given, it escaped my observation. I incline to believe that the victim destined to be eaten first always is stabbed three times, but that economy causes the others only to be struck twice.

The last cricket being finished, the larva sets to work to spin a cocoon. In less than forty-eight hours the work is completed, and henceforward the skilful worker may yield within an impenetrable shelter to the overpowering lethargy which is stealing over it – a state of being which is neither sleeping nor waking, death nor life, whence it will issue transfigured ten months later.

Moths

Who does not know the magnificent Moth, the largest in Europe, clad in maroon velvet with a necktie of white fur? The wings, with their sprinkling of grey and brown, crossed by a faint zigzag and edged with smoky white, have in the centre a round patch, a great eye with a black pupil and a variegated iris containing successive black, white, chestnut, and purple arcs.

Well, on the morning of the 6th of May [1897], a female emerges from her cocoon in my presence, on the table of my insect laboratory. I forthwith cloister her, still damp with the humours of the hatching, under a wire-gauze bell-jar. For the rest, I cherish no particular plans. I incarcerate her from mere habit, the habit of the observer always on the look-out for what may happen.

It was a lucky thought. At nine o'clock in the evening, just as the household is going to bed, there is a great stir in the room next to mine. Little Paul, half-undressed, is rushing about, jumping and stamping, knocking the chairs over like a mad thing. I hear him call me:

'Come quick!' he screams. 'Come and see these Moths, big as birds! The room is full of them!'

I hurry in. There is enough to justify the child's enthusiastic and hyperbolical exclamations, an invasion as yet unprecedented in our house, a raid of giant Moths. Four are already caught and lodged in a bird-cage. Others, more numerous, are fluttering on the ceiling.

At this sight, the prisoner of the morning is recalled to my mind.

'Put on your things, laddie,' I say to my son. 'Leave your cage and come with me. We shall see something interesting.'

We run downstairs to go to my study, which occupies the right wing of the house. In the kitchen I find the servant, who is also bewildered by what is happening and stands flicking her apron at great Moths whom she took at first for Bats.

The Great Peacock, it would seem, has taken possession of pretty well every part of the house. What will it be around my prisoner, the cause of this incursion? Luckily, one of the two windows of the study had been left open. The approach is not blocked.

We enter the room, candle in hand. What we see is unforgettable. With a soft flick-flack the great Moths fly around the bell-jar, alight, set off again, come back, fly up to the ceiling and down. They rush at the candle, putting it out with a stroke of their wings; they descend on our shoulders, clinging to our clothes, grazing our faces. The scene suggests a wizard's cave, with its whirl of Bats. Little Paul holds my hand tighter than usual, to keep up his courage.

How many of them are there? About a score. Add to these the number that have strayed into the kitchen, the nursery, and the other rooms of the house; and the total of those who have arrived from the outside cannot fall far short of forty. As I said, it was a memorable evening, this Great Peacock evening. Coming from every direction and apprised I know not how, here are forty lovers eager to pay their respects to the marriageable bride born that morning amid the mysteries of my study.

Fossils

My very window-ledge, the confidant of bygone ages, talks to me of a vanished world. It is, literally speaking, an ossuary, each particle of which retains the imprint of past lives. That block of stone has lived. Spines of sea-urchins, teeth and vertebræ of fish, broken pieces of shells, shivers of madrepores form a pulp of dead existences. Examined ashlar by ashlar, my house would resolve itself into a reliquary, a rag-fair of things that were alive in the days of old.

The rocky layer from which building-materials are derived in these parts covers, with its mighty shell, the greater portion of the neighbouring upland. Here the quarry-man has dug for none knows how many centuries, since the time when Agrippa hewed cyclopean flags to form the stages and façade of the Orange theatre. And here, daily, the pick-axe uncovers curious fossils. The most remarkable of these are teeth, wonderfully polished in the heart of their rough veinstone, bright with enamel as though still in a fresh state. Some of them are most formidable, triangular, finely jagged at the edges, almost as large as one's hand. What an insatiable abyss, a jaw armed with such a set of teeth in manifold rows, placed stepwise almost to the back of the gullet; what mouthfuls, snapped up and lacerated by those serrate shears! You are seized with a shiver merely at the imaginary reconstruction of that awful implement of destruction!

The monster thus equipped as a prince of death belonged to the order of Squalidæ. Paleontology calls him Carcharodon Megalodon. The shark of to-day, the terror of the seas, gives an approximate idea of him, in so far as the dwarf can give an idea of the giant.

Other Squali abound in the same stone, all fierce gullets. It contains Oxyrhinæ (Oxyrhina Xiphodon, Agass.), with teeth shaped like pointed cleavers; Hemipristes (Hemipristis Serra, Agass.), whose mouths bristle with flexuous, steeled daggers, flattened on one side, convex on the other; Notidani (Notidanus Primigenius, Agass.), whose sunk teeth are crowned with radiate indentations.

This dental arsenal tells me how extermination came at all times to lop off the surplus of life; it says:

'On the very spot where you stand meditating upon a shiver of stone, an arm of the sea once stretched, filled with truculent devourers and peaceable victims. A long gulf occupied the future site of the Rhône Valley. Its billows broke at no great distance from your dwelling.'

Here, in fact, are the cliffs of the bank, in such a state of preservation that, on concentrating my thoughts, I seem to hear the thunder of the waves. Sea-urchins, Lithodomi, Petricolæ, Pholaidids have left their signatures upon the rock: hemispherical recesses large enough to contain one's fist, round cells, cabins with a narrow conduit-pipe through which the recluse received the incoming water, constantly renewed and laden with nourishment. Sometimes, the erstwhile occupant is there, mineralized, intact to the tiniest details of his striæ and scales, a frail ornament; more often, he has disappeared, dissolved, and his house has filled with a fine sea mud, hardened into a chalky kernel.

In this quiet inlet, some eddy has collected and drowned at the bottom of the mire, now turned into marl, enormous heaps of shells, of every shape and size. It is a molluscs' burying-ground, with hills for tumuli. I dig up oysters a cubit long and weighing five or six pounds apiece. One could shovel up, in the immense pile, Scallops, Cones, Cytheridæ, Mactridæ, Murices, Turritellidæ, Mitridæ and others too numerous, too innumerable to mention. You stand stupefied before the vital ardour of the days of old, which was able to supply such a pile of relics in a mere nook of earth.

The necropolis of shells tells us, besides, that time, that patient renewer of the order of things, has mown down not only the individual, a precarious being, but also the species. Nowadays, the neighbouring sea, the Mediterranean, has almost nothing identical with the population of the vanished gulf. To find a few features of similarity between the present and the past, we should have to seek them in the tropical seas. The climate, therefore, has become colder; the sun is slowly becoming extinguished; the species are dying out. Thus speak the numismatics of the stones on my window-ledge.

Without leaving my field of observation, so modest, so limited and yet so rich, let us once more consult the stone and, this time, on the subject of the insect. The country round Apt abounds in a strange rock that breaks off in thin plates, similar to sheets of whitish cardboard. It burns with a sooty flame and a bituminous smell; and it was deposited at the bottom of great lakes haunted by crocodiles and giant tortoises. Those lakes no human eye has ever seen. Their basins have been replaced by the ridges of the hills; their muds, peacefully deposited in thin courses, have become mighty banks of rock.

Let us break off a slab and subdivide it into sheets with the point of a knife, a work as easy as separating the superposed layers of a piece of paste- or mill-board. In so doing, we are examining a volume taken from the library of the mountains, we are turning the pages of a magnificently illustrated book. It is a manuscript of nature, far superior to the Egyptian papyrus. On almost every page are diagrams; nay, better: realities converted into pictures.

On this page are fish, grouped at random. One might take them for a dish fried in oil. Back-bones, fins, vertebral links, bones of the head, crystal of the eye turned to a black globule, everything is there, in its natural arrangement. One thing alone is absent: the flesh. No matter: our dish of gudgeons looks so good that we feel an inclination to scratch off a bit with our finger and taste this supramillenary preserve. Let us indulge our fancy and put between our teeth a morsel of this mineral fry seasoned with petroleum.

There is no inscription to the picture. Reflection makes good the deficiency. It says to us:

'These fishes lived here, in large numbers, in peaceful waters. Suddenly, swells came and asphyxiated them in their mud-thickened waves. Buried forthwith in the mire and thus rescued from the agents of destruction, they have passed through time, will pass through it indefinitely, under the cover of their winding-sheet.'

The same swells brought from the adjacent rain-swept shores a host of refuse, both vegetable and animal, so much so that the lacustrian deposit talks to us also of things on land. It is a general record of the life of the time.

Let us turn a page of our slab, or rather our album. Here are winged seeds, leaves drawn in brown prints. The stone herbal vies in botanical accuracy with a normal herbal. It repeats what the shells had already told us: the world is changing, the sun is losing its strength. The vegetation of modern Provence is not what it was in former days; it no longer includes palm-trees, camphor-yielding laurels, tufted araucarias and many other trees and shrubs whose equivalents belong to the torrid regions.

Continue to turn the pages. We now come to the insects. The most frequent are the Diptera, of middling size, often very humble flies and gnats. The teeth of the great Squali astonished us by their soft polish amid the roughness of their chalky veinstone. What shall we say of these frail midges preserved intact in their marly shrine? The frail creature, which our fingers could not grasp without crushing it, lies undeformed beneath the weight of the mountains!

The six slender legs, which the least thing is enough to disjoint, here lie spread upon the stone, correct in shape and arrangement, in the attitude of the insect at rest. There is nothing lacking, not even the tiny double claws of the extremities. Here are the two wings, unfurled. The fine net-work of their nervures can be studied under the lens as clearly as in the Dipteron of the collections, stuck upon its pin. The antennary tufts have lost none of their subtle elegance; the belly gives us the number of the rings, edged with a row of atoms that were cilia.

The carcase of a mastodont, defying time in its sandy bed, already astonishes us: a gnat of exquisite delicacy, preserved intact in the thickness of the rock, staggers our imagination.

Certainly, the Mosquito, carried by the rising swells, did not come from far away. Before his arrival, the hurly-burly of a thread of water must have reduced him to that annihilation to which he was so near. He lived on the shores of the lake. Killed by the joys of a morning – the old age of gnats – he fell from the top of his reed, was forthwith drowned and disappeared in the muddy catacombs.

Who are those others, those dumpy ones, with hard, convex elytra, the most numerous next to the Diptera? Their small heads, prolonged into a snout, tell us plainly. They are proboscidian Coleoptera, Rhynchopora, or, in less hard terms, Weevils. There are small ones, middling ones, large ones, similar in dimensions to their counterparts of to-day.

Their attitudes on the chalky slab are not as correct as those of the Mosquito. The legs are entangled anyhow; the beak, the rostrum is at one time hidden under the chest, at another projects forward. Some show it in profile; others – more frequent these – stretch it to one side, as the result of a twist in the neck.

These dislocated, contorted insects did not receive the swift and peaceful burial of the Dipteron. Though sundry of them may have lived on the plants on the banks, the others, the majority, come from the surrounding neighbourhood, brought by the rains, which warped their joints in crossing such obstacles as branches and stones. A stout armour has kept the body unscathed, but the delicate articulations of the members have given way to some extent; and the miry winding-sheet received the drowned Beetles as the disorder of the passage left them.

These strangers, come perhaps from afar, supply us with precious information. They tell us that, whereas the banks of the lake had the Mosquito as the chief representative of the insect class, the woods had the Weevil.

Outside the snout-carrying family, the sheets of my Apt rock show me hardly anything more, especially in the order of the Coleoptera. Where are the other terrestrial groups, the Carabus, the Dung-beetle, the Capricorn, which the wash of the rains, indifferent as to its harvests, would have brought to the lake even as it did the Weevil? There is not the least vestige of those tribes, so prosperous to-day.

Where are the Hydrophilus, the Gyrinus, the Dytiscus, all inhabitants of the water? These lacustrians had a great chance of coming down to us mummified between two sheets of marl. If there were any in those days, they lived in the lake, whose muds would have preserved these horn-clad insects even more perfectly than the little fishes and especially than the Dipteron. Well, of those aquatic Coleoptera there is no trace either.

Where were they, where were those missing from the geological reliquary? Where were they of the thickets, of the green-sward, of the worm-eaten trunks: Capricorns, borers of wood; Sacred Beetles, workers in dung; Carabi, disembowellers of game? One and all were in the limbo of the time to come. The present of that period did not possess them: the future awaited them. The Weevil, therefore, if I may credit the modest records which I am free to consult, is the oldest of the Coleoptera.

Life, at the start, fashioned oddities which would be screaming discords in the present harmony of things. When it invented the Saurian, it revelled at first in monsters fifteen and twenty yards long. It placed horns on their noses and eyes, paved their backs with fantastic scales, hollowed their necks into spiny wallets, wherein their heads withdrew as into a hood. It even tried, though not with great success, to give them wings. After these horrors, the procreating ardour calmed down and produced the charming green Lizard of our hedges.

When it invented the bird, it filled its beak with the pointed teeth of the reptile and appended a long, feathered tail unto its rump. These undetermined and revoltingly ugly creatures were the distant prelude to the Robin Redbreast and the Dove.

All these primitives are noted for a very small skull, an idiot's brain. The brute of antiquity is, first and foremost, an atrocious machine for snapping, with a stomach for digesting. The intellect does not count as yet. That will come later.

The Weevil, in his fashion, to a certain extent, repeats these aberrations. See the extravagant appendage to his little head. It is here a short, thick snout; there a sturdy beak, round or cut four-square; elsewhere a crazy reed, thin as a hair, long as the body and longer. At the tip of this egregious instrument, in the terminal mouthpiece, are the fine shears of the mandibles; on the sides, the antennæ, with their first joints set in a groove.

What is the use of this beak, this snout, this caricature of a nose? Where did the insect find the model? Nowhere. The Weevil is its inventor and retains the monopoly. Outside his family, no Coleopteron indulges in these buccal eccentricities.

Observe, also the smallness of the head, a bulb that hardly swells beyond the base of the snout. What can it have inside? A very poor nervous equipment, the sign of exceedingly limited instincts. Before seeing them at work, we make small account of these microcephali, in respect of intelligence; we class them among the obtuse, among creatures bereft of working capacity. These surmises will not be very largely upset.

Though the Curculio be but little glorified by his talents, this is no reason for scorning him. As we learn from the lacustrian schists, he was in the van of the insects with the armoured wing-cases; he was long stages ahead of the workers in incubation within the limits of possibility. He speaks to us of primitive forms, sometimes so quaint; he is, in his own little world, what the bird with the toothed jaws and the Saurian with the horned eyebrows are in a higher world.

In ever-thriving legions, he has been handed down to us without changing his characteristics. He is to-day as he was in the old times of the continents: the prints in the chalky slates proclaim the fact aloud. Under any such print, I would venture to write the name of the genus, sometimes even of the species.

Permanence of instinct must go with permanence of form. By consulting the modern Curculionid, therefore, we shall obtain a very approximate chapter upon the biology of his predecessors, at the time when Provence had great lakes filled with crocodiles and palm-trees on their banks wherewith to shade them. The history of the present will teach us the history of the past.

Sources: J.-H. Fabre, *Insect Life. Souvenirs of a Naturalist*, trans. from the French by the author of *Mademoiselle Mori*, with a Preface by David Sharp, MA, PRS, and edited by F. Merrifield, London, Macmillan, 1901; *The Life of Jean-Henri Fabre*, by the Abbé Augustin Fabre, trans. Bernard Miall, London, Hodder and Stoughton, 1921; J.-H. Fabre, *The Life and Love of the Insect*, trans. Alexander Teixeira de Mattos, London, Adam and Charles Black, 1911.

The Massacre of the Males

Maurice Maeterlinck (1862–1949), the Belgian Symbolist poet and playwright, was a keen bee-keeper. In *The Life of the Bee* (1901) he describes the bees' year – the departure of the old queen from the hive, accompanied by a swarm, in spring or early summer; the hatching of a new queen; her nuptial flight and her impregnation by one of the hundreds of drones that follow her; her return to the hive to begin her life of egg-laying; and – in the following extract – the extermination of the drones in autumn.

If skies remain clear, the air warm, and pollen and nectar abound in the flowers, the workers, through a kind of forgetful indulgence, or over-scrupulous prudence perhaps, will for a short time longer endure the importunate, disastrous presence of the males. These comport themselves in the hive as did Penelope's suitors in the house of Ulysses. Indelicate and wasteful, sleek and corpulent, fully content with their idle existence as honorary lovers, they feast and carouse, throng the alleys, obstruct the passages, and hinder the work; jostling and jostled, fatuously pompous, swelled with foolish, good-natured contempt; harbouring never a suspicion of the deep and calculating scorn wherewith the workers regard them, of the constantly growing hatred to which they give rise, or of the destiny that awaits them. For their pleasant slumbers they select the snuggest corners of the hive; then, rising carelessly, they flock to the open cells where the honey smells sweetest, and soil with their excrements the combs they frequent. The patient workers, their eyes steadily fixed on the future, will silently set things right. From noon till three, when the purple country trembles in blissful lassitude beneath the invincible gaze of a July or August sun, the drones will appear on the threshold. They have a helmet made of enormous black pearls, two lofty, quivering plumes, a doublet of iridescent, yellowish velvet, an heroic tuft, and a four-fold mantle, translucent and rigid. They create a prodigious stir, brush the sentry aside, overturn the cleaners, and collide with the foragers as these return, laden with their humble spoil. They have the busy air, the extravagant, contemptuous gait of indispensable gods who should be simultaneously venturing towards some destiny unknown to the vulgar. One by one they sail off into space, irresistible, glorious, and tranquilly make for the nearest flowers, where they sleep till the afternoon freshness awake them. Then, with the same majestic pomp, and still overflowing with magnificent schemes, they return to the hive, go straight to the cells, plunge their head to the neck in the vats of honey, and fill themselves tight as a drum to repair their exhausted strength; whereupon, with heavy steps, they go forth to meet the good, dreamless, and careless slumber that shall fold them in its embrace till the time for the next repast.

But the patience of the bees is not equal to that of men. One morning the long-expected word of command goes through the hive; and the peaceful workers turn into judges and executioners. Whence this word issues we know not; it would seem to emanate suddenly from the cold, deliberate, indignation of the workers; and no sooner has it been uttered than every heart throbs with it, inspired with the genius of the unanimous republic. One part of the people renounce their foraging duties to devote themselves to the work of justice. The great idle drones, asleep in unconscious groups on the melliferous walls, are rudely torn from their slumbers by an army of wrathful virgins. They wake, in pious wonder; they cannot believe their eyes; and their astonishment struggles through their sloth as a moonbeam through marshy water. They stare amazedly round them, convinced that they must be victims of some mistake; and the mother-idea of their life being first to assert itself in their dull brain, they take a step towards the vats of honey to seek comfort there. But ended for them are the days of May honey, the wine-flower of lime-trees and fragrant ambrosia of thyme and sage, of marjoram and white clover. Where the path once lay open to the kindly, abundant reservoirs, that so invitingly offered their waxen and sugary mouths, there stands now a burning-bush all alive with poisonous, bristling stings. The atmosphere of the city is changed; in lieu of the friendly perfume of honey the acrid odour of poison prevails; thousands of tiny drops glisten at the end of the stings, and diffuse rancour and hatred. Before the bewildered parasites are able to realize that the happy laws of the city have crumbled, dragging down in most inconceivable fashion their own plentiful destiny, each one is assailed by three or four envoys of justice; and these vigorously proceed to cut off his wings, saw through the petiole that connects the abdomen with the thorax, amputate the feverish antennæ, and seek an opening between the rings of his cuirass through which to pass their sword. No defence is attempted by the enormous, but unarmed, creatures; they try to escape, or oppose their mere bulk to the blows that rain down upon them. Forced on to their back, with their relentless enemies clinging doggedly to them, they will use their powerful claws to shift them from side to side; or, turning on themselves, they will drag the whole group round and round in wild circles, which exhaustion soon brings to an end. And, in a very brief space, their appearance becomes so deplorable, that pity, never far from justice in the depths of our heart, quickly returns, and would seek forgiveness, though vainly, of the stern workers who recognise only Nature's harsh and profound laws. The wings of the wretched creatures are torn, their antennæ bitten, the segments of their legs wrenched off; and their magnificent eyes, mirrors once of the exuberant flowers, flashing back the blue light and the innocent pride of summer, now, softened by suffering, reflect only the anguish and distress of their end. Some succumb to their wounds, and are at once borne away to distant cemeteries by two or three of their executioners. Others, whose injuries are less, succeed in sheltering themselves in some corner, where they lie, all huddled together, surrounded by an inexorable guard, until they perish of want. Many will reach the door and escape into space, dragging their adversaries with them; but, towards evening, impelled by hunger and cold, they return in crowds to the entrance of the hive to beg for shelter. But there they encounter another pitiless guard. The next morning, before setting forth on their journey, the workers will clear the threshold, strewn with the corpses of the useless giants; and all recollections of the idle race disappear till the following spring.

Source: Maurice Maeterlinck, *The Life of the Bee*, trans. Alfred Sutro, London, George Allen, 1901.

Freud on Perversion

As the creator of psychoanalysis Sigmund Freud (1856–1939) gave mankind the first scientific instrument for examining the human mind – or so it has been claimed. Others have denied Freud's theories scientific status, dismissing them as myths or at best unprovable hypotheses. Despite these criticisms, Freud's major 'discoveries' – the unconscious mind, infant sexuality, the almost universal prevalence of the Oedipus complex, the tripartite division of the psyche into ego, super-ego and id – have taken their place among our commonest cultural assumptions. He was among the first to expose to reasoned investigation areas of experience that had previously been shrouded in guilt, shame and prejudice. The following extracts from *Three Essays on Sexuality* show this rationality at its best.

Deviations in Respect of the Sexual Aim

The normal sexual aim is regarded as being the union of the genitals in the act known as copulation, which leads to a release of the sexual tension and a temporary extinction of the sexual instinct – a satisfaction analogous to the sating of hunger. But even in the most normal sexual process we may detect rudiments which, if they had developed, would have led to the deviations described as 'perversions'. For there are certain intermediate relations to the sexual object, such as touching and looking at it, which lie on the road towards copulation and are recognized as being preliminary sexual aims. On the one hand these activities are themselves accompanied by pleasure, and on the other hand they intensify the excitation, which should persist until the final sexual aim is attained. Moreover, the kiss, one particular contact of this kind, between the mucous membrane of the lips of the two people concerned, is held in high sexual esteem among many nations (including the most highly civilized ones), in spite of the fact that the parts of the body involved do not form part of the sexual apparatus but constitute the entrance to the digestive tract. Here, then, are factors which provide a point of contact between the perversions and normal sexual life and which can also serve as a basis for their classification.

Perversions are sexual activities which either (a) extend, in an anatomical sense, beyond the regions of the body that are designed for sexual union, or (b) linger over the immediate relations to the sexual object which should normally be traversed rapidly on the path towards the final sexual aim . . .

Sexual Use of the Mucous Membrane of the Lips and Mouth

The use of the mouth as a sexual organ is regarded as a perversion if the lips (or tongue) of one person are brought into contact with the genitals of another, but not if the mucous membranes of the lips of both of them come together. This exception is the point of contact with what is normal. Those who condemn the other practices (which have no doubt been common among mankind from primaeval times) as being perversions, are giving way to an unmistakable feeling of *disgust*, which protects them from accepting sexual aims of the kind. The limits of such disgust are, however, often purely conventional: a man who will kiss a pretty girl's lips passionately, may perhaps be disgusted at the idea of using her toothbrush, though there are no grounds for supposing that his own oral cavity, for which he feels no disgust, is any cleaner than the girl's. Here, then, our attention is drawn to the factor of disgust, which interferes with the libidinal over-valuation of the sexual object but can in turn be overridden by libido. Disgust seems to be one of the forces which have led to a restriction of the sexual aim. These forces do not as a rule extend to the genitals themselves. But there is no doubt that the genitals of the opposite sex can in themselves be an object of disgust and that such an attitude is one of the characteristics of all hysterics, and especially of hysterical women. The sexual instinct in its strength enjoys overriding this disgust.

Sexual Use of the Anal Orifice

Where the anus is concerned it becomes still clearer that it is disgust which stamps that sexual aim as a perversion. I hope, however, I shall not be accused of partisanship when I assert that people who try to account for this disgust by saying that the organ in question serves the function of excretion and comes in contact with excrement – a thing which is disgusting in itself – are not much more to the point than hysterical girls who account for their disgust at the male genital by saying that it serves to void urine.

The playing of a sexual part by the mucous membrane of the anus is by no means limited to intercourse between men: preference for it is in no way characteristic of inverted feeling. On the contrary, it seems that *paedicatio* with a male owes its origin to an analogy with a similar act performed with a woman; while mutual masturbation is the sexual aim most often found in intercourse between inverts . . .

Sadism and masochism occupy a special position among the perversions, since the contrast between activity and passivity which lies behind them is among the universal characteristics of sexual life.

The history of human civilization shows beyond any doubt that there is an intimate connection between cruelty and the sexual instinct; but nothing has been done towards explaining the connection, apart from laying emphasis on the aggressive factor in the libido. According to some authorities this aggressive element of the sexual instinct is in reality a relic of cannibalistic desires – that is, it is a contribution derived from the apparatus for obtaining mastery, which is concerned with the satisfaction of the other and, ontogenetically, the older of the great instinctual needs. It has also been maintained that every pain contains in itself the possibility of a feeling of pleasure. All that need be said is that no satisfactory explanation of this perversion has been put forward and that it seems possible that a number of mental impulses are combined in it to produce a single resultant.

But the most remarkable feature of this perversion is that its active and passive forms are habitually found to occur together in the same individual. A person who feels pleasure in producing pain in someone else in a sexual relationship is also capable of enjoying as pleasure any pain which he may himself derive from sexual relations. A sadist is always at the same time a masochist, although the active or the passive aspect of the perversion may be the more strongly developed in him and may represent his predominant sexual activity . . .

Variation and Disease

It is natural that medical men, who first studied perversions in outstanding examples and under special conditions, should have been inclined to regard them, like inversion, as indications of degeneracy or disease. Nevertheless, it is even easier to dispose of that view in this case than in that of inversion. Everyday experience has shown that most of these extensions, or at any rate the less severe of them, are constituents which are rarely absent from the sexual life of healthy people, and are judged by them no differently from other intimate events. If circumstances favour such an occurrence, normal people too can substitute a perversion of this kind for the normal sexual aim for quite a time, or can find place for the one alongside the other. No healthy person, it appears, can fail to make some addition that might be called perverse to the normal sexual aim; and the universality of this finding is in itself enough to show how inappropriate it is to use the word perversion as a term of reproach. In the sphere of sexual life we are brought up against peculiar and, indeed, insoluble difficulties as soon as we try to draw a sharp line to distinguish mere variations within the range of what is physiological from pathological symptoms.

In his poem 'In Memory of Sigmund Freud', W. H. Auden praised Freud's rationality, and placed recognition of the repressed and the unconscious among the greatest of his achievements:

> . . . he would have us remember most of all
> to be enthusiastic over the night,
> not only for the sense of wonder
> it alone has to offer, but also
>
> because it needs our love. With large sad eyes
> its delectable creatures look up and beg
> us dumbly to ask them to follow:
> they are exiles who long for the future
>
> that lies in our power, they too would rejoice
> if allowed to serve enlightenment like him,
> even to bear our cry of 'Judas',
> as he did and all must bear who serve it.
>
> One rational voice is dumb. Over his grave
> the household of Impulse mourns one dearly loved:
> sad is Eros, builder of cities,
> and weeping anarchic Aphrodite.

Sources: *The Standard Edition of the Complete Psychological Works of Sigmund Freud. Translated from the German under the General Editorship of James Strachey*, volume VII (1901–5), London, The Hogarth Press and the Institute of Psychoanalysis, 1953. W. H. Auden, *Collected Shorter Poems, 1927–1957*, London, Faber and Faber, 1966.

Kitty Hawk

The Wright brothers, Wilbur (1867–1912) and Orville (1871–1948), sons of a bishop of the United Brethren Church, had a keen interest in mechanical inventions from boyhood. Their thoughts were turned towards flying machines in June 1878 when their father gave them a toy helicopter designed by the Frenchman Alphonse Penaud, who first used rubber bands to power model aircraft. As young men, they experimented with kites and gliders, while running a business repairing and building bicycles. It was Orville who had the idea of constructing an aircraft wing with movable sections (ailerons), so that the pilot could vary their inclination. This was the original Wright Brothers patent. Their first powered machine was a 40ft-wingspan biplane with a 16 h.p. four-cylinder motor. To reduce the risk of its falling on the pilot, the motor was mounted on the lower wing right of centre, and the pilot lay flat, left of centre, to balance it. It had two propellers, and sledge-like runners instead of wheels. For take-off it was put on a truck with wheels that fitted into the groove of a monorail track. It first flew on 14 December 1902 at Kill Devil Hill, a few miles south of the remote coastal hamlet of Kitty Hawk, North Carolina, watched by five locals from the Kill Devil Life-Saving Station, two small boys and a dog. The brothers tossed a coin to decide who should be pilot first, and Wilbur won. This is Orville's account, written in 1913.

I took a position at one of the wings, intending to help balance the machine as it ran down the track. But when the restraining wire was slipped, the machine started off so quickly I could stay with it only a few feet. After a 35- to 40-foot run, it lifted from the rail.

But it was allowed to turn up too much. It climbed a few feet, stalled, and then settled to the ground near the foot of the hill, 105 feet below. My stop-watch showed that it had been in the air just three and a half seconds. In landing, the left wing touched first. The machine swung around, dug the skids into the sand and broke one of them. Several other parts were also broken, but the damage to the machine was not serious. While the tests had shown nothing as to whether the power of the motor was sufficient to keep the machine up, since the landing was made many feet below the starting point, the experiment had demonstrated that the method adopted for launching the machine was a safe and practical one. On the whole, we were much pleased.

Two days were consumed in making repairs, and the machine was not ready again till late in the afternoon of the sixteenth. While we had it out on the track in front of the building, making the final adjustments, a stranger came along. After looking at the machine a few seconds he inquired what it was. When we told him it was a flying-machine he asked whether we intended to fly it. We said we did, as soon as we had a suitable wind. He looked at it several minutes longer and then, wishing to be courteous, remarked that it looked as if it would fly, if it had a 'suitable wind.' We were much amused, for, no doubt, he had in mind the recent 75-mile gale when he repeated our words, 'a suitable wind'!

During the night of December 16th a strong cold wind blew from the north. When we arose on the morning of the seventeenth, the puddles of water, which had been standing about the camp since the recent rains, were covered with ice. The wind had a velocity of 10 to 12 metres per second (22 to 27 miles an hour). We thought it would die down before long, and so remained indoors the early part of the morning. But when ten o'clock arrived, and the wind was as brisk as ever, we decided that we had better get the machine out and attempt a flight. We hung out the signal for the men of the Life-saving Station. We thought that by facing the flyer into a strong wind, there ought to be no trouble in launching it from the level ground about camp. We realized the difficulties of flying in so high a wind, but estimated that the added dangers in flight would be partly compensated for by the slower speed in landing.

We laid the track on a smooth stretch of ground about one hundred feet west of the new building. The biting cold wind made work difficult, and we had to warm up frequently in our living-room, where we had a good fire in an improvised stove made of a large carbide can. By the time all was ready J. T. Daniels, W. S. Dough and A. D. Etheridge, members of the Kill Devil Life-saving Station, W. C. Brinkley of Manteo, and Johnny Moore, a boy from Nag's Head, had arrived.

We had a 'Richard' hand anemometer with which we measured the velocity of the wind. Measurements made just before starting the first flight showed velocities of 11 to 12 metres per second, or 24 to 27 miles per hour. Measurements made just before the last flight gave between 9 and 10 metres per second. One made just afterwards showed a little over 8 metres. The record of the Government Weather Bureau at Kitty Hawk gave the velocity of the wind between the hours of ten-thirty and twelve o'clock, the time during which the four flights were made, as averaging 27 miles at the time of the first flight and 24 miles at the time of the last.

With all the knowledge and skill acquired in thousands of flights in the last ten years, I would hardly think today of making my first flight on a strange machine in a 27-mile wind, even if I knew that the machine had already been flown and was safe. After these years of experience, I look with amazement upon our audacity in attempting flights with a new and untried machine under such circumstances. Yet faith in our calculations and the design of the first machine, based upon our tables of air pressure, obtained by months of careful laboratory work, and confidence in our system of control developed by three years of actual experiences in balancing gliders in the air had convinced us that the machine was capable of lifting and maintaining itself in the air, and that, with a little practice, it could be safely flown.

Wilbur having used his turn in the unsuccessful attempt on the fourteenth, the right to the first trial now belonged to me. After running the motor a few minutes to heat it up, I released the wire that held the machine to the track, and the machine started forward into the wind. Wilbur ran at the side of the machine, holding the wing to balance it on the track. Unlike the start on the fourteenth, made in a calm, the machine, facing a 27-mile wind, started very slowly. Wilbur was able to stay with it till it lifted from the track after a forty-foot run. One of the Life-saving men snapped the camera for us, taking a picture just as the machine had reached the end of the track and had risen to a height of about two feet. The slow forward speed of the machine over the ground is clearly shown in the picture by Wilbur's attitude. He stayed along beside the machine without any effort.

The course of the flight up and down was exceedingly erratic, partly due to the irregularity of the air and partly to lack of experience in handling this machine. The control of the front rudder was difficult on account of its being balanced too near the centre. This gave it a tendency to turn itself when started, so that it turned too far on one side and then too far on the other. As a result, the machine would rise suddenly to about ten feet, and then as suddenly dart for the ground. A sudden dart when a little over a hundred feet from the end of the track, or a little over 120 feet from the point at which it rose into the air, ended the flight. As the velocity of the wind was over 35 feet per second and the speed of the machine over the ground against this wind ten feet per second, the speed of the machine relative to the air was over 45 feet per second, and the length of the flight was equivalent to a flight of 540 feet made in calm air.

This flight lasted only 12 seconds, but it was nevertheless the first in the history of the world in which a machine carrying a man had raised itself by its own power into the air in full flight, had sailed forward without reduction of speed, and had finally landed at a point as high as that from which it started.

With the assistance of our visitors we carried the machine back to the track and prepared for another flight. The wind, however, had chilled us all through, so that before attempting a second flight we all went to the building again to warm up. Johnny Moore, seeing under the table a box filled with eggs, asked one of the Station men where we got so many of them. The people of the neighbourhood eke out a bare existence by catching fish during the short fishing season, and their supplies of other articles of food are limited. He probably never had seen so many eggs at one time in his whole life.

The one addressed jokingly asked him whether he hadn't noticed the small hen running about the outside of the building. 'That chicken lays eight to ten eggs a day!' Moore, having just seen a piece of machinery lift itself from the ground and fly, a thing at that time considered as impossible as perpetual motion, was ready to believe nearly anything. But after going out and having a good look at the wonderful fowl, he returned with the remark, 'It's only a common-looking chicken!'

At twenty minutes after eleven Wilbur started on the second flight. The course of this flight was much like that of the first, very much up and down. The speed over the ground was somewhat faster than that of the first flight, due to the lesser wind. The duration of the flight was less than a second longer than the first, but the distance covered was about seventy-five feet greater.

Twenty minutes later, the third flight started. This one was steadier than the first one an hour before. I was proceeding along pretty well when a sudden gust from the right lifted the machine up twelve to fifteen feet and turned it up sidewise in an alarming manner. It began a lively sidling off to the left. I warped the wings to try to recover the lateral balance and at the same time pointed the machine down to reach the ground as quickly as possible. The lateral control was more effective than I had imagined and before I reached the ground the right wing was lower than the left and struck first. The time of this flight was 15 seconds and the distance over the ground a little over 200 feet.

Wilbur started the fourth and last flight at just twelve o'clock. The first few hundred feet were up and down, as before, but by the time three hundred feet had been covered, the machine was under much better control. The course for the next four or five hundred feet had but little undulation. However, when out about eight hundred feet the machine began pitching again, and, in one of its darts downward, struck the ground. The distance over the ground was measured and found to be 852 feet; the time of the flight 59 seconds. The frame supporting the front rudder was badly broken, but the main part of the machine was not injured at all. We estimated that the machine could be put in condition for flight again in a day or two.

While we were standing about discussing this last flight a sudden strong gust of wind struck the machine and began to turn it over. Everybody made a rush for it. Wilbur, who was at one end, seized it in front. Mr Daniels and I, who were behind, tried to stop it by holding to the rear uprights.

All our efforts were in vain. The machine rolled over and over. Daniels, who had retained his grip, was carried along with it, and was thrown about, head over heels, inside of the machine. Fortunately he was not seriously injured, though badly bruised in falling about against the motor, chain guides, etc. The ribs in the surfaces of the machine were broken, the motor injured and the chain guides badly bent, so that all possibility of further flights with it for that year were at an end.

Source: Orville Wright, account published in *Flying* magazine December 1913.

A Cuckoo in a Robin's Nest

W. H. Hudson (1841–1922), one of the greatest natural history writers in English, grew up on his parents' ranch in Argentina. His childhood, spent roaming the pampas, watching birds and snakes, is described in his autobiography *Far Away and Long Ago* (1918). Settling in England in 1874, he endured poverty and ill health, achieving fame late in life with his novel *Green Mansions* and the countryside books such as *Afoot in England* (1909) and *A Shepherd's Life* (1910), which helped to foster the back-to-nature movement. He left nearly all he had to the Royal Society for the Protection of Birds, of which he was a founder. His gentle, romantic realism is quite individual. Joseph Conrad, a friend and admirer, said of him, 'One can't tell how this fellow gets his effects; he writes as the grass grows.' This excerpt is from *Hampshire Days* (1903).

A robin's nest with three robin's eggs and one of the cuckoo was found in a low bank at the side of the small orchard on 19th May, 1900. The bird was incubating, and on the afternoon of 27th May the cuckoo hatched out. Unfortunately I did not know how long incubation had been going on before the 19th, but from the fact that the cuckoo was first out, it seems probable that the parasite has this further advantage of coming first from the shell. Long ago I found that this was so in the case of the parasitical troupials of the genus *Molothrus* in South America.

I kept a close watch on the nest for the rest of that afternoon and the whole of the following day (the 28th), during which the young cuckoo was lying in the bottom of the nest, helpless as a piece of jelly with a little life in it, and with just strength enough in his neck to lift his head and open his mouth; and then, after a second or two, the wavering head would drop again. At eight o'clock next morning (29th) I found that one robin had come out of the shell, and one egg had been ejected and was lying a few inches below the nest on the sloping bank. Yet the young cuckoo still appeared a weak, helpless, jelly-like creature, as on the previous day. But he had increased greatly in size. I believe that in forty-eight hours from the time of hatching he had quite doubled his bulk, and had grown darker, his naked skin being of a bluish-black colour. The robin, thirty or more hours younger, was little more than half his size, and had a pale, pinkish-yellow skin, thinly clothed with a long black down. The cuckoo occupied the middle of the deep, cup-shaped nest, and his broad back, hollow in the middle, formed a sort of false bottom; but there was a small space between the bird's sides and the nest, and in this space or interstice the one unhatched egg that still remained and the young robin were lying.

During this day (29th) I observed that the pressure of the egg and young robin against his sides irritated the cuckoo: he was continually moving, jerking and wriggling his lumpish body this way and that, as if to get away from the contact. At intervals this irritation would reach its culminating point, and a series of mechanical movements would begin, all working blindly but as surely towards the end as if some devilish intelligence animated the seemingly helpless infant parasite.

Of the two objects in the nest the unhatched egg irritated him the most. The young robin was soft, it yielded when pressed, and could be made somehow to fit into the interstice; but the hard, round shell, pressing against him like a pebble, was torture to him, and at intervals became unendurable. Then would come that magical change in him, when he seemed all at once to become possessed of a preternatural power and intelligence, and then the blind struggle down in the nest would begin. And after each struggle – each round it might be called – the cuckoo would fall back again and lie in a state of collapse, as if the mysterious virtue had gone out of him. But in a very short time the pressure on his side would begin again to annoy him, then to torment him, and at last he would be wrought up to a fresh effort. Thus in a space of eight minutes I saw him struggle four separate times, with a period of collapse after each, to get rid of the robin's egg; and each struggle involved a long series of movements on his part. On each of these occasions the egg was pushed or carried up to the wrong or upper side of the nest, with the result that when the bird jerked the egg from him it rolled back into the bottom of the nest. The statement is therefore erroneous that the cuckoo knows at which side to throw the egg out. Of course he *knows* nothing, and, as a fact, he tries to throw the egg up as often as down the slope.

The process in each case was as follows: The pressure of the egg against the cuckoo's side, as I have said, was a constant irritation; but the irritability varied in degree in different parts of the body. On the under parts it scarcely existed; its seat was chiefly on the upper surface, beginning at the sides and increasing towards the centre, and was greatest in the hollow of the back. When, in moving, the egg got pushed up to the upper edge of his side, he would begin to fidget more and more, and this would cause it to move round, and so to increase the irritation by touching and pressing against other parts. When all the bird's efforts to get away from the object had only made matters worse, he would cease wriggling and squat down lower and lower in the bottom of the nest, and the egg, forced up, would finally roll right into the cavity in his back – the most irritable part of all. Whenever this occurred, a sudden change that was like a fit would seize the bird; he would stiffen, rise in the nest, his flabby muscles made rigid, and stand erect, his back in a horizontal position, the head hanging down, the little naked wings held up over the back. In that position he looked an ugly, lumpish negro mannikin, standing on thinnest dwarf legs, his back bent, and elbows stuck up above the hollow flat back.

Once up on his small stiffened legs he would move backwards, firmly grasping the hairs and hair-like fibres of the nest-lining, and never swerving, until the rim of the cup-like structure was reached; and then standing, with feet sometimes below and in some cases on the rim, he would jerk his body, throwing the egg off or causing it to roll off. After that he would fall back into the nest and lie quite exhausted for some time, his jelly-like body rising and falling with his breathing.

These changes in the bird strongly reminded me of a person with an epileptic fit, as I had been accustomed to see it on the pampas, where, among the gauchos, epilepsy is one of the commonest maladies: the sudden rigidity of muscle in some weak, sickly, flabby-looking person, the powerful grip of the hand, the strength in struggling, exceeding that of a man in perfect health, and finally, when this state is over, the weakness of complete exhaustion.

I witnessed several struggles with the egg, but at last, in spite of my watchfulness, I did not see it ejected. On returning after a very short absence I found the egg had been thrown out and had rolled down the bank, a distance of fourteen inches from the nest.

The young cuckoo appeared to rest more quietly in the nest now, but after a couple of hours the old fidgeting began again, and increased, until he was in the same restless state as before. The rapid growth of the birds made the position more and more miserable for the cuckoo, since the robin, thrust against the side of the nest, would throw his head and neck across the cuckoo's back, and he could not endure being touched there. And now a fresh succession of struggles began, the whole process being just the same as when the egg was struggled with. But it was not so easy with the young bird, not because of its greater weight, but because it did not roll like the egg and settle in the middle of the back; it would fall partly on to the cuckoo's back and then slip off into the nest again. But success came at last, after many failures. The robin was lying partly across the cuckoo's neck, when, in moving its head, its little curved beak came down and rested on the very centre of that irritable hollow in the back of its foster-brother. Instantly the cuckoo pressed down into the nest, shrinking away as if hot needles had pricked him, as far as possible from the side where the robin was lying against him, and this movement of course brought the robin more and more over him, until he was thrown right upon the cuckoo's back.

Instantly the rigid fit came on, and up rose the cuckoo as if the robin weighed no more than a feather on him; and away backwards he went, right up the nest, without a pause, and standing actually on the rim, jerked his body, causing the robin to fall off, clean away from the nest. It fell, in fact, on to a large dock leaf five inches below the rim of the nest, and rested there.

After getting rid of his burden the cuckoo continued in the same position, perfectly rigid, for a space of five or six seconds, during which it again and again violently jerked its body, as if it had the feeling of the burden on it still. Then, the fit over, it fell back, exhausted as usual.

The end of the little history – the fate of the ejected nestling and the attitude of the parent robins – remains to be told. When the young cuckoo throws out the nestlings from nests in trees, hedges, bushes, and reeds, the victims, as a rule, fall some distance to the ground, or in the water, and are no more seen by the old birds. Here the young robin, when ejected, fell a distance of but five or six inches, and rested on a broad, bright green leaf, where it was an exceedingly conspicuous object; and when the mother robin was on the nest – and at this stage she was on it a greater part of the time – warming that black-skinned, toad-like spurious babe of hers, her bright, intelligent eyes were looking full at the other one, just beneath her, which she had grown in her body and had hatched with her warmth, and was her very own. I watched her for hours; watched her when warming the cuckoo, when she left the nest and when she returned with food, and warmed it again, and never once did she pay the least attention to the outcast lying there so close to her. There, on its green leaf, it remained, growing colder by degrees, hour by hour, motionless, except when it lifted its head as if to receive food, then dropped it again, and when, at intervals, it twitched its body as if trying to move. During the evening even these slight motions ceased, though that feeblest flame of life was not yet extinguished; but in the morning it was dead and cold and stiff; and just above it, her bright eyes on it, the mother robin sat on the nest as before, warming her cuckoo.

How amazing and almost incredible it seems that a being such as a robin, intelligent above most birds as we are apt to think, should prove in this instance to be a mere automaton! The case would, I think, have been different if the ejected one had made a sound, since there is nothing which more excites the parent bird, or which is more instantly responded to, than the cry of hunger or distress of the young. But at this early stage the nestling is voiceless – another point in favour of the parasite. The sight of its young, we see, slowly and dumbly dying, touches no chord in the parent: there is, in fact, no recognition; once out of the nest it is no more than a coloured leaf, or a bird-shaped pebble, or fragment of clay.

Source: W. H. Hudson, *Hampshire Days*, London, Longmans Green, 1903.

Was the World Made for Man?

The great naturalist Alfred Russel Wallace (1823–1913) was co-founder, with Darwin of the theory of evolution (see p. 114) and inaugurated the modern study of biogeography (the geographical distribution of plants and animals). In *Man's Place in the Universe* (1903) he argued that, since only a small difference in the structure of the universe would have prevented organic life from developing, it proves that the universe was designed by an intelligent being with the purpose of generating organic life, 'culminating in man'. Mark Twain's sardonic riposte to this popular and fallacious argument was written in 1903 but not published in Twain's lifetime. The Lord Kelvin Twain refers to was the British physicist (1824–1907), and the authority Twain mistakenly cites as 'Herbert Spencer' was actually the Canadian geologist Joseph W. W. Spencer (1851–92).

Alfred Russel Wallace's revival of the theory that this earth is at the centre of the stellar universe, and is the only habitable globe, has aroused great interest in the world. – Literary Digest

For ourselves we do thoroughly believe that man, as he lives just here on this tiny earth, is in essence and possibilities the most sublime existence in all the range of non-divine being – the chief love and delight of God. – Chicago 'Interior' *(Presb.)*

I seem to be the only scientist and theologian still remaining to be heard from on this important matter of whether the world was made for man or not. I feel that it is time for me to speak.

I stand almost with the others. They believe the world was made for man, I believe it likely that it was made for man; they think there is proof, astronomical mainly, that it was made for man, I think there is evidence only, not proof, that it was made for him. It is too early, yet, to arrange the verdict, the returns are not all in. When they are all in, I think that they will show that the world was made for man; but we must not hurry, we must patiently wait till they are all in.

Now as far as we have got, astronomy is on our side. Mr Wallace has clearly shown this. He has clearly shown two things: that the world was made for man, and that the universe was made for the world – to stiddy it, you know. The astonomy part is settled, and cannot be challenged.

We come to the geological part. This is the one where the evidence is not all in, yet. It is coming in, hourly, daily, coming in all the time, but naturally it comes with geological carefulness and deliberation, and we must not be impatient, we must not get excited, we must be calm, and wait. To lose our tranquillity will not hurry geology; nothing hurries geology.

It takes a long time to prepare a world for man, such a thing is not done in a day. Some of the great scientists, carefully ciphering the evidences furnished by geology, have arrived at the conviction that our world is prodigiously old, and they may be right, but Lord Kelvin is not of their opinion. He takes a cautious, conservative view, in order to be on the safe side, and feels sure it is not so old as they think. As Lord Kelvin is the highest authority in science now living, I think we must yield to him and accept his view. He does not concede that the world is more than a hundred million years old. He believes it is that old, but not older. Lyell believed that our race was introduced into the world 31,000 years ago. Herbert Spencer makes it 32,000. Lord Kelvin agrees with Spencer.

Very well. According to these figures it took 99,968,000 years to prepare the world for man, impatient as the Creator doubtless was to see him and admire him. But a large enterprise like this has to be conducted warily, painstakingly, logically. It was foreseen that man would have to have the oyster. Therefore the first preparation was made for the oyster. Very well, you cannot make an oyster out of whole cloth, you must make the oyster's ancestor first. This is not done in a day. You must make a vast variety of invertebrates, to start with – belemnites, trilobites, jebusites, amalekites, and that sort of fry, and put them to soak in a primary sea, and wait and see what will happen. Some will be a disappointment – the belemnites, the ammonites and such; they will be failures, they will die out and become extinct, in the course of the 19,000,000 years covered by the experiment, but all is not lost, for the amalekites will fetch the home-stake, they will develop gradually into encrinites, and stalactites, and blatherskites, and one thing and another as the mighty ages creep on and the Archaean and the Cambrian Periods pile their lofty crags in the primordial seas, and at last the first grand stage in the preparation of the world for man stands completed, the Oyster is done. An oyster has hardly any more reasoning power than a scientist has; and so it is reasonably certain that this one jumped to the conclusion that the nineteen-million years was a preparation for *him*; but that would be just like an oyster, which is the most conceited animal there is, except man. And anyway, this one could not know, at that early date, that he was only an incident in a scheme, and that there was some more to the scheme, yet.

The oyster being achieved, the next thing to be arranged for in the preparation of the world for man, was fish. Fish, and coal – to fry it with. So the Old Silurian seas were opened up to breed the fish in, and at the same time the great work of building Old Red Sandstone mountains 80,000 feet high to cold-storage their fossils in was begun. This latter was quite indispensable, for there would be no end of failures again, no end of extinctions – millions of them – and it would be cheaper and less trouble to can them in the rocks than keep tally of them in a book. One does not build the coal beds and 80,000 feet of perpendicular Old Red Sandstone in a brief time – no, it took twenty million years. In the first place, a coal bed is a slow and troublesome and tiresome thing to construct. You have to grow prodigious forests of tree-ferns and reeds and calamites and such things in a marshy region; then you have to sink them under out of sight and let them rot; then you have to turn the streams on them, so as to bury them under several feet of sediment, and the sediment must have time to harden and turn to rock; next you must grow another forest on top, then sink it and put on another layer of sediment and harden it; then more forest and more rock, layer upon layer, three miles deep – ah, indeed it is a sickening slow job to build a coal-measure and do it right!

So the millions of years drag on; and meantime the fish-culture is lazying along and frazzling out in a way to make a person tired. You have developed ten thousand kinds of fishes from the oyster; and come to look, you have raised nothing but fossils, nothing but extinctions. There is nothing left alive and progressive but a ganoid or two and perhaps half a dozen asteroids. Even the cat wouldn't eat such.

Still, it is no great matter; there is plenty of time, yet, and they will develop into something tasty before man is ready for them. Even a ganoid can be depended on for that, when he is not going to be called on for sixty million years.

The Palaeozoic time-limit having now been reached, it was necessary to begin the next stage in the preparation of the world for man, by opening up the Mesozoic Age and instituting some reptiles. For man would need reptiles. Not to eat, but to develop himself from. This being the most important detail of the scheme, a spacious liberality of time was set apart for it – thirty million years. What wonders followed! From the remaining ganoids and asteroids and alkaloids were developed by slow and steady and pains-taking culture those stupendous saurians that used to prowl about the steamy world in those remote ages, with their snaky heads reared forty feet in the air and sixty feet of body and tail racing and thrashing after. All gone, now, alas – all extinct, except the little handful of Arkansawrians left stranded and lonely with us here upon this far-flung verge and fringe of time.

Yes, it took thirty million years and twenty million reptiles to get one that would stick long enough to develop into something else and let the scheme proceed to the next step.

Then the Pterodactyl burst upon the world in all his impressive solemnity and grandeur, and all Nature recognized that the Cainozoic threshold was crossed and a new Period open for business, a new stage begun in the preparation of the globe for man. It may be that the Pterodactyl thought the thirty million years had been intended as a preparation for himself, for there was nothing too foolish for a Pterodactyl to imagine, but he was in error, the preparation was for man. Without doubt the Pterodactyl attracted great attention, for even the least observant could see that there was the making of a bird in him. And so it turned out. Also the makings of a mammal, in time. One thing we have to say to his credit, that in the matter of picturesqueness he was the triumph of his Period; he wore wings and had teeth, and was a starchy and wonderful mixture altogether, a kind of long-distance premonitory symptom of Kipling's marine:

> 'E isn't one o' the reg'lar Line, nor 'e isn't one of the crew,
> 'E's a kind of a giddy harumfrodite – soldier an' sailor too!

From this time onward for nearly another thirty million years the preparation moved briskly. From the Pterodactyl was developed the bird; from the bird the kangaroo, from the kangaroo the other marsupials; from these the mastodon, the megatherium, the giant sloth, the Irish elk, and all that crowd that you make useful and instructive fossils out of – then came the first great Ice Sheet, and they all retreated before it and crossed over the bridge at Behring's strait and wandered around over Europe and Asia and died. All except a few, to carry on the preparation with. Six Glacial Periods with two million years between Periods chased these poor orphans up and down and about the earth, from weather to weather – from tropic swelter at the poles to Arctic frost at the equator and back again and to and fro, they never knowing what kind of weather was going to turn up next; and if ever they settled down anywhere the whole continent suddenly sank under them without the least notice and they had to trade places with the fishes and scramble off to where the seas had been, and scarcely a dry rag on them; and when there was nothing else doing a volcano would let go and fire them out from wherever they had located. They led this unsettled and irritating life for twenty-five million years, half the time afloat, half the time aground, and always wondering what it was all for, they never suspecting, of course, that it was a preparation for man and had to be done just so or it wouldn't be any proper and harmonious place for him when he arrived.

And at last came the monkey, and anybody could see that man wasn't far off, now. And in truth that was so. The monkey went on developing for close upon 5,000,000 years, and then turned into a man – to all appearances.

Such is the history of it. Man has been here 32,000 years. That it took a hundred million years to prepare the world for him is proof that that is what it was done for. I suppose it is. I dunno. If the Eiffel tower were now representing the world's age, the skin of paint on the pinnacle-knob at its summit would represent man's share of that age; and anybody would perceive that that skin was what the tower was built for. I reckon they would, I dunno.

Source: *The Works of Mark Twain. What Is Man? and Other Philosophical Writings*, edited with an introduction by Paul Baender, published for the Iowa Center for Textual Studies by the University of California Press, Berkeley, Los Angeles, London, 1973.

Drawing the Nerves

The greatest Spanish scientist, and the virtual founder of neuroscience, Santiago Ramón y Cajal (1852–1934) was born in one of the poorest regions of Aragon, the son of a country surgeon. A failure at school, where he resisted attempts to teach him Latin and Greek, he was apprenticed to a shoemaker, but developed an interest in human anatomy, encouraged by his father. They used to rob churchyards together, carrying bones home for inspection. After medical school at Zaragoza, Cajal taught anatomy at Valencia, where he learnt about the technique, developed by the Italian neurologist Camillo Golgi (1844–1926), of staining nervous tissue with silver nitrate, in an attempt to reveal its fine structures. As he explains in this extract from his autobiography, Cajal decided to apply the technique, in the first instance, to the brain and sensory centres of embryos and young animals, in order to simplify the problem. Using similar nerve-specific stains, he was able to determine the fine structure of the retina of the eye, and to trace the connections of nerve cells in grey matter and the spinal cord. With Golgi, he received the 1906 Nobel Prize for Medicine for establishing the neuron or nerve cell as the basic unit of nervous structure.

In my systematic explorations through the realms of microscopic anatomy, there came the turn of the nervous system, that masterpiece of life. It is important to remember that the technical resources of those times were quite inadequate for attacking the great and alluring problem effectively. Colouring agents capable of staining selectively the processes of the nerve cells so that they could be followed with some certainty across the formidable tangle of the gray matter were as yet unknown.

Nevertheless, in spite of the weakness of our methods of analysis, the problem attracted us irresistibly. We saw that an exact knowledge of the structure of the brain was of supreme interest for the building up of a rational psychology. To know the brain, we said, is equivalent to ascertaining the material course of thought and will, to discovering the intimate history of life in its perpetual duel with external forces; a history summarized, and in a way engraved in the defensive neuronal coordinations of the reflex, of instinct, and of the association of ideas.

Unfortunately, we lacked a weapon sufficiently powerful to pierce the impenetrable thicket of the gray matter, that constellation of unknowns.

In spite of all this, my pessimism was exaggerated, as we are about to see. Obviously, the *desideratum* referred to was and is even today an unattainable ideal, but some progress towards it could be made by taking advantage of the technique of the time. As a matter of fact, the instrument of revelation already existed; only I, isolated in my corner, was not acquainted with it, nor had it yet become known to any extent among scientists, in spite of having been made public in the years 1880 and 1885. It was discovered by C. Golgi, the famous histologist of Pavia, through the favour of chance, the muse who inspires great discoveries. In his staining experiments, this savant noticed that the protoplasm of the nerve cells, which is so refractory to artificial staining, possesses the valuable attribute of attracting strongly a precipitate of silver chromate when this precipitate is produced right within the thickness of the pieces of tissue. The *modus operandi*, which is of the simplest, consists essentially of impregnating fragments of gray matter for several days in solutions of potassium bichromate (or of Müller's fluid), or better still, in a mixture of bichromate and 1 per cent osmic acid solution, and treating them afterwards with dilute solutions (0.75 per cent) of crystalline silver nitrate. In this way there is formed a deposit of silver bichromate which, by a happy peculiarity that has not yet been explained, picks out certain nerve cells to the absolute exclusion of the others. When one examines the preparation, the granules of the gray matter appear coloured brownish black even to their finest branchlets, which stand out with unsurpassable clarity upon a transparent yellow background, formed by the elements which are not impregnated. Thanks to such a valuable reaction, Golgi succeeded during several years of labour in clarifying not a few points of importance in the morphology of the nerve cells and processes. As I have already mentioned, however, the admirable method of Golgi was then (1887–8) unknown to the immense majority of neurologists or was undervalued by those who had the requisite information about it . . . I decided to employ the method of Golgi on a large scale and to study it with all the patience of which I was capable. Innumerable tests by Bartual and myself in many parts of the central nervous system and many species of animals convinced us that the new method of analysis had before it a brilliant future, especially if there could be found some way of overcoming its highly capricious and uncertain character.* The procuring of a good preparation constituted a delightful surprise and gave rise to jubilant hopes.

Up to that time, our preparations of the cerebrum, the cerebellum, the spinal cord, etc., confirmed fully the discoveries of the celebrated histologist of Pavia, but nothing new of any importance arose out of them. I did not, however, lose faith in the method on that account. I was fully convinced that, in order to make a significant advance in the knowledge of the structure of the nerve centres, it was absolutely necessary to make use of procedures capable of showing the most delicate rootlets of the nerve fibres vigorously and selectively coloured upon a clear background. It is well known that the gray matter is formed by something like a very dense felt of excessively fine threads; and for following these filaments thin sections or completely stained preparations are worthless. What is required for this purpose is very intense reactions which, nevertheless, permit the use of very thick almost macroscopic sections (the processes from nerve cells are sometimes many millimeters or even centimeters long), the transparency of which, in spite of their unusual thickness, is made possible by the exclusive colouration of some few cells or fibres which stand out in the midst of extensive masses of cells that are uncoloured. Only thus does the undertaking to follow a nervous conductor from its origin to its termination become possible.

In any case, we were now in possession of the required instrument. It remained only to determine carefully the conditions of the chrome-silver reaction, and to regulate it so as to adapt it to each particular case. And if the brain and other adult central organs of man and other vertebrates are too complex to permit of scrutinizing their structural plan by the method referred to, why not apply the method systematically to lower vertebrates or to the early stages of ontogenetic

*It was due, no doubt, to these inconstancies of chrome-silver impregnation that Simarro, the introducer of the methods and discoveries of Golgi into Spain, abandoned his efforts in discouragement. In a letter to me in 1889 he said: 'I received your last publication on the structure of the spinal cord, which seems to me an important work but not *convincing*, because of the method of Golgi, which, even in your hands, who have perfected it so much, is a method which *suggests* rather than demonstrates.' Unfortunately, Simarro, who was endowed with great talent, lacked perseverence, the virtue of the less brilliant.

development, in which the nervous system should present a simple and, so to speak, diagrammatic organization?

Such was the programme of work which I laid out for myself. It was commenced in Valencia, but only after I had removed to Barcelona was it completed, with a perseverence, an enthusiasm, and a success which surpassed my expectations.

The year 1888 arrived, my greatest year, my year of fortune. For during this year, which rises in my memory with the rosy hues of dawn, there emerged at last those interesting discoveries so eagerly hoped and longed for. Had it not been for them, I should have vegetated sadly in a provincial university without passing in the scientific order beyond the category of more or less estimable delvers after details. As a result of them I attained the enjoyment of the sour flattery of celebrity; my humble surname, pronounced in the German manner (Cayal), crossed the frontiers; and my ideas, made known among scientific men, were discussed hotly. From that time on, the trench of science had one more recognized digger.

How did it happen? The reader will, I hope, forgive me if I devote a few remarks and explanations here to an occurrence so decisive for my career. I declare, in the first place, that the *new truth*, laboriously sought and so elusive during two years of vain efforts, rose up suddenly in my mind like a revelation. The laws governing the morphology and connections of the nerve cells in the gray matter, which became patent first in my studies of the cerebellum, were confirmed in all the organs which I successively explored. I may be permitted to formulate them at once:

1. The collateral and terminal ramifications of every axis cylinder* end in the gray matter, not in a diffuse network as maintained by Gerlach and Golgi, and most other neurologists, but by free arborizations arranged in a variety of ways (pericellular, baskets or nests, climbing branches, etc.).

2. These ramifications are applied very closely to the bodies and dendrites [tree-like outgrowths from the cell body] of the nerve cells, a contact or articulation being established between the receptive protoplasm and the ultimate axonic branchlets.

From the anatomical laws stated spring two physiological corollaries:

*The axis cylinder, or axon, is the fibre which conducts the nerve impulse away from the cell body. A nerve is a bundle of many such fibres. (Translator's note.)

3. Since the final rootlets of the axis cylinders are applied closely to the bodies and dendrites of the neurons,* it must be admitted that the cell bodies and their protoplasmic processes enter into the chain of conduction, that is to say, that they receive and propagate the nervous impulse, contrary to the opinion of Golgi, according to whom these parts of the cell perform a merely nutritive rôle.

4. The continuity of substance between cell and cell being excluded, the view that the nerve impulse is transmitted by contact, as in the junctions of electric conductors, or by an induction effect, as in induction coils, becomes inescapable.

The laws mentioned, a purely inductive outcome of the structural analysis of the cerebellum, were afterwards confirmed in all the nervous structures examined (retina, olfactory bulb, sensory and sympathetic ganglia, cerebrum, spinal cord, medulla oblongata, etc.). Later studies by myself and by others revealed that these structural and physiological standards apply equally, without modification, to the nervous system of vertebrates and to that of invertebrates. As happens with all legitimate conceptions, mine become more thoroughly established and gained progressively in dignity as the circle of confirmatory studies was extended.

However, in my eagerness to condense the essentials of the results obtained in brief propositions, I have not replied as yet to the question formulated in preceding paragraphs.

How were these laws discovered? Why did my work suddenly acquire surprising originality and broad importance?

I wish to be frank with the reader. To my successes of those days there contributed, without doubt, some improvements of the chrome silver method, particularly the modification designated the *procedure of double impregnation*; but the principal thing, the really efficacious cause, was – who would have thought it? – *the application to the solution of the problem of the gray matter of the dictates of the most ordinary common sense*. Instead of taking the bull by the horns, as the saying is, I permitted myself some strategic subterfuges. This demands explanation.

I have already pointed out that the great enigma in the organization

*The neuron is one complete nerve element, the architectural unit of the nervous system, and typically comprises a cell body, one or more dendrites, *i.e.*, processes which conduct towards the cell body, and an axon. (Translator's note.)

of the brain was the way in which the nervous ramifications ended and in which the neurons were mutually connected. Repeating a simile already used, it was a case of finding out how the roots and branches of these trees in the gray matter terminate, in that forest so dense that, by a refinement of complexity, there are no spaces in it, so that the trunks, branches, and leaves touch everywhere.

Two methods come to mind for investigating adequately the true form of the elements in this inextricable thicket. The most natural and simple apparently, but really the most difficult, consists of exploring the full-grown forest intrepidly, clearing the ground of shrubs and parasitic plants, and eventually isolating each species of tree, as well from its parasites as from its relatives. Such was the approach employed in neurology by most authors from the time of Stilling, Deiters, and Schültze (mechanical and chemical dissociations) to that of Weigert and Golgi, in which the isolation of each form of cell or of each fibre is procured optically, that is by the disappearance or absence of colour or the majority of the interlacing elements in the gray matter. Such tactics, however, to which Golgi and Weigert owed important discoveries, are inappropriate for the elucidation of the problem proposed, by reason of the enormous length and extraordinary luxuriance of the nervous ramifications, which inevitably appear mutilated and almost indecipherable in each section.

The second path open to reason is what, in biological terms, is designated the ontogenetic or embryological method. Since the full grown forest turns out to be impenetrable and indefinable, why not revert to the study of the young wood, in the nursery stage, as we might say? Such was the very simple idea which inspired my repeated trials of the silver method upon embryos of birds and mammals. If the stage of development is well chosen, or, more specifically, if the method is applied before the appearance of the myelin sheaths upon the axons (these forming an almost insuperable obstacle to the reaction), the nerve cells, which are still relatively small, stand out complete in each section; the terminal ramifications of the axis cylinder are depicted with the utmost clearness and perfectly free; the pericellular nests, that is the interneuronal articulations, appear simple, gradually acquiring intricacy and extension; in sum, the fundamental plan of the histological composition of the gray matter rises before our eyes with admirable clarity and precision. As a crowning piece of good fortune, the chrome silver reaction, which is so incomplete and uncertain in the adult, gives in embryos splendid colourations, singularly extensive and constant.

How is it, one may ask, that scientists did not hit upon so obvious a step? Certainly the idea must have occurred to many. In after years I learned that Golgi himself had already applied his method to embryos and young animals and obtained some excellent results; but he did not persist in his efforts, perhaps not thinking that he could progress by such a path in the elucidation of the problem of the structure of the centres. So little importance did he evidently attach to such experiments that in his greatest work the observations described have reference exclusively to the adult nervous system of man and mammals. In any case, my easy success proves once more that ideas do not show themselves productive with those who suggest them or apply them for the first time, but with those persevering workers who feel them strongly and put all their faith and love in their efficacy. From this point of view, it may be affirmed that scientific accomplishments are creations of the will and rewards of ardour.

Realizing that I had discovered a rich field, I proceeded to take advantage of it, dedicating myself to work, no longer merely with earnestness, but with fury. In proportion as new facts appeared in my preparations, ideas boiled up and jostled each other in my mind. A fever for publication devoured me. In order to make known my thoughts, I made use chiefly of a certain professional medical review, the *Gaceta Médica Catalana*. The tide of ideas and impatience for publication rising rapidly, however, this outlet became too narrow for me. I was much annoyed by the slowness of the press and the lateness of the dates of appearance. To extricate myself once and for all from such fetters, I decided to publish upon my own account a new review, the *Revista trimestral de Histología normal y patológica*. The first number saw the light in May, 1888, and the second appeared in the month of August of the same year. Naturally, all the articles, six in number, sprang from my own pen. From my hands emerged also the six lithographic plates which were included. Financial considerations obliged me not to print more than sixty copies altogether at the time and these were distributed almost entirely among foreign scientists.

Needless to say the vortex of publication entirely swallowed up my income, both ordinary and supplementary. Before that desolating cyclone of expenditure, my poor wife, taken up with caring for and watching five little demons (during the first year of my residence in

Barcelona, another son was born to me), determined to get along without a servant. She divined no doubt that there was maturing in my brain something unusual and of decisive importance for the future of the family, and, discreetly and self-sacrificingly, avoided any suggestion of rivalry or competition between the children of the flesh and the creatures of the mind.

As a distraction for the reader, who, I suppose, will be surfeited with the foregoing lucubrations, I should like to tell here how I freed myself from a tenacious and inveterate vice, the game of chess, which seriously menaced my evenings.

Knowing my fondness for the noble game of Ruy López y Philidor, various members of the *Casino Militar* invited me to join it.

I was weak enough to do so; I made my debut with varying success, measuring myself against players of considerable skill; and soon my skill increased and with it the morbid eagerness to overcome my adversaries. In my foolish vanity, I reached the point of playing four games simultaneously, against separate combatants, besides numerous onlookers who discussed at length the consequences of every move. There was one game that lasted two or three days. In my desire to shine at all costs and my confidence in my rather good visual memory, I even played without looking at the board.

Needless to say, I acquired as many books on the aristocratic pastime as I could lay my hands on and I even fell into the folly of sending solutions of problems to foreign illustrated papers. Carried away by the growing passion, I found my sleep broken by dreams and nightmares, in which pawns, knights, queens, and bishops were jumbled together in a frenzied dance. After being defeated the evening before in one or several games, it often happened that I wakened with a sudden start in the early hours of the morning, with my brain burning and in a whirl, breaking out in phrases of irritation and despair and exclaiming: 'I am a fool! I had a checkmate at the fourth move and did not see it.' In fact, putting the board on the table, I proved with sorrow the delayed clairvoyance of my *unconscious mind*, which had been working within me during the few hours of repose.

This could not continue. The almost permanent fatigue and cerebral congestion weakened me. If one does not lose money in playing chess, one loses time and brain energy, which are worth infinitely more, and one's will is turned aside and runs through the wrong channels. In my opinion, far from exercising the intelligence, as many claim, chess warps it and wears it out. Conscious of the danger of my position, I trembled before the distressing prospect of becoming converted into one of those amorphous types, sedentary and corpulent, who grow old unproductively and insensibly, seated at a card table or a chess table, without arousing any sincere affection or exciting, when the inevitable apoplexy or the terrible uraemia comes, more than a feeling of cold and formal commiseration. 'Too bad about Pérez! He was a good player! We shall have to look about for someone to take his place.' – For the player at a club or casino is no more than a table leg, something like the common picture which occupies a place in the room simply to balance the others.

But how was I to cure myself thoroughly? Feeling myself incapable of an inexorable, 'I do not play any more,' the possession of a will of iron; constantly excited by the eagerness for revenge, the evil genius of every player; the only supreme remedy which occurred to me was the *similia similibus* of the homeopathists: to study the works upon chess thoroughly and reproduce the most celebrated plays; and besides to discipline my rather sensitive nerves, augmenting the imaginative and reflex tension to the utmost. It was indispensable, also, to abandon my usual style of play, with consistently romantic and audacious attacks, and stick to the rules of the most cautious prudence.

In this way, expending my whole inhibitory capacity in the undertaking, I finally attained my desired end. This consisted, as the reader will have guessed, in flattering and lulling to sleep my insatiable self-love by defeating my skilful and cunning competitors for a whole week. Having demonstrated my superiority, eventually or by chance, the devil of pride smiled and was satisfied. Fearful of a relapse, I abandoned my place in the casino and did not move a pawn again for more than twenty-five years. Thanks to my psychological stratagem, I emancipated my modest intellect, which had been sequestrated by such stupid and sterile competitions, and was now able to devote it, fully and without distraction, to the noble worship of science.

Source: Santiago Ramón y Cajal, *Recollections of My Life*, trans. E. Horne Craigie and Juan Cano, New York, American Philosophical Society, 1937.

Discovering the Nucleus

Ernest Rutherford (1871–1937) was a New Zealander, the son of an odd-job man, who won a scholarship to university and came to England to work under J. J. Thomson at Cambridge. His famous gold foil experiment was carried out in 1909, when he was Professor at Manchester. This extract is from C. P. Snow's *The Physicists* (1981). Snow, a scientist turned novelist, caused a furore in 1959 with his Rede Lecture *The Two Cultures*, which argued that scientists 'have the future in their bones', whereas 'intellectuals, in particular literary intellectuals, are natural Luddites'.

If any scientist had a nose for, to use Medawar's phrase, 'the solution of the possible', Rutherford had. His attack was simple and direct, or rather he saw his way, through the hedges of complication, to a method which was the simplest and most direct.

An example is the most dramatic event of his career, the experiments by which he proved the existence of the atomic nucleus. The Curies had shown that radium emits various kinds of 'radiation', and one of these was now known to consist of a stream of electrically charged particles. These 'alpha particles' were identical to helium atoms with their electrons removed; but they originated not from helium gas but sprang spontaneously from the radium atoms as they disintegrated.

Even though atomic disintegration was still little understood, Rutherford saw these high-speed alpha particles as useful projectiles. He intercepted them with a thin sheet of gold foil, to see what happened as they passed through. If atoms were diffuse spheres of electrical charge, as Thomson had imagined, then most of the alpha particles should have gone straight through; a few should be deflected slightly. But some of the alpha particles bounced straight back again. It was like firing artillery shells at a piece of tissue paper, and getting some of them returning in the direction of the gun.

Rutherford could only explain this by postulating that these alpha particles were hitting small, massive concentrations within the atoms.

He thus concluded that most of an atom's mass resided in a minute, positively charged nucleus at the centre, while the electrons went around the outside – very much like the planets orbiting the massive sun. Most of the atom was just empty space. If an atom were expanded to the size of the dome of St Paul's Cathedral, virtually all its mass would lie within a central nucleus no larger than an orange. The large majority of alpha particles passed the atoms' emptiness and carried on through the foil; but just occasionally one would hit a nucleus head-on, and rebound along the way it had come.

Positive, like all Rutherford's physics. He said that he knew it was convincing, and maintained that he was completely surprised. One wonders if he hadn't had a secret inkling. He was superlatively good at making predictions about nature.

In 1919, he started firing alpha particles at nitrogen atoms. Nothing much should have happened. A great deal did.

As in his earlier experiments, the alpha particles came from radium. This time he was directing them down a tube filled with nitrogen gas. At the far end, he found he was detecting not just alpha particles, but also particles with all the properties of hydrogen nuclei. There was, however, no hydrogen in the tube. With his high-speed alpha-particle projectiles, Rutherford had actually broken them off the nuclei of the nitrogen atoms.

The discovery of radioactivity had earlier shown that certain, rare types of atom could spontaneously disintegrate. Now Rutherford had shown that ordinary atoms were not indestructible. By knocking out a hydrogen nucleus (later called a proton) from the nucleus of nitrogen he had converted it into another element, oxygen. Rutherford had, to a limited extent, achieved the dream of the alchemists and changed one element to another.

Snow's image of the artillery shells was Rutherford's own. He said that to find the alpha particles bouncing back 'was almost as incredible as if you fired a 15-inch shell at a piece of tissue paper and it came back and hit you'. Despite his skill at prediction, remarked on by Snow, he did not believe that the energy of the atomic nucleus could ever be released. He said this quite explicitly in 1933, four years before his death. Nine years later, in Chicago, the first atomic pile began to run (see p. 324).

Source: C. P. Snow, *The Physicists*, with an introduction by William Cooper, London, Macmillan, 1981.

Death of a Naturalist

Bruce Frederick Cummings (1889–1919), who wrote under the pseudonym W. N. P. Barbellion, was a self-taught naturalist, the son of a journalist from Barnstaple, Devon. At the age of 22, in competition with university-trained candidates, he won a place on the staff of the Natural History Museum, South Kensington, where he worked on lice. By this time, though he did not realize it, he was already suffering from multiple sclerosis. His *Journal of a Disappointed Man*, published in the year of his death, records his passionate thirst for life and charts the progress of his disease.

22 June 1910

How I hate the man who talks about the 'brute creation', with an ugly emphasis on *brute*. Only Christians are capable of it. As for me, I am proud of my close kinship with other animals. I take a jealous pride in my Simian ancestry. I like to think that I was once a magnificent hairy fellow living in the trees and that my frame has come down through geological time *via* sea jelly and worms and Amphioxus, Fish, Dinosaurs, and Apes. Who would exchange these for the pallid couple in the Garden of Eden? . . .

22 December 1912

Palæontology has its comfortable words too. I have revelled in my littleness and irresponsibility. It has relieved me of the harassing desire to live, I feel content to live dangerously, indifferent to my fate; I have discovered I am a fly, that we are all flies, that nothing matters. It's a great load off my life, for I don't mind being such a micro-organism – to me the honour is sufficient of belonging to the universe – such a great universe, so grand a scheme of things. Not even Death can rob me of that honour. For nothing can alter the fact that I *have* lived; *I have been I*, if for ever so short a time. And when I am dead, the matter which composes my body is indestructible – and eternal, so that come what may to my 'Soul', my dust will always be going on, each separate atom of me playing its separate part – I shall still have some sort of a finger in the Pie. When I am dead, you can boil me, burn me, drown me, scatter me – but you cannot destroy me: my little atoms would merely deride such heavy vengeance. Death can do no more than kill you . . .

16 August 1915

I probably know more about Lice than was ever before stored together within the compass of a single human mind! I know the Greek for Louse, the Latin, the French, the German, the Italian. I can reel off all the best remedies for Pediculosis [infestation with lice]: I am acquainted with the measures adopted for dealing with a nuisance in the field by the German Imperial Board of Health, by the British R.A.M.C., by the armies of the Russians, the French, the Austrians, the Italians. I know its life history and structure, how many eggs it lays and how often, the anatomy of its brain and stomach and the physiology of all its little parts. I have even pursued the Louse into ancient literature and have read old medical treatises about it, as, for example, the *De Phthiriasi* of Gilbert de Frankenau. Mucius the lawgiver died of this disease, so also did the Dictator Sylla, Antiochus Epiphanes, the Emperor Maximilian, the philosopher Pherecydes, Philip II of Spain, the fugitive Ennius, Callisthenes, Alcman and many other distinguished people including the Emperor Arnauld in 899. In 955, the Bishop of Noyon had to be sewn up in a leather sack before he could be buried. (See *Des Insectes reputés venimeux*, par M. Amoureux Fils, Doctor of Medicine in the University of Montpellier, Paris, 1789.) In Mexico and Peru, a poll-tax of Lice was exacted and bags of these treasures were found in the Palace of Montezuma (see Bingley, *Animal Biog.*, first edition, iii). In the *United Service Magazine* for 1842 (clix, 169) is an account of the wreck of the *Wager*, a vessel found adrift, the crew in dire straits and Captain Cheap lying on the deck – 'like an ant-hill'.

So that as an ancient writer puts it, 'you must own that for the quelling of human pride and to pull down the high conceits of mortal man, this most loathesome of all maladies (Pediculosis) has been the inheritance of the rich, the wise, the noble and the mighty – poets, philosophers, prelates, princes, Kings and Emperors'.

In his well-known *Bridgewater Treatise*, the Rev. Dr Kirby, the Father of English Entomology, asked: 'Can we believe that man in his pristine state of glory and beauty and dignity could be the receptacle of prey so loathsome as these unclean and disgusting creatures?' (Vol. I, p. 13). He therefore dated their creation *after* the Fall.

The other day a member of the staff of the Lister Institute called to see me on a lousy matter, and presently drew some live Lice from his waistcoat pocket for me to see. They were contained in pill boxes with little bits of muslin stretched across the open end thro' which the Lice could thrust their little hypodermic needles when placed near the skin. He feeds them by putting these boxes into a specially constructed belt and at night ties the belt around his waist and all night sleeps in Elysium. He is not married.

In this fashion he has bred hundreds from the egg upwards and even hybridized the two different species!

In the enfranchised mind of the scientific naturalist, the usual feelings of repugnance simply do not exist. Curiosity conquers prejudice . . .

20 January 1917

I am over 6 feet high and as thin as a skeleton; every bone in my body, even the neck vertebrae, creak at odd intervals when I move. So that I am not only a skeleton but a badly articulated one to boot. If to this is coupled the fact of the creeping paralysis, you have the complete horror. Even as I sit and write, millions of bacteria are gnawing away my precious spinal cord, and if you put your ear to my back the sound of the gnawing I dare say could be heard . . .

8 March 1917

As, for all practical purposes, I have done with life, and my own existence is often a burden to me and is like to become a burden also to others, I wish I possessed the wherewithal to end it at my will. With two or three tabloids in my waistcoat pocket, and my secret locked in my heart, how serenely I would move about among my friends and fellows, conscious that at some specially selected moment – at midnight or high noon – just when the spirit moved me, I could quietly slip out to sea on this Great Adventure. It would be well to be able to control this: the time, the place, and the manner of one's exit. For what disturbs me in particular is how I shall conduct myself; I am afraid lest I become afraid, it is a fear of fear. By means of my tabloids, I could arrange my death in an artistic setting, say underneath a big tree on a summer's day, with an open Homer in my hand, or more appropriately, a magnifying glass and Miall and Denny's *Cockroach*. It would be stage-managing my own demise and surely the last thing in self-conscious elegance! . . .

1 June 1917

We discuss post mortem affairs quite genially and without restraint. It is the contempt bred of familiarity, I suppose, Eleanor [his wife] says widows' weeds have been so vulgarised by the war widows that she won't go into deep mourning. 'But you'll wear just one weed or two for me?' I plead, and then we laugh . . .

7 August 1917

I become dreadfully emaciated. This morning, before getting off the bed I lifted my leg and gazed wistfully along all its length. My flabby *gastrocnemius* [calf muscle] swung suspended from the *tibia* like a gondola from a Zeppelin. I touched it gently with the tip of my index finger and it oscillated . . .

3 September 1917

My bedroom is on the ground floor as I cannot mount the stairs. But the other day when they were all out, I determined to clamber upstairs if possible, and search in the bedrooms for a half-bottle of laudanum, which Mrs — told me she found the other day in a box, a relic of the time when — had to take it to relieve pain.

I got off the bed on to the floor and crawled around on hands and knees to the door, where I knelt up straight, reached the handle and turned it. Then I crawled across the hall to the foot of the stairs, where I sat down on the bottom step and rested. It is a short flight of only 12 steps and I soon reached the top by sitting down on each and raising myself up to the next one with my hands.

Arrived at the top, I quickly decided on the most likely room to reach first, and painfully crawled along the passage and thro' the bathroom by the easiest route to the small door – there are two. The handles of all the doors in the house are fixed some way up above the middle, so that only by kneeling with a straight back could I reach them from the floor. This door in addition was at the top of a high but narrow step, and I had to climb on to this, balance myself carefully, and then carefully pull myself up towards the handle by means of a towel hung on the handle. After three attempts I reached the handle and found the door locked on the inside.

I collapsed on the floor and could have cried. I lay on the floor of the bathroom resting with head on my arm, then set my teeth and crawled around the passage along two sides of a square, up three more steps to the other door which I opened and then entered. I had only examined two drawers containing only clothes, when a key turned in the front door lock and Eleanor entered with — and gave her usual whistle.

I closed the drawers and crawled out of the room in time to hear Eleanor say in a startled voice to her mother: 'Who's that upstairs?' I whistled, and said that being bored I had come up to see the cot: which passed at that time all right.

Next morning my darling asked me why I went upstairs. I did not answer, and I think she knows.

4 September 1917

I am getting ill again, and can scarcely hold the pen. So good-bye Journal – only for a time perhaps.

Source: W. N. P. Barbellion, *Journal of a Disappointed Man*, London, Chatto & Windus, 1919.

Relating Relativity

Albert Einstein (1879–1955) became a cultural icon in his own lifetime. While still a young man he provided physicists with entirely new ways of thinking about space, time and gravitation. In later years his appearance and personality – the thistledown hair-do, the sad chimpanzee face, the pacifism, the simple tastes – gave the west its paramount symbol of benevolent scientific genius.

Born in Ulm, Germany, Einstein went to school in Munich, where his father ran a small engineering works. The pedantic educational regime daunted him, and he was credited with little academic ability. Two of his uncles encouraged his interest in physics and maths, though, and he studied these subjects as an undergraduate at Zurich Polytechnic. Becoming a Swiss citizen, he taught maths, and acted as an examiner at the Swiss Patent Office in Berne. In 1905, at the age of 26, he published four epoch-making scientific papers, including his Special Theory of Relativity. The General Theory followed in 1916.

In his *Popular Exposition* of the two theories (1920), Einstein explains for non-scientific readers how his ideas differ from those of orthodox physics. In orthodox physics the distance between two points on a rigid body is assumed to be always the same, irrespective of whether the body is in motion. Measurement of the time interval between two events is also assumed to be unaffected by motion. These assumptions accord with the dictates of Newton, who had proclaimed (in *Mathematical Principles*, 1686) that space and time were 'absolute':

> Absolute Space, in its own nature, without regard to anything external, remains always similar and immovable . . . Absolute, True and Mathematical Time, of itself, and from its own nature, flows equably without regard to anything external.

To illustrate how his own theory differs from these classic views, Einstein asks us to imagine a train speeding along an embankment. He also asks us to imagine two observers: one a passenger on the train, the other a man standing on the embankment watching the train go by. In orthodox physics the train would, of course, be the same length to both observers. But in Einstein's theory it is not. To the watcher on the embankment, it is shorter. This shortening is not due to optical illusion, but to a change in the nature of space itself, caused by motion.

The same principle holds whatever moving body we substitute for the train. Suppose that the passenger is holding a metre rod, and pointing it in the direction of the train's motion, then the length of the rod seen by the watcher on the embankment will, Einstein asserts, be less than a metre, by a mathematically calculable fraction:

> The rigid rod is shorter when in motion than when at rest, and the more quickly it is moving the shorter is the rod.

The rod will shrink only in the direction of its motion. Its length will diminish, but its thickness will remain identical to what it was at rest.

As with space, so with time. Einstein asks us to imagine a clock on the train. The time that elapses between two successive ticks, when it is stationary, is exactly one second, and to the passenger on the train, seated by the clock, the interval will still be one second when the train is moving. But to the watcher on the embankment the interval, Einstein states, will be greater:

> As a consequence of its motion the clock goes more slowly than when at rest.

The same will apply whatever measured interval we substitute for clock ticks. Heartbeats, for example, or biological growth, or the process of human ageing, will all be slower as observed from the embankment.

These effects of foreshortening and slowing down are all reciprocal. That is, if the passenger on the train looked out at the watcher on the embankment, and if the watcher had a metre rod, which he held parallel to the rails, and a clock identical with the passenger's, then the watcher's metre rod would be shorter than the passenger's and the watcher's clock would go slower. This reciprocity follows from the fact that in Einstein's theory the watcher on the embankment is no less in motion than the passenger:

> Every motion must only be considered as a relative motion. Returning to the illustration we have frequently used of the embankment and the railway carriage, we can express the fact of motion here taking place in the following two forms, both of which are equally justifiable:
>
> (a) The carriage is in motion relative to the embankment.
>
> (b) The embankment is in motion relative to the carriage.

Einstein emphasizes that the effects he describes become discernible only at very high speeds, approaching the speed of light:

We have experience of such rapid motions only in the case of electrons and ions; for other motions the variation from the laws of classical mechanics are too small to make themselves evident in practice.

Nevertheless the many popular accounts of Relativity Theory published in the 1920s continued to draw their illustrations from everyday life, so as to give the public graphic means of imagining the new concepts. Bertrand Russell's *The ABC of Relativity* (1926) retained Einstein's railway train but, in keeping with the ethos of an upper-middle-class pastoral England, gave it a dining car and made the watcher on the embankment a fisherman:

Let us suppose that you are in a train on a long straight railway, and that you are travelling at three-fifths of the velocity of light. Suppose that you measure the length of your train, and find that it is a hundred yards. Suppose that the people who catch a glimpse of you as you pass succeed by skilful scientific methods, in taking observations which enable them to calculate the length of your train. If they do their work correctly, they will find that it is eighty yards long. Everything in the train will seem to them shorter in the direction of the train than it does to you. Dinner plates, which you see as ordinary circular plates, will look to the outsider as if they were oval: they will seem only four-fifths as broad in the direction in which the train is moving as in the direction of the breadth of the train. And all this is reciprocal. Suppose you see out of the window a man carrying a fishing-rod which by his measurement, is fifteen feet long. If he is holding it upright, you will see it as he does; so you will if he is holding it horizontally at right angles to the railway. But if he is pointing it along the railway, it will seem to you to be only twelve feet long. All lengths in the direction of motion are diminished by twenty per cent, both for those who look into the train from outside and for those who look out of the train from inside.

The astronomer A. S. Eddington in his *Space, Time and Gravitation* (1920) substituted for Einstein's locomotive an advanced type of aircraft with an airman lying full length on the floor.

Suppose that by development in the powers of aviation, a man flies past us at the rate of 161,000 miles a second. We shall suppose that he is in a comfortable travelling conveyance in which he can move about, and act normally, and that his length is in the direction of the flight. If we could catch an instantaneous glimpse as he passed, we should see a figure about three feet high, but with the breadth and girth of a normal human being. And the strange thing is that he would be sublimely unconscious of his own undignified appearance. If he looks in a mirror in his conveyance, he sees his usual proportions . . . But when he looks down on us, he sees a strange race of men who have apparently gone through some flattening-out process: one man looks hardly ten inches across the shoulders; another standing at right angles is almost 'length and breadth without thickness'. As they turn about they change appearance like the figures seen in the old-fashioned convex mirrors. If the reader has watched a cricket match through a pair of prismatic binoculars, he will have seen this effect exactly.

It is the reciprocity of these appearances – that each party should think the other has contracted – that is so difficult to realize. Here is a paradox beyond even the imagination of Dean Swift. Gulliver regarded the Lilliputians as a race of dwarfs; and the Lilliputians regarded Gulliver as a giant. That is natural. If the Lilliputians had appeared dwarfs to Gulliver, and Gulliver had appeared a dwarf to the Lilliputians – but no! that is too absurd for fiction, and is an idea only to be found in the sober pages of science.

It is not only in space, but in time that these strange variations occur. If we observed the aviator carefully we should infer that he was unusually slow in his movements; and events in the conveyance moving with him would be similarly retarded – as though time had forgotten to go on. His cigar lasts twice as long as ours.

But here again reciprocity comes in, because in the aviator's opinion it is we who are travelling at 161,000 miles a second past him; and when he has made all allowances, he finds that it is we who are sluggish. Our cigar lasts twice as long as his.

Whereas in Newtonian physics space and time were absolute, in Einstein's theory, as these examples illustrate, they become fluid. Space is involved in time, and time in space. For Einstein space and time cease to exist as separate concepts. All events occur in a new (and unimaginable) 'continuum' called space-time. Unlike the old geometry, the geometry of this continuum does not contain any straight lines, but is curved. Left to themselves, objects move in curves within it. But large masses, such as the sun, cause 'puckers' in space-time, affecting the motion of any object near them. According to Einstein, gravity is the result of these puckers, rather than (as Newton had believed) of a force operating between masses.

Working on this assumption Einstein predicted that the light from a distant star would be 'bent' by the sun's gravitational field as it passed close to the sun.

The occurrence of a solar eclipse in 1919 gave scientists a chance of testing this prediction (it is only during a solar eclipse that stars near the sun can be observed), and accordingly two British expeditions were sent to photograph the eclipse, one, under Eddington, to the Isle of Principe in West Africa, the other to Brazil. They reported that the apparent displacement of stars near the sun was as Einstein had predicted.

When this news broke in November 1919 it projected relativity theory into the world's headlines. 'Light All Askew In The Heavens: Einstein's Theory Triumphs', announced the *New York Times*. The London *Times* carried an article headed 'The Fabric Of The Universe' hailing Einstein's theory as 'a new philosophy that will sweep away nearly all that has been hitherto accepted as the axiomatic basis of physical thought'. The effects of the new theory on time and space also inspired the cartoonists and limerick-writers, like Arthur Buller:

> There was a young lady named Bright,
> Who travelled much faster than light.
> She started one day
> In the relative way
> And returned on the previous night.

The *Scientific American* ran a competition for the clearest 3,000-word account of Einstein's theory. Several world-famous scientists entered, but the prize ($5,000) went to Lyndon Bolton, an Irish-born schoolmaster working in the London Patent Office (appropriately, since Einstein had been a school-master working in the Swiss Patent Office when he formulated the theory).

Having explained the Special Theory (much as Russell and Eddington do above) Bolton's prize-winning essay (printed in the *Scientific American*, 5 February 1921) goes on to tackle the more difficult General Theory which, unlike the Special Theory, takes account of gravitational fields. Developing an illustration used by Einstein himself, Bolton asks us to imagine a simulated gravitational field, consisting of a large revolving disk with a man standing on it. Since the disk is isolated in space, the man is not aware it is revolving (any more than we are aware the earth is). This is (as Bolton concedes) not a perfect model of a gravitational field, because the man will feel himself thrown away from the centre by centrifugal force, whereas in a gravitational field he would feel himself pulled towards the centre. However, in other respects it is, Bolton shows, a useful model for understanding what happens to conventional geometry when it is put into a gravitational field:

Let us note the experiences of an observer on a rotating disk which is isolated so that the observer has no direct means of perceiving the rotation . . .

He will notice as he walks about on the disk that he himself and all the objects on it, whatever their constitution or state, are acted upon by a force directed away from a certain point upon it and increasing with the distance from that point. This point is actually the center of rotation, though the observer does not recognize it as such. The space on the disk in fact presents the characteristic properties of a gravitational field. The force differs from gravity as we know it by the fact that it is directed away from instead of toward a center, and it obeys a different law of distance, but this does not affect the characteristic properties that it acts on all bodies alike, and cannot be screened from one body by the interposition of another. An observer aware of the rotation of the disk would say that the force was centrifugal force; that is, the force due to inertia which a body always exerts when it is accelerated.

Next suppose the observer to stand at the point of the disk where he feels no force, and to watch someone else comparing, by repeated applications of a small measuring rod, the circumference of a circle having its center at that point, with its diameter. The measuring rod when laid along the circumference is moving lengthwise relatively to the observer, and is therefore subject to contraction by his reckoning. When laid radially to measure the diameter this contraction does not occur. The rod will therefore require a greater proportional number of applications to the circumference than to the diameter, and the number representing the ratio of the circumference of the circle to the diameter thus measured will therefore be greater than 3.14159+, which is its normal value. Moreover the relative velocity decreases as the center is approached, so that the contraction of the measuring rod is less when applied to a smaller circle; and the ratio of the circumference to the diameter, while still greater than the normal, will be nearer to it than before, and the smaller the circle the less the difference from the normal. For circles whose centers are not at the point of zero force the confusion is still greater, since the velocities relative to the observer of points on them now change from point to point. The whole scheme of geometry as we know it is thus disorganized. Rigidity becomes an unmeaning term since the standards by which alone rigidity can be tested are themselves subject to alteration . . .

The same confusion arises in regard to clocks. No two clocks will in general go at the same rate, and the same clock will alter its rate when moved about . . .

The region therefore requires a space-time geometry of its own, and be it noted that with this special geometry is associated a definite gravitational field, and if the gravitational field ceases to exist, for example if the disk were brought to rest, all the irregularities of measurement disappear . . .

Gravitational fields arise in the presence of matter. Matter is therefore presumed to be accompanied by a special geometry, as though it imparted some peculiar kink or twist to space.

Relativity theory created shock-waves in literature and the visual arts as well as in popular thought. The painter Juan Gris had studied physics and maths before coming to Paris in 1906 where with Braque and Picasso he led the Cubist movement. Cubism, showing objects as seen from more than one position at once, could be seen as a crude attempt to represent four-dimensional space-time, and Gris explained it in terms of the new science: 'The mathematics of picture-making leads me to the physics of representation.'

The earliest poetic celebration of Einstein seems to be 'St Francis Einstein of the Daffodils', by the American poet William Carlos Williams. Published in April 1921, it sees Einstein as affirming love, youth and springtime against the dead values of the First World War:

In March's black boat
Einstein and April
have come at the time in fashion
up out of the sea
through the rippling daffodils
in the foreyard of
the dead Statue of Liberty
whose stonearms
are powerless against them
the Venusremembering wavelets
breaking into laughter . . .

Study of relativity theory prompted Williams to metrical experiment – a 'variable' poetic foot and a poetic line that was only 'relatively stable'. Taking up these ideas other American poets furthered developments in free verse – Louis Zukofsky (who translated a biography of Einstein), and the 'Black Mountain' poets, Charles Olson (who demanded 'What is measure when the universe flips?'), Robert Creeley and Robert Duncan. Poetic equivalents to relativity theory were at best vague, however, for lines of verse do not move relative to the observer at speeds approaching that of light, and if they did their lengths would become variable without any help from the poet.

Similar difficulties hindered the adoption of relativity theory in novels. Major novelists such as James Joyce and William Faulkner were aware of the new ideas and made efforts to incorporate them. But language inevitably reflects human perceptions of space and time, and cannot represent space-time. Virginia Woolf's *The Waves* (1931) merges spatial and temporal images and modishly adopts Einstein's contractable train. 'The train slows and lengthens, as we approach London,' one character remarks. But since the speaker is here on the train he would not be aware of any lengthening, even supposing Einstein's effects applied perceptibly to railway trains. *The Waves*, with its six characters all talking like Virginia Woolf, has really nothing to do with relativity.

Einstein was often misunderstood as saying that everything was relative, including truth, or that all observations were subjective. Some writers imagined they were being Einsteinian if they presented events from a number of different viewpoints. Lawrence Durrell made prominent claims, on these grounds, for his *Alexandria Quartet* (1957–60) – 'Three sides of space and one of time constitute the soup-mix of a continuum. The four novels follow this pattern.' In fact switches of viewpoint had been common practice since the epistolary novels of the eighteenth century, and Durrell's understanding of mathematics was so limited that he could not even master the game of chess.

Far from condoning subjectivity, Einstein believed in mathematical certainty, arguing that measurements within any reference-frame are definite and unalterable, and that measurements within other reference-frames can be accurately predicted. He told the *Saturday Evening Post* on 26 October 1929:

> Everything is determined, the beginning as well as the end, by forces over which we have no control. It is determined for the insect as well as the star. Human beings, vegetables, or cosmic dust, we all dance to a mysterious tune, intoned in the distance by an invisible piper.

Einstein's most famous equation ($E = mc^2$) did not belong to the original theory of relativity, but was the subject of a supplementary paper in 1905. As he later explained to readers of *Science Illustrated*, the huge amounts of energy the formula attributes to any given mass are not apparent in ordinary life since they are locked up in atoms of the mass, and can be released only by atomic fission:

It is customary to express the equivalence of mass and energy (though somewhat inexactly) by the formula $E = mc^2$, in which c represents the velocity of light, about 186,000 miles per second. E is the energy that is contained in a stationary body; m is its mass. The energy that belongs to the mass m is equal to this mass, multiplied by the square of the enormous speed of light – which is to say, a vast amount of energy for every unit of mass.

But if every gram of material contains this tremendous energy, why did it go so long unnoticed? The answer is simple enough: so long as none of the energy is given off externally, it cannot be observed. It is as though a man who is fabulously rich should never spend or give away a cent; no one could tell how rich he was. . . . We know of only one sphere in which such amounts of energy per mass unit are released: namely, radioactive disintegration. . . .

Now, we cannot actually weigh the atoms individually. However, there are indirect methods for measuring their weights exactly. . . . Thus it has become possible to test and confirm the equivalence formula. Also, the law permits us to calculate in advance, from precisely determined atom weights, just how much energy will be released with any atom disintegration we have in mind. The law says nothing, of course, as to whether – or how – the disintegration reaction can be brought about.

Einstein's equation states that mass is convertible into energy, and vice versa. Mass becomes, as it were, simply very, very concentrated energy. The formula extends the law of the conservation of energy (that energy can neither be created nor destroyed) into a law of the conservation of energy and mass (energy and mass can be neither created nor destroyed, though one form of energy or matter can be converted into another form of matter or energy).

Einstein had not anticipated, in his 1905 paper, that his formula could have any practical use. When the atomic age dawned, however, he found himself portrayed as a sinister prophet. The cover of *Time* magazine for 1 July 1946 carried a portrait of Einstein, with the mushroom-cloud of an atomic explosion in the background, and the caption 'Cosmoclast'. As Alan Friedman and Carol Donley put it in their book *Einstein as Myth and Muse* (1983), the benign scientist had become a modern Prometheus.

Sources: Albert Einstein, *Relativity: The Special and General Theory, A Popular Exposition*, trans. Robert W. Lawson, London, Methuen, 1920. Bertrand Russell, *The ABC of Relativity*, London, Kegan Paul, Trench, Trubner, 1925. A. S. Eddington, *Space, Time and Gravitation: an Outline of the General Theory of Relativity*, Cambridge University Press, 1920. Lyndon Bolton's essay from *Scientific American*, 5 February 1921, pp. 106–7. Einstein on $E = mc^2$ from *Science Illustrated* (April 1946), copyright McGraw Hill Publishing Co Inc: Philosophical Library Inc. William Carlos Williams, *The Complete Collected Poems 1906–38*, Norfolk, Conn., New Directions, 1938.

Uncertainty and Other Worlds

The development of the quantum theory of matter at the beginning of the twentieth century drastically altered conventional scientific wisdom. The conviction that the world was understandable had been science's most important gift to civilization. It had redeemed mankind from centuries of superstition. The new physics destroyed this cherished certainty. It found that the subatomic world was random and ultimately unintelligible. Electrons and other subatomic particles do not move along predictable paths, and they behave, incomprehensibly, like waves as well as like particles. It seems that, though they are the basic components of our material world, and of us, they are not 'things' at all, in the sense of having an independent identity, but remain in a suspended state until someone observes or measures them, whereupon they 'collapse' into one of many possible versions of reality. Thus the observer effectively creates the universe, or his version of it, by his observations. As the Danish physicist Niels Bohr (1885–1962) – one of the founders, with Max Planck (1858–1947), of quantum theory – puts it, there are:

> fundamental limitations, met with in atomic physics, of the objective existence of phenomena independent of their means of observation.

The 'uncertainty principle', formulated in 1927 by the German Werner Heisenberg (1901–76), established this indeterminacy as an inherent feature of subatomic matter. Heisenberg declared that:

> For the first time in history man, on this planet, is discovering that he is alone with himself . . . The conventional division of the world into subject and object, into inner and outer world, into body and soul, is no longer applicable.

Even before Heisenberg propounded his principle, alert and imaginative writers had seized on the idea that the new physics put reality back inside the human head, and made the material world unreal, just as mystics and religious thinkers had always insisted it was. The apprentice-mystic Mr Calamy, in Aldous Huxley's novel *Those Barren Leaves* (1925), explains that:

> The human mind . . . has invented space, time and matter, picking them

> out of reality in a quite arbitrary fashion . . . Everything that seems real is in fact entirely illusory – *maya*, in fact, the cosmic illusion.

Calamy (and Huxley) found this release from scientific 'fact' exhilarating. But others were dismayed. In March 1929 F. W. Bridgman, Professor of Mathematics and Philosophy at Harvard, told readers of *Harper's Magazine* that, though it was 'enormously upsetting', they must accept that:

> Nature is intrinsically and in its elements neither understandable nor subject to law . . . The physical properties of the electron are not absolutely inherent in it, but involve also the choice of the observer . . . This means nothing more nor less than that the law of cause and effect must be given up . . . The world is not a world of reason, understandable by the intellect of man . . . It is probable that new methods of education will have to be painfully developed and applied to very young children in order to inculcate the instinctive and successful use of habits of thought so contrary to those which have been naturally acquired.

All human languages would have to be remodelled too, Bridgman predicted, since our present verbal habits are based on assumptions about cause and effect that have been proved to be no longer valid.

According to the conventional 'Copenhagen' interpretation of quantum theory only our world 'really' exists. The potential alternative worlds not brought into being at any moment of observation remain only potential. However, the 'many-universe' interpretation of quantum theory argues that the other possible worlds we do not select when we make an observation do really exist. Paul Davies, Professor of Mathematical Physics at the University of Adelaide, envisages the results of this in his book *Other Worlds: Space, Superspace and the Quantum Universe*:

> Taking the widest possible view of superspace, it seems that every situation that can be reached along some convoluted path of development will occur in at least one of these other worlds. Every atom is offered billions of trajectories by the quantum randomization, and in the many-worlds theory it accepts them all, so every conceivable atomic arrangement will come about somewhere. There will be worlds that have no Earth, no sun, even no Milky Way. Others may differ so much from ours that no stars or galaxies of any kind exist. Some universes will be all darkness and chaos, with black holes

roaming about swallowing up haphazardly strewn material, while others will be seared with radiation.

Universes will exist that look superficially like ours but have different stars and planets. Even those with essentially the same astronomical arrangement will have very different life forms: in many, there will be no life on Earth, but in others life will have progressed more rapidly and there will be Utopian societies. Still others will have suffered total destruction from war, while in some the whole Milky Way will be colonized by aliens, Earth included.

Heinsenberg's uncertainty principle states that the position and momentum of a particle cannot be specified simultaneously: the greater the precision with which one property is specified, the less will be the precision of the other measurement. Not all scientists regard this as a fatal blow to understanding the world, however, as P. W. Atkins explains:

Although most people appear to consider that the uncertainty principle abolishes any chance we once might have thought we had to comprehend the world, it is more optimistic (and perhaps more correct) to consider that the uncertainty principle is an indication that our classically inspired template for understanding the world is over-elaborate. Classical physics, the physics of the farmyard of everyday experience, forged a template that led us to expect that we should be able to describe the world using the language of speed and location simultaneously. Quantum mechanics takes its axe to this naive, superficial view. It reminds us of what should be obvious: that farmyard-inspired theories may be too gross and unsophisticated, too covered in the dung of their own origin. It provides a template that in effect requires us to choose one language or another. It tells us to speak in terms *either* of location *or* of speed, and never to mix the two. It tells us to speak German for complete sentences or to speak English for complete sentences. It warns us not to start a sentence in German and then end it in English. Quantum mechanics tells us that the mathematization of Nature should be done using formulas drawn from the language of position or from the language of speed. It instructs us to separate the muddled classical template into two sheets and to use either one sheet or the other. Quantum mechanics clarifies our vision of the world and in so doing exposes more sharply its mathematical structure. That is just one example. In the end, if there is an end, we shall possess a mathematical theory of the universe that matches it in every test: the fit of reality to the template will be exact and we shall have a theory of everything.

Sources: F. W. Bridgman, 'The New Vision of Science', *Harper's Magazine*, March, 1929, pp. 443–51. Paul Davies, *Other Worlds: Space, Superspace and the Quantum Universe*, London, Penguin Books, 1990; first published by J. M. Dent and Sons, 1980. Peter Atkins, *Creation Revisited: The Origin of Space, Time and the Universe*, London, Penguin Books, 1994.

Quantum Mechanics: Mines and Machine-Guns

The German physicist Max Born (1882–1970) collaborated with his pupil Werner Heisenberg (see p. 277) in developing the mathematical formulation of quantum theory. Born fled the Nazis in 1933 and came to Britain, becoming Professor of Natural Philosophy at Edinburgh in 1936. This extract is from his inaugural lecture, where he tries to explain the problems underlying quantum theory to a non-specialist audience – using as illustration the weapons of the coming conflict, mines and machine-guns.

Let us start with the old problem of the constitution of light. At the beginning of the scientific epoch two rival theories were proposed: the corpuscular theory by Newton, the wave theory by Huygens. About a hundred years elapsed before experiments were found deciding in favour of one of them, the wave theory, by the discovery of interference. When two trains of waves are superposed, and a crest of one wave coincides with a valley of the other, they annihilate one another; this effect creates the well-known patterns which you can observe on any pond on which swimming ducks or gulls excite water-waves. Exactly the same kind of pattern can be observed when two beams of light cross one another, the only difference being that you need a magnifying-lens to see them; the inference is that a beam of light is a train of waves of short wave-length. This conclusion has been supported by innumerable experiments.

But about a hundred years later, during my student days, another set of observations began to indicate with equal cogency that light consists of corpuscles. This type of evidence can best be explained by analogy with two types of instruments of war, mines and guns. When a mine explodes you will be killed if you are near it, by the energy transferred to you as a wave of compressed air. But if you are some hundred yards away you are absolutely safe; the explosion-wave has lost its dangerous energy by continuously spreading out over a large area. Now imagine that the same amount of explosive is used as the propellant in a machine-gun which is rapidly fired, turning round in all directions. If you are near it you will almost certainly be shot, unless you hastily run away. When you have reached a distance of some hundred yards you will feel much safer, but certainly not quite safe. The probability of being hit has dropped enormously, but if you are hit the effect is just as fatal as before.

Here you have the difference between energy spread out from a centre in the form of a continuous wave-motion, and a discontinuous rain of particles. Planck discovered, in 1900, the first indication of this discontinuity of light in the laws governing the heat radiated from hot bodies. In his celebrated paper of 1905, Einstein pointed out that experiments on the energetic effect of light, the so-called photoelectric effect, could be interpreted in the way indicated as showing unambiguously the corpuscular constitution of light. These corpuscles are called quanta of light or photons.

This dual aspect of the luminous phenomenon has been confirmed by many observations of various types. The most important step was made by Bohr, who showed that the enormous amount of observations on spectra collected by the experimentalists could be interpreted and understood with the help of the conception of light quanta. For this purpose he had also to apply the idea of discontinuous behaviour to the motion of material particles, the atoms, which are the source of light.

I cannot follow out here the historical development of the quantum idea which led step by step to the recognition that we have here to do with a much more general conception. Light is not the only 'radiation' we know; I may remind you of the cathode rays which appear when electric currents pass through evacuated bulbs, or the rays emitted by radium and other radioactive substances. These rays are certainly not light. They are beams of fast-moving electrons, i.e. atoms of electricity, or ordinary atoms of matter like helium. In the latter case this has been proved directly by Rutherford, who caught the beam (a so-called α-ray of radium) in an evacuated glass vessel and showed that it was finally filled with helium gas. Today one can actually photograph the tracks of these particles of radiating matter in their passage through other substances.

In this case the corpuscular evidence was primary. But in 1924 de Broglie, from theoretical reasoning, suggested the idea that these radiations should show interference and behave like waves under proper conditions. This idea was actually confirmed by experiments a short time later. Not only electrons, but real atoms of ordinary matter like hydrogen or helium have all the properties of waves if brought into the form of rays by giving them a rapid motion.

This is a most exciting result, revolutionizing all our ideas of matter and motion. But when it became known, theoretical physics was already prepared to treat it by proper mathematical methods, the so-called quantum mechanics, initiated by Heisenberg, worked out in collaboration with Jordan and myself, and quite independently by Dirac; and another form of the same theory, the wave-mechanics, worked out by Schrödinger in close connection with de Broglie's suggestion. The mathematical formalism is a wonderful invention for describing complicated things. But it does not help much towards a real understanding. It took several years before this understanding was reached, even to a limited extent. But it leads right amidst philosophy, and this is the point about which I have to speak.

The difficulty arises if we consider the fundamental discrepancy in describing one and the same process sometimes as a rain of particles, and at other times as a wave. One is bound to ask, what is it really? You see here the question of reality appears. The reason why it appears is that we are talking about particles or waves, things considered as well known; but which expression is adequate depends on the method of observation. We thus meet a situation similar to that in relativity, but much more complicated. For here the two representations of the same phenomenon are not only different but contradictory. I think everyone feels that a wave and a particle are two types of motion which cannot easily be reconciled. But if we take into account the simple quantitative law relating energy and frequency already discovered by Planck, the case becomes very serious. It is clear that the properties of a given ray when appearing as a rain of particles must be connected with its properties when appearing as a train of waves. This is indeed the case, and the connecting law is extremely simple when all the particles of the beam have exactly the same velocity. Experiment then shows that the corresponding train of waves has the simplest form possible, which is called harmonic, and is characterized by a definite sharp frequency and wave-length. The law of Planck states that the kinetic energy of the particles is exactly proportional to the frequency of vibration of the wave; the factor of proportionality, called Planck's constant, and denoted by the letter h, has a definite numerical value which is known from experiment with fair accuracy.

There you have the logical difficulty: a particle with a given velocity is, *qua* particle, a point, existing at any instant without extension in space. A train of waves is by definition harmonic only if it fills the whole of space and lasts from eternity to eternity! (The latter point may not appear so evident; but a mathematical analysis made by Fourier more than a hundred years ago has clearly shown that every train of waves finite in space and time has to be considered as a superposition of many infinite harmonic waves of different frequencies and wave-lengths which are arranged in such a way that the outer parts destroy one another by interference; and it can be shown that every finite wave can be decomposed into its harmonic components.) Bohr has emphasized this point by saying that Planck's principle introduces an irrational feature into the description of nature.

Indeed the difficulty cannot be solved unless we are prepared to sacrifice one or other of those principles which were assumed as fundamental for science. The principle to be abandoned now is that of causality as it has been understood ever since it could be formulated exactly. I can indicate this point only very shortly. The laws of mechanics as developed by Galileo and Newton allow us to predict the future motion of a particle if we know its position and velocity at a given instant. More generally, the future behaviour of a system can be predicted from a knowledge of proper initial conditions. The world from the standpoint of mechanics is an automaton, without any freedom, determined from the beginning. I never liked this extreme determinism, and I am glad that modern physics has abandoned it. But other people do not share this view.

To understand how the quantum idea and causality are connected, we must explain the second fundamental law relating particles and waves. This can be readily understood with the help of our example of the exploding mine and the machine-gun. If the latter fires not only horizontally but equally in all directions, the number of bullets, and therefore the probability of being hit, will decrease with distance in exactly the same ratio as the surface of the concentric spheres, over which the bullets are equally distributed, increases. But this corresponds exactly to the decrease of energy of the expanding wave of the exploding mine. If we now consider light spreading out from a small source, we see immediately that in the corpuscular aspect the number of photons will decrease with the distance in exactly the same way as does the energy of the wave in the undulatory aspect. I have generalized this idea for electrons and any other kind of particles by the statement that we have to do with 'waves of probability' guiding the particles in such a way that the intensity of the wave at a point is always proportional to the probability of finding a particle at that point. This suggestion has been confirmed by a great number of direct and indirect experiments. It has to be modified if the particles do not move independently, but act on one another; for our purpose, however, the simple case is sufficient.

Now we can analyse the connection between the quantum laws and causality.

Determining the position of a particle means restricting it physically to a small part of space. The corresponding probability wave must also be restricted to this small part of space, according to our second quantum law. But we have seen that by Fourier's analysis such a wave is a superposition of a great number of simple harmonic waves with wave-lengths and frequencies spread over a wide region. Using now the first quantum law stating the proportionality of frequency and energy, we see that this geometrically well-defined state must contain a wide range of energies. The opposite holds just as well. We have derived qualitatively the celebrated uncertainty law of Heisenberg: exact determination of position and velocity exclude one another; if one is determined accurately the other becomes indefinite.

The quantitative law found by Heisenberg states that for each direction in space the product of the uncertainty interval of space and that of momentum (equal to mass times velocity) is always the same, being given by Planck's quantum constant h.

Here we have the real meaning of this constant as an absolute limit of simultaneous measurement of position and velocity. For more complicated systems there are other pairs or groups of physical quantities which are not measurable at the same instant.

Now we remember that the knowledge of position and velocity at one given time was the supposition of classical mechanics for determining the future motion. The quantum laws contradict this supposition, and this means the break-down of causality and determinism. We may say that these propositions are not just wrong, but empty: the premise is never fulfilled.

Source: Max Born, *Physics in My Generation*, Oxford, Pergamon Press, 1956.

Why Light Travels in Straight Lines

One of the most challenging and well-written of modern science-books, P. W. Atkins's *Creation Revisited* (1992), undertakes to explain how the universe came into existence by chance. Atkins (a physical chemist at Oxford) adopts the hypothesis of an 'infinitely lazy creator', and then shows that even such a minimal supernatural agency is unnecessary, because the behaviour of things is determined by their nature, without any need to impose rules. (Atkins's notes, which elaborate particular points and suggest further reading, have been omitted from this extract).

Light, we all know, travels in straight lines. If it could bend round corners, the world would be harder to discern. It would be like listening to it instead of seeing it. We would be immersed in a symphony of colour from objects that could be vaguely located but only hazily scrutinized. There would be no night; the symphony would be endless.

But saying that light travels in straight lines is not quite right. It conflicts with observation. Light bends at the junction of different media. Your leg in your bath looks broken even if it isn't. A lens bends light, and is shaped to focus the image on a film or on an eye. We therefore have to find a rule that captures both the straightness of the path when the medium is uniform and its bending when it passes from one medium to another.

The rule that captures both turns out to be elegantly simple (like all acceptable rules prior to their elimination): light travels by the path that takes the least time.

This succinct rule obviously accounts for the motion of light through air or any other uniform medium, because a straight line is then also the briefest path for anything travelling with a uniform speed. The rule also accounts for light's bending at the junction of media. Light travels at different speeds in different substances; the briefest path is then no longer the straightest, as can be understood by thinking about drowning.

Suppose the victim is out to sea, and you are on the shore. What path brings you to him in the shortest time, bearing in mind that you can run faster than you can swim? One possibility is for you to select a geometrically straight path from your deckchair to where he is sinking: that involves a certain amount of running and swimming. Alternatively you could run to a point on the water's edge directly opposite him and swim out straight from there. That is greater in distance but it may be briefer in duration if you can run very much faster than you can swim. By trial and error, or trigonometry, you would find that the path involving the least time is one where you run at some angle across the beach, then change direction and swim at another angle in a straight line towards your target (if it is not too late by now). This is exactly the behaviour of light passing into a denser medium.

But how does light know, apparently in advance, which is the briefest path? And, anyway, why should it care? The only way of discovering the briefest path appears to be to try them all, and then to eliminate all traces of having done so. There must be something about the nature of light which entails that it naturally tries all paths, and then eliminates all but the briefest.

The essential property is that light travels as a wave. Once that is realized, its other properties fall into place: light cannot help travelling by the briefest path.

A wave is an undulation, a series of peaks and troughs. Two or more waves of disturbance may spread into the same region. If it happens that the peaks of one coincide with the troughs of the other, then they tend to cancel, and an observer sees less disturbance, and perhaps no disturbance at all if they happen to cancel completely. That is basically all the information we need in order to see how light's character determines its destiny.

We are taking the view that things happen if they are not expressly forbidden, and that an infinitely lazy creator does not trouble to forbid. Think, then, of a light ray that happens to travel from A to B along some meandering path. *We* know that light doesn't travel like that, but light doesn't. If that path is permissible, then so too, as far as the light is concerned, is one that lies very close to it. So the light also travels by that path. Whereas the light that snaked by the first path may have reached B with a peak, the light that snakes by the second might reach B with a trough, or something in between. There are very many paths lying close to the first, and an observer at B sees the total disturbance arising from the waves that explore them all: many are troughs at B, many are peaks, and many are all the possibilities in between. The total disturbance at B is consequently zero, because there is always a neighbour to wash its neighbour out. In other words, by letting light travel by any path, it appears to be unable to travel at all. But light does travel.

One step was too hasty. Think of a ray that happens to go straight from A to B. Now think of a neighbouring path and the ray that takes it. If that path lies close to the first, it will have a trough at B if the first had a trough, and a peak if the first had a peak. There are very many almost-straight lines from A to B, and they all give disturbances at B differing only slightly from the disturbance due to the straight path. These paths therefore do not wash each other out, and an observer at B sees the light. He observes that the light travelled to him by lines that are straight, or very nearly straight.

The extent to which the nearly but not quite straight rays contribute to the overall disturbance at B depends on their wavelength (the separation of successive peaks). If the wavelength is short, then only rays correspondingly close to the straight line survive, all the others having sufficiently destructive neighbours. As the wavelength increases the waves get out of step less quickly, and the eliminating power of neighbours declines. Then even quite bent paths survive and can deliver their disturbance. That is the reason why radio transmissions (which use long-wavelength waves) can circumvent houses, and why we cannot see round corners. We can hear round corners: sound waves' wavelengths are long.

The wave nature of light accounts for the inevitability of its selection of straight lines. That, though, is true of uniform media, as in air. When light passes from one medium into something denser it travels more slowly. As a result, the positions of its peaks and troughs are modified. Still it explores all possible paths, but no longer is the geometrically straight line the one without neighbours that annihilate. Now the surviving path, because of the shift of peaks and troughs, is the one that bends at the junction. The surviving path also happens to be the briefest path. That rule therefore turns out to be merely a distant commentary on deeper purposelessness. Light automatically discovers briefest paths by trying all paths, and automatically eradicates all traces of its explorations; this presents itself to us as a behaviour, which we summarize as a rule.

We see in this example how perfect freedom generates its own constraint. As well as accounting for observed behaviour, everything we have said accords with the common-sense view that inanimate things are innately simple. That is one more step along the path to the view that animate things, being innately inanimate, are innately simple too.

The next step in the development involves noticing a similar observation about another thing. Since the behaviour is similar, we can suspect that the explanation is similar too. I should like you to notice that particles of matter also travel in straight lines unless subject to a force. Why?

According to the view we are taking, they do so because it is their intrinsic nature. But what can be this intrinsic nature that determines such behaviour? It must be that particles are distributed as waves.

In a single leap, impelled by common sense, we have gone from the old-fashioned original physics of Newton to the modern theory of matter, quantum theory, which regards the qualities of 'particle' and 'wave' as inseparable. Many feel at home with classical physics and regard quantum theory, being less familiar, as contrary to common sense. In my view, though, common sense drives us to accept quantum theory in place of classical physics as more consistent with common sense. I hold that the mind-shutting familiarities of classical physics actually conceal its incomprehensibility, except as a commentary and a mode of calculation. When they are inspected, the explanations of classical physics fall apart, and are seen to be mere superficial delusions, like film-sets.

There is much more to quantum theory than the assertion that particles are intrinsically wavelike, but that remark is at its core, and is what we develop here, first by seeking the actual rule that appears to govern the classical mechanics of particles, and then by looking for an explanation.

The rule that appears to govern the propagation of particles is remarkably, and therefore suspiciously, like the rule that appears to govern the propagation of light: particles follow trajectories between A and B that involve least action. Never mind the technical meaning of action; it is good enough, and truthful enough, to think of action as having its everyday meaning. In particular, if the particle is not subject to a force then the path that involves least action – no meandering and no acceleration – is uniform and straight.

Now, how does a particle know, before it tries, which of the infinity of possible paths from A to B corresponds to the one of least action? And why should it care?

As soon as we take the view that particles are distributed like waves, both questions are eliminated by the same reasoning that eliminates them in the case of light. The intrinsic nature of particles, their wavelike character, ensures that they travel in straight lines of least action, because all other paths, which they are perfectly free to explore, are eliminated automatically. The reason why particles like pigs and people do not normally seem to be waves is simply that their wavelengths are normally so short as to be undetectable. Nevertheless, distributed as waves they are, and that attribute provides explanations which are totally beyond the reach of classical physics.

This picture accounts for motion in straight lines, because in the absence of forces such paths are survivors; the wave nature lets them survive.

Source: Peter Atkins, *Creation Revisited: The Origin of Space, Time and the Universe*, London, Penguin Books, 1994.

Puzzle Interest

William Empson (1906–84) went up to Cambridge to study mathematics, but changed to English and wrote, while still at university, *Seven Types of Ambiguity* (1930), which revolutionized the practice of literary criticism. His poems frequently embody concepts from physics and mathematics, and offer, he said, 'a sort of puzzle interest'. 'Camping Out' is from *Poems* (1935). Empson's notes, printed below it, help in solving the puzzle.

Camping Out

And now she cleans her teeth into the lake:
Gives it (God's grace) for her own bounty's sake
What morning's pale and the crisp mist debars:
Its glass of the divine (that Will could break)
Restores, beyond Nature: or lets Heaven take
(Itself being dimmed) her pattern, who half awake
Milks between rocks a straddled sky of stars.

Soap tension the star pattern magnifies.
Smoothly Madonna through-assumes the skies
Whose vaults are opened to achieve the Lord.
No, it is we soaring explore galaxies,
Our bullet boat light's speed by thousands flies.
Who moves so among stars their frame unties;
See where they blur, and die, and are outsoared.

CAMPING OUT. The intention behind the oddness of the theme, however much it may fail, was not to be satirical but to show indifference to satire from outside. She gives the lake its pattern of reflected stars, now made of toothpaste, as God's grace allows man virtues that nature wouldn't; the mist and pale (pale light or boundary) of morning have made it unable to reflect real stars any longer. *Soap tension* is meant to stand for the action of surface tension between more and less concentrated soap solutions which makes the specks fly apart. *Their frame unties*: if any particle of matter got a speed greater than that of light it would have infinite mass and might be supposed to crumple up round itself the whole of space-time – 'a great enough ecstasy makes the common world unreal.'

Source: William Empson, *Collected Poems*, London, Chatto & Windus, 1955.

Submarine Blue

William Beebe and Otis Barton set a new record for ocean diving when they reached 3,028 feet (923 metres) off Bermuda on 15 August 1934. They dived in a 4-foot 9-inch diameter steel 'bathysphere' (a word Beebe coined), with two windows of fused quartz, and a 14-inch entrance which had a steel cover bolted over it. The sphere was winched down on a steel cable from the deck of a barge, the *Ready*. During three seasons of diving Beebe, who was Director of Tropical Research at the New York Zoological Society, observed an immense variety of marine life, including many previously unknown species, at depths that had been thought virtually lifeless. In this extract from his classic account, *Half Mile Down* (1934), he describes the bathysphere's first manned descent, on 6 June 1930. The 'hose' he refers to carried electricity and telephone lines.

We were all ready and I looked around at the sea and sky, the boats and my friends, and not being able to think of any pithy saying which might echo down the ages, I said nothing, crawled painfully over the steel bolts, fell inside and curled up on the cold, hard bottom of the sphere. This aroused me to speech and I called for a cushion only to find that we had none on hand. Otis Barton climbed in after me, and we disentangled our legs and got set. I had no idea that there was so much room in the inside of a sphere only four and a half feet in diameter, and although the longer we were in it the smaller it seemed to get, yet, thanks to our adequate physique, we had room and to spare. At Barton's suggestion I took up my position at the windows, while he hitched himself over to the side of the door, where he could keep watch on the various instruments. He also put on the ear-phones.

Miss Hollister on deck took charge of the other end of the telephone and arranged the duplicate control electric light so that she could watch it. Mr Tee-Van assumed control of the deck crew.

At our signal, the four-hundred-pound door was hoisted and clanged into place, sliding snugly over the ten great steel bolts. Then the huge nuts were screwed on. If either of us had had time to be nervous, this would have been an excellent opportunity – carrying out Poe's idea of being sealed up, not all at once, but little by little. For after the door was securely fastened, there remained a four-inch round opening in the center, through which we could see and talk and just slip a hand. Then this mighty bolt was screwed in place, and there began the most infernal racket I have ever heard. It was necessary, not only to screw the nuts down hard, but to pound the wrenches with hammers to take up all possible slack. I was sure the windows would be cracked, but having forcibly expressed our feelings through the telephone we gradually got used to the ear-shattering reverberations. Then utter silence settled down.

I turned my attention to the windows, cleaned them thoroughly and tested the visual angles which I could attain by pressing my face close to the surface. I could see a narrow sector of the deck with much scurrying about, and as we rolled I caught sight of the ultramarine sea and the *Gladisfen* [the sea-going tug that towed the *Ready*] dipping at the end of the slack tow rope. Faint scuffling sounds reached us now and then, and an occasional hollow beating. Then it seemed as if the steel walls fell away, and we were again free among our fellows, for a voice came down the half mile of hose coiled on the deck, and such is the human mind, that slender vocal connection seemed to restore physical as well as mental contact. While waiting for the take-off, Barton readjusted the phone, tested the searchlight, and opened the delicate oxygen valve. He turned it until we both verified the flow as two litres a minute – that being the amount suggested to us for two people. I remembered what I had read of Houdini's method of remaining in a closed coffin for a long time, and we both began conscientiously regulating our breathing, and conversing in low tones.

Another glance through my porthole showed Tee-Van looking for a signal from old Captain Millet. I knew that now it was actually a propitious wave or rather a propitious lack of one for which they waited. Soon Millet waved his hand, and exactly at one o'clock the winch grumbled, the wire on the deck tightened, and we felt our circular home tremble, lean over, and lift clear. Up we went to the yard-arm, then a half-score of the crew pulled with all their might and swung us out over the side. This all between two, big, heaving swells. We were dangling in mid-air and slowly we revolved until I was facing in toward the side of the *Ready*. And now our quartz windows played a trick on us. Twice already, in an experimental test submergence, we had not gauged correctly the roll of the ship or the distance outboard and the sphere had crashed into the half-rotten bulwarks. Now as I watched, I saw us begin to swing and my eyes told me that we were much too close, and that a slightly heavier roll would crash us, windows first, into the side of the vessel. Barton could not see the imminent danger, and the next message I got was 'Gloria wants to know why the Director is swearing so.' By this time we had swung far out, and I realized that every word which we spoke to each other in our tiny hollow chamber was clearly audible at the other end of the wire. I sent up word that any language was justifiable at such gross neglect as to allow our window to swing back and forth only a yard from the boat. And very decisively the word came back that fifteen feet was the nearest it had ever been, and we were now twenty-five feet away. Barton looked out with me and we could not believe our eyes. Fused quartz, as I have said, is the clearest, the most transparent material in the world, and the side of the *Ready* seemed only a yard away. My apologies must have cost us several litres of good oxygen.

To avoid any further comment on our part, profane or otherwise, we were lowered 20 feet. I sensed the weight and sturdy resistance of the bathysphere more at this moment than at any other time. We were lowered gently but we struck the surface with a splash which would have crushed a rowboat like an eggshell. Yet within we hardly noticed the impact, until a froth of foam and bubbles surged up over the glass and our chamber was dimmed to a pleasant green. We swung quietly while the first hose clamp was put on the cable. At the end of the first revolution the great hull of the barge came into view. This was a familiar landscape, which I had often seen from the diving helmet – a transitory, swaying reef with waving banners of seaweed, long tubular sponges, jet black blobs of ascidians and tissue-thin plates of rough-spined pearl shells. Then the keel passed slowly upward, becoming one with the green water overhead.

With this passed our last visible link with the upper world; from now on we had to depend on distant spoken words for knowledge of our depth, or speed, or the weather, or the sunlight, or anything having to do with the world of air on the surface of the Earth.

A few seconds after we lost sight of the hull of the *Ready*, word came down the hose that we were at 50 feet, and I looked out at the brilliant bluish-green haze and could not realize that this was almost my limit in the diving helmet. Then '100 feet' was called out, and still the only change was a slight twilighting and chilling of the green. As we sank slowly I knew that we must be passing the 132-foot level, the depth where Commander Ellsberg labored so gallantly to free the men in the Submarine S-57. '200 feet' came and we stopped with the slightest possible jerk and hung suspended while a clamp was attached – a double gripping bit of brass which bound the cable and hose together to prevent the latter from breaking by its own weight. Then the call came that all was clear and again I knew that we were sinking, although only by the upward passing of small motes of life in the water.

We were now very far from any touch of Mother Earth; ten miles south of the shore of Bermuda, and one and a half miles from the sea bottom far beneath us. At 300 feet, Barton gave a sudden exclamation and I turned the flash on the door and saw a slow trickle of water beneath it. About a pint had already collected in the bottom of the sphere. I wiped away the meandering stream and still it came. There flashed across my mind the memory of gentle rain falling on a window pane, and the first drops finding their way with difficulty over the dry surface of the glass. Then I looked out through the crystal clear quartz at the pale blue, and the contrast closed in on my mind like the ever deepening twilight.

We watched the trickle. I knew the door was solid enough – a mass of four hundred pounds of steel – and I knew the inward pressure would increase with every foot of depth. So I gave the signal to descend quickly. After that, the flashlight was turned on the door-sill a dozen times during our descent, but the stream did not increase.

Two minutes more and '400 feet' was called out; 500 and 600 feet came and passed overhead, then 700 feet where we remained for a while.

Ever since the beginnings of human history, when first the Phœnicians dared to sail the open sea, thousands upon thousands of human beings had reached the depth at which we were now suspended, and had passed on to lower levels. But all of these were dead, drowned victims of war, tempest, or other Acts of God. We were the first living men to look out at the strange illumination: And it was stranger than any imagination could have conceived. It was of an indefinable translucent blue quite unlike anything I have ever seen in the upper world, and it excited our optic nerves in a most confusing manner. We kept thinking and calling it brilliant, and again and again I picked up a book to read the type, only to find that I could not tell the difference between a blank page and a colored plate. I brought all my logic to bear, I put out of mind the excitement of our position in watery space and tried to think sanely of comparative color, and I failed utterly. I flashed on the searchlight, which seemed the yellowest thing I have ever seen, and let it soak into my eyes, yet the moment it was switched off, it was like the long vanished sunlight – it was as though it had never been – and the blueness of the blue, both outside and inside our sphere, seemed to pass materially through the eye into our very beings. This is all very unscientific; quite worthy of being jeered at by optician or physicist, but there it was. I was excited by the fishes that I was seeing perhaps more than I have ever been by other organisms, but it was only an intensification of my surface and laboratory interest: I have seen strange fluorescence and ultra-violet illumination in the laboratories of physicists: I recall the weird effects of color shifting through distant snow crystals on the high Himalayas, and I have been impressed by the eerie illumination, or lack of it, during a full eclipse of the sun. But this was beyond and outside all or any of these. I think we both experienced a wholly new kind of mental reception of color impression. I felt I was dealing with something too different to be classified in usual terms.

All our remarks were recorded by Miss Hollister and when I read them later, the repetition of our insistence upon the brilliance, which yet was not brilliance, was almost absurd. Yet I find that I must continue to write about it, if only to prove how utterly inadequate language is to translate vividly, feeling and sensations under a condition as unique as submersion at this depth.

The electric searchlight now became visible. Heretofore we could see no change whatever in the outside water when it was turned on, but now a pale shaft of yellow – intensely yellow – light shot out through the blue, very faint but serving to illuminate anything which crossed it. Most of the time I chose to have it cut off, for I wanted more than anything else to see all that I could of the luminescence of the living creatures.

After a few minutes I sent up an order, and I knew that we were again sinking. The twilight (the word had become absurd, but I could coin no other) deepened, but we still spoke of its brilliance. It seemed to me that it must be like the last terrific upflare of a flame before it is quenched. I found we were both expecting at any moment to have it blown out, and to enter a zone of absolute darkness. But only by shutting my eyes and opening them again could I realize the terrible *slowness* of the change from dark blue to blacker blue. On the earth at night in moonlight I can always imagine the yellow of sunshine, the scarlet of invisible blossoms, but here, when the searchlight was off, yellow and orange and red were unthinkable. The blue which filled all space admitted no thought of other colors.

We spoke very seldom now. Barton examined the dripping floor, took the temperatures, watched and adjusted the oxygen tank, and now and then asked, 'What depth now?' 'Yes, we're all right.' 'No, the leak's not increasing.' 'It's as brilliant as ever.'

And we both knew it was not as brilliant as ever, but our eyes kept telling us to say so. It actually seemed to me to have a brilliance and intensity which the sunshine lacked; sunshine, that is, as I remembered it in what seemed ages ago.

'800 feet' now came down the wire and I called a halt. There seemed no reason why we should not go on to a thousand; the leak was no worse, our palm-leaf fan kept the oxygen circulating so that we had no sense of stuffiness and yet some hunch – some mental warning which I have had at half a dozen critical times in my life – spelled *bottom* for this trip. This settled, I concentrated on the window for five minutes.

The three exciting internal events which marked this first trip were, first, the discovery of the slight leak through the door at 300 feet, which lessened as we went down; next, the sudden short-circuiting of the electric light switch, with attendant splutterings and sparks, which was soon remedied. The third was absurd, for it was only Barton pulling out his palm-leaf fan from between the wall of the sphere and the wire lining of the chemical rack. I was wholly absorbed at the time in watching some small fish, when the sudden shrieking rasp in the confines of our tiny cell gave me all the reactions which we might imagine from the simultaneous caving in of both windows and door! After that, out of regard for each other's nerves, we squirmed about and carried on our various duties silently.

Coming up to the surface and through it was like hitting a hard ceiling – I unconsicously ducked, ready for the impact, but there followed only a slather of foam and bubbles, and the rest was sky.

We reached the deck again just one hour after our start, and sat quietly while the middle bolt was slowly unscrewed. We could hear our compressed air hissing outward through the threads until finally the bolt popped off, and our ear-drums vibrated very slightly. After a piece of boiler-factory pounding the big door finally swung off. I started to follow and suddenly realized how the human body could be completely subordinated to the mind. For a full hour I had sat in almost the same position with no thought either of comfort or discomfort, and now I had severally to untwist my feet and legs and bring them to life. The sweater which was to have served as cushion, I found reposing on one of the chemical racks, while I had sat on the hard cold steel in a good-sized puddle of greasy water. I also bore the distinct imprint of a monkey wrench for several days. I followed Barton out on deck into the glaring sunshine, whose yellowness can never heareafter be as wonderful as blue can be.

While still upside down, creeping painfully, sea-lion-wise, over the protruding circle of bolts, I fancied that I heard a strange, inexplicable ringing in my ears. When I stood up, I found it was the screeching whistles on the boilers and the deeper toned siren of the *Gladisfen* giving us, all to ourselves, a little celebration in mid-ocean. The wind was right and my staff on Nonsuch ten miles away saw the escaping steam through the telescope binoculars, later heard the sound faintly, and knew that we had made our dive and ascent in safety.

Source: William Beebe, *Half Mile Down*, New York, Harcourt Brace, 1934.

Sea-Cucumbers

In March 1940 John Steinbeck and his marine biologist friend Ed Ricketts hired a boat, *Western Flyer*, and set off on a trip across the Gulf of Mexico to collect specimens of marine life. This is from Steinbeck's log for 25 March. The creatures he describes are *Euapta godeffroyi*, worm-like sea-cucumbers – one of a group of spiny-skinned animals, some varieties of which under the commercial name *bêche-de-mer* are used by the Chinese for food.

As soon as the anchor was out, we dropped the fishing lines and immediately hooked several hammer-head sharks and a large red snapper. The air here was hot and filled with the smell of mangrove flowers. The little outer bay was our first collecting station, a shallow warm cove with a mud bottom and edged with small boulders, smooth and unencrusted with algae. On the bottom we could see long snake-like animals, gray with black markings, with purplish-orange floriate heads like chrysanthemums. They were about three feet long and new to us. Wading in rubber boots, we captured some of them and they proved to be giant synaptids. They were strange and frightening to handle, for they stuck to anything they touched, not with slime but as though they were coated with innumerable suction-cells. On being taken from the water, they collapsed to skin, for their bodily shape is maintained by the current of water which they draw through themselves. When lifted out, this water escapes and they hang as limp as unfilled sausage skins. Since they were new and fascinating to us, we took many specimens, maneuvering them gently to the surface and then sliding them into submerged wooden collecting buckets to prevent them from dropping their water. On the bottom they crawled about, their flower-heads moving gently, while the current of water passing through their bodies drew food into their stomachs. When we took them on board, we found they had to a high degree the habit of a number of holothurians: eviscerating. These *Euapta* were a nervous lot. We tried to relax them with Epsom salts so that we might kill them with their floriate heads extended, but the salts, no matter how carefully administered, caused the heads to retract, and soon afterwards they threw their stomachs out into the water. The word 'stomach' is used here inadvisedly, for what they actually disgorge is the intestinal tract and respiratory tree.

We intoxicated them with pure oxygen and then tried the salts, but with the same result. Finally, by administering the salts in minute quantities and very slowly, we were able to preserve some uneviscerated specimens, but none with the head extended.

Source: John Steinbeck, *The Log from the Sea of Cortez*, London, Heinemann, 1958.

Telling the Workers about Science

J. B. S. Haldane (1892–1965) was a child prodigy. At three, it is claimed, he asked, on looking at blood from his cut forehead, 'Is it oxyhaemoglobin or carboxyhaemoglobin?' Because of his precociousness he was savagely bullied at Eton, which turned him against authority for life. In the 1930s he became an ardent Communist, visiting Spain during the Civil War to observe the effects of Fascist aerial bombardment.

As a student at Oxford he switched from maths to Classics, in which he got a first class degree. He had a wide knowledge of literature and a huge memory for English poetry. In the First World War he was commissioned into the Black Watch, earning the nickname 'Bombo' for his experiments with bombs and grenades.

After the war, as a don at New College, Oxford, he became a member of Lady Ottoline Morrell's Garsington Manor set, meeting writers of the day – Katherine Mansfield, Lytton Strachey and Aldous Huxley, who satirized him as Shearwater in *Antic Hay*. Encouraged by his first wife, a journalist, he began writing the popular scientific articles for the *Manchester Guardian* and the *Daily Worker* for which he is best known. As a scientist, his main interest was in applying genetics to evolution theory, to show how evolution worked mathematically, and he became the first Professor of Biometry at University College London in 1937.

Of the two articles that follow, the first is from Haldane's first collection, *Possible Worlds* (1927) and the other explains nuclear fission to readers of the *Daily Worker* in 1939.

On Being the Right Size

The most obvious differences between different animals are differences of size, but for some reason the zoologists have paid singularly little attention to them. In a large textbook of zoology before me I find no indication that the eagle is larger than the sparrow, or the hippopotamus bigger than the hare, though some grudging admissions are made in the case of the mouse and the whale. But yet it is easy to show that a hare could not be as large as a hippopotamus, or a whale as small as a herring. For every type of animal there is a most convenient size, and a large change in size inevitably carries with it a change of form.

Let us take the most obvious of possible cases, and consider a giant man sixty feet high – about the height of Giant Pope and Giant Pagan in the illustrated *Pilgrim's Progress* of my childhood. These monsters were not only ten times as high as Christian, but ten times as wide and ten times as thick, so that their total weight was a thousand times his, or about eighty to ninety tons. Unfortunately the cross sections of their bones were only a hundred times those of Christian, so that every square inch of giant bone had to support ten times the weight borne by a square inch of human bone. As the human thigh-bone breaks under about ten times the human weight, Pope and Pagan would have broken their thighs every time they took a step. This was doubtless why they were sitting down in the picture I remember. But it lessens one's respect for Christian and Jack the Giant Killer.

To turn to zoology, suppose that a gazelle, a graceful little creature with long thin legs, is to become large, it will break its bones unless it does one of two things. It may make its legs short and thick, like the rhinoceros, so that every pound of weight has still about the same area of bone to support it. Or it can compress its body and stretch out its legs obliquely to gain stability, like the giraffe. I mention these two beasts because they happen to belong to the same order as the gazelle, and both are quite successful mechanically, being remarkably fast runners.

Gravity, a mere nuisance to Christian, was a terror to Pope, Pagan, and Despair. To the mouse and any smaller animal it presents practically no dangers. You can drop a mouse down a thousand-yard mine shaft; and, on arriving at the bottom, it gets a slight shock and walks away. A rat is killed, a man is broken, a horse splashes. For the resistance presented to movement by the air is proportional to the surface of the moving object. Divide an animal's length, breadth, and height each by ten; its weight is reduced to a thousandth, but its surface only to a hundredth. So the resistance to falling in the case of the small animal is relatively ten times greater than the driving force.

An insect, therefore, is not afraid of gravity; it can fall without danger, and can cling to the ceiling with remarkably little trouble. It can go in for elegant and fantastic forms of support like that of the daddy-long-legs. But there is a force which is as formidable to an insect as gravitation to a mammal. This is surface tension. A man coming out of a bath carries with him a film of water of about one-fiftieth of an inch in thickness. This weighs roughly a pound. A wet mouse has to carry about its own weight of water. A wet fly has to lift many times its own weight and, as every one knows, a fly once wetted by water or any other liquid is in a very serious position indeed. An insect going for a drink is in as great danger as a man leaning out over a precipice in search of food. If it once falls into the grip of the surface tension of the water – that is to say, gets wet – it is likely to remain so until it drowns. A few insects, such as water-beetles, contrive to be unwettable, the majority keep well away from their drinks by means of a long proboscis.

Of course tall land animals have other difficulties. They have to pump their blood to greater heights than a man and, therefore, require a larger blood pressure and tougher blood-vessels. A great many men die from burst arteries, especially in the brain, and this danger is presumably still greater for an elephant or a giraffe. But animals of all kinds find difficulties in size for the following reason. A typical small animal, say a microscopic worm or rotifer, has a smooth skin through which all the oxygen it requires can soak in, a straight gut with sufficient surface to absorb its food, and a simple kidney. Increase its dimensions tenfold in every direction, and its weight is increased a thousand times, so that if it is to use its muscles as efficiently as its miniature counterpart, it will need a thousand times as much food and oxygen per day and will excrete a thousand times as much of waste products.

Now if its shape is unaltered its surface will be increased only a hundredfold, and ten times as much oxygen must enter per minute through each square millimetre of skin, ten times as much food through each square millimetre of intestine. When a limit is reached to their absorptive powers their surface has to be increased by some special device. For example, a part of the skin may be drawn out into tufts to make gills or pushed in to make lungs, thus increasing the oxygen-absorbing surface in proportion to the animal's bulk. A man, for example, has a hundred square yards of lung. Similarly, the gut, instead of being smooth and straight, becomes coiled and develops a velvety surface, and other organs increase in complication. The higher animals are not larger than the lower because they are more complicated. They are more complicated because they are larger. Just the same is true of plants. The simplest plants, such as the green algae growing in stagnant water or on the bark of trees, are mere round cells. The higher plants increase their surface by putting out leaves and roots. Comparative anatomy is largely the story of the struggle to increase surface in proportion to volume.

Some of the methods of increasing the surface are useful up to a point, but not capable of a very wide adaptation. For example, while vertebrates carry the oxygen from the gills or lungs all over the body in the blood, insects take air directly to every part of their body by tiny blind tubes called tracheae which open to the surface at many different points. Now, although by their breathing movements they can renew the air in the outer part of the tracheal system, the oxygen has to penetrate the finer branches by means of diffusion. Gases can diffuse easily through very small distances, not many times larger than the average length travelled by a gas molecule between collisions with other molecules. But when such vast journeys – from the point of view of a molecule – as a quarter of an inch have to be made, the process becomes slow. So the portions of an insect's body more than a quarter of an inch from the air would always be short of oxygen. In consequence hardly any insects are much more than half an inch thick. Land crabs are built on the same general plan as insects, but are much clumsier. Yet like ourselves they carry oxygen around in their blood, and are therefore able to grow far larger than any insects. If the insects had hit on a plan for driving air through their tissues instead of letting it soak in, they might well have become as large as lobsters, though other considerations would have prevented them from becoming as large as man.

Exactly the same difficulties attach to flying. It is an elementary principle of aeronautics that the minimum speed needed to keep an aeroplane of a given shape in the air varies as the square root of its length. If its linear dimensions are increased four times, it must fly twice as fast. Now the power needed for the minimum speed increases more rapidly than the weight of the machine. So the larger aeroplane, which weighs sixty-four times as much as the smaller, needs one hundred and twenty-eight times its horsepower to keep up. Applying the same principles to the birds, we find that the limit to their size is soon reached. An angel whose muscles developed no more power weight for weight than those of an eagle or a pigeon would require a breast projecting for about four feet to house the muscles engaged in working its wings, while to economize in weight, its legs would have to be reduced to mere stilts. Actually a large bird such as an eagle or kite does not keep in the air mainly by moving its wings. It is generally to be seen soaring, that is to say balanced on a rising column of air. And even soaring becomes more and more difficult with increasing size. Were this not the case eagles might be as large as tigers and as formidable to man as hostile aeroplanes.

But it is time that we passed to some of the advantages of size. One of the most obvious is that it enables one to keep warm. All warm-blooded animals at rest lose the same amount of heat from a unit area of skin, for which purpose they need a food-supply proportional to their surface and not to their weight. Five thousand mice weigh as much as a man. Their combined surface and food or oxygen consumption are about seventeen times a man's. In fact a mouse eats about one quarter of its own weight in food every day, which is mainly used in keeping it warm. For the same reason small animals cannot live in cold countries. In the arctic regions there are no reptiles or amphibians, and no small mammals. The smallest mammal in Spitzbergen is the fox. The small birds fly away in the winter, while the insects die, though their eggs can survive six months or more of frost. The most successful mammals are bears, seals, and walruses.

Similarly, the eye is a rather inefficient organ until it reaches a large size. The back of the human eye on which an image of the outside world is thrown, and which corresponds to the film of a camera, is composed of a mosaic of 'rods and cones' whose diameter is little more than a length of an average light wave. Each eye has about half a million, and for two objects to be distinguishable their images must fall on separate rods or cones. It is obvious that with fewer but larger rods and cones we should see less distinctly. If they were twice as broad two points would have to be twice as far apart before we could distinguish them at a given distance. But if their size were diminished and their number increased we should see no better. For it is impossible to form a definite image smaller than a wave-length of light. Hence a mouse's eye is not a small-scale model of a human eye. Its rods and cones are not much smaller than yours, and therefore there are far fewer of them. A mouse could not distinguish one human face from another six feet away. In order that they should be of any use at all the eyes of small animals have to be much larger in proportion to their bodies than our own. Large animals on the other hand only require relatively small eyes, and those of the whale and elephant are little larger than our own.

For rather more recondite reasons the same general principle holds true of the brain. If we compare the brain-weights of a set of very similar animals such as the cat, cheetah, leopard, and tiger, we find that as we quadruple the body-weight the brain-weight is only doubled. The larger animal with proportionately larger bones can economize on brain, eyes, and certain other organs.

Such are a very few of the considerations which show that for every type of animal there is an optimum size. Yet although Galileo demonstrated the contrary more than three hundred years ago, people still believe that if a flea were as large as a man it could jump a thousand feet into the air. As a matter of fact the height to which an animal can jump is more nearly independent of its size than proportional to it. A flea can jump about two feet, a man about five. To jump a given height, if we neglect the resistance of the air, requires an expenditure of energy proportional to the jumper's weight. But if the jumping muscles form a constant fraction of the animal's body, the energy developed per ounce of muscle is independent of the size, provided it can be developed quickly enough in the small animal. As a matter of fact an insect's muscles, although they can contract more quickly than our own, appear to be less efficient; as otherwise a flea or grasshopper could rise six feet into the air.

Atom-Splitting

This is a sensational article. I am sorry. In these articles I try to keep to facts. But occasionally facts are sensational. A discovery has just been made which may revolutionize human life as completely as the steam engine, and much more quickly. The odds are against its doing so, but not more than ten to one, if so much. So it is worth writing about it.

In the *Daily Worker* of March 30th, 1939, I described the recent work on splitting the nuclei of uranium atoms. A certain number of them explode when neutrons collide with them. Neutrons are among the so-called elementary particles – that is to say, particles which have not yet been broken up, such as electrons, protons, and perhaps a few others. This does not mean that they will never be broken up.

Ordinary atoms hold together when they collide at a speed of about a mile a second, as they do in air. When the temperature is raised and the speed of collisions goes up to ten miles or so a second, they cannot hold together, but electrons – that is to say, elementary particles with a negative charge – are torn off them. That is why a flame conducts electricity.

But at moderate speeds – say, a few thousand miles per second – collisions only break up the atoms temporarily. They soon pick up their lost electrons. When the speed rises to tens or hundreds of thousands of miles per second, the nuclei, or cores of the atoms, are sometimes broken up.

When a current is passed through the heavy variety of hydrogen at a voltage of half a million or so, the atomic nuclei become formidable projectiles, and if they hit a light metal called lithium they break up its atomic nuclei and let neutrons loose. Neutrons can penetrate the nuclei of many atoms even when moving slowly and cause still further changes.

Generally they only chip a piece off. But when they attack uranium, an element which is unstable, anyway, and produces radium, though very slowly, when left to itself, the uranium nuclei split up. The new fact, first discovered by Joliot and his colleagues in Paris, is that when the uranium nucleus splits, it produces neutrons also. In the experiments so far made, very small pieces of uranium were used.

So most of the neutrons, which can penetrate even metals for some distance, get out. But if the neutrons are liberated in the middle of a sufficiently large lump of uranium, they will cause further nuclei to break up, and the process will spread. The principle involved is quite simple. A single stick burns with difficulty, because most of the heat gets away. But a large pile of sticks will blaze, even if most of them are damp.

Nobody knows how large a lump of uranium is needed before it begins to set itself alight, so to say. But experiments are already under way in two British and one German laboratory to my knowledge, and doubtless in others in America, the Soviet Union and elsewhere.

In the current number [May 13th, 1939] of *Nature* Joliot and Halban, a French and a German physicist working together in Paris, published an S.O.S. letter suggesting means for slowing the process down, so as to avoid disaster.

If the experiment succeeds several things may happen. The change may take place slowly, the metal gradually warming up. It may occur fairly quickly, in which case there will be a mild explosion, and the lump will fly apart into vapour before one atom in a million has been affected. Or there may be a really big explosion. For if about one four-hundredth of the mass of the exploding uranium is converted into energy, as seems to be probable, an ounce would produce enough heat to boil about 1,000 tons of water. So 1 oz. of uranium, if it exploded suddenly, would be equivalent to over 100,000 tons of high explosive.

Of course, no one will begin with an ounce. Still, they may do a good deal of damage. Most probably, however, nothing much will happen. It may be, for example, that the majority of uranium atoms are stable, and only one of the several isotopes (as the different sorts of atom of the same element are called) is explosive. If so it will take several years to separate the isotopes.

Nevertheless, the next few months may see the problem solved in principle. If so, power will be available in vast quantities. There will be a colossal economic crisis in capitalist countries. There is plenty of uranium in different parts of the world, notably northern Canada, the Belgian Congo, Czechoslovakia and in several parts of the Soviet Union.

So the owners of uranium ores will make vast fortunes and millions of coalminers will be thrown out of work. The Soviet Union will adopt the new energy source on a vast scale, but the rest of the world will have a much tougher job to do so. Fortunately, uranium bombs cannot at once be adapted for war, as the apparatus needed is very heavy and also very delicate, so it cannot at present be dropped from an aeroplane. But doubtless uranium will be used for killing in some way.

An intelligent reader may well ask why, if uranium is so explosive, under certain conditions, explosions do not occur in Nature. The answer is that uranium does not occur in Nature in a pure state. It is generally found combined with oxygen, and neutrons would be stopped by the oxygen atoms to such an extent that an explosion could not possibly spread.

I repeat that this article is highly speculative. I am prepared to bet against immediate 'success' in these experiments. Nevertheless, some of the world's ablest physicists are hard at work on the problem. And the time has gone past when the ordinary man and woman can neglect what they are doing.

Sources: 'On Being the Right Size', from J. B. S. Haldane, *Possible Worlds and Other Essays*, London, Chatto & Windus, 1927; 'Atom-Splitting' from J. B. S. Haldane, *Science in Peace and War*, London, Lawrence & Wishart, 1940.

The Making of the Eye

The English physiologist Sir Charles Sherrington (1857–1952) won a Nobel Prize in 1932 for his work on the nervous system of mammals. Experimenting on cats, dogs, monkeys and apes that had had their cerebral hemispheres removed, he showed that reflexes must be regarded as integrated activities of the whole organism. He coined the words 'neuron' and 'synapse' to mean the nerve cell and the point at which the nervous impulse is transmitted from one cell to another. His book *Man and His Nature* (1940) presents man – both mind and body – as the product of natural forces acting upon the materials of our planet. This extract is from the second edition, 1951.

Can then physics and chemistry out of themselves explain that a pin's-head ball of cells in the course of so many weeks becomes a child? They more than hint that they can. A highly competent observer, after watching a motion-film photo-record taken with the microscope of a cell-mass in the process of making bone, writes: 'Team-work by the cell-masses. Chalky spicules of bone-in-the-making shot across the screen, as if labourers were raising scaffold-poles. The scene suggested purposive behaviour by individual cells, and still more by colonies of cells arranged as tissues and organs.'* That impression of concerted endeavour comes, it is no exaggeration to say, with the force of a self-evident truth. The story of the making of the eye carries a like inference.

The eye's parts are familiar even apart from technical knowledge and have evident fitness for their special uses. The likeness to an optical camera is plain beyond seeking. If a craftsman sought to construct an optical camera, let us say for photography, he would turn for his materials to wood and metal and glass. He would not expect to have to provide the actual motor power adjusting the focal length or the size of the aperture admitting light. He would leave the motor

*E. G. Drury, '*Psyche and the Physiologists' and other Essays on Sensation* (London, 1938), p. 4.

power out. If told to relinquish wood and metal and glass and to use instead some albumen, salt and water, he certainly would not proceed even to begin. Yet this is what that little pin's-head bud of multiplying cells, the starting embryo, proceeds to do. And in a number of weeks it will have all ready. I call it a bud, but it is a system separate from that of its parent, although feeding itself on juices from its mother. And the eye it is going to make will be made out of those juices. Its whole self is at its setting out not one ten-thousandth part the size of the eye-ball it sets about to produce. Indeed it will make two eyeballs built and finished to one standard so that the mind can read their two pictures together as one. The magic in those juices goes by the chemical names, protein, sugar, fat, salts, water. Of them 80% is water.

Water is a great menstruum of 'life'. It makes life possible. It was part of the plot by which our planet engendered life. Every egg-cell is mostly water, and water is its first habitat. Water it turns to endless purposes; mechanical support and bed for its membranous sheets as they form and shape and fold. The early embryo is largely membranes. Here a particular piece grows fast because its cells do so. There it bulges or dips, to do this or that or simply to find room for itself. At some other centre of special activity the sheet will thicken. Again at some other place it will thin and form a hole. That is how the mouth, which at first leads nowhere, presently opens into the stomach. In the doing of all this, water is a main means.

The eye-ball is a little camera. Its smallness is part of its perfection. A spheroid camera. There are not many anatomical organs where exact shape counts for so much as with the eye. Light which will enter the eye will traverse a lens placed in the right position there. *Will* traverse; all this making of the eye which *will* see in the light is carried out in the dark. It is a preparing in darkness for use in light. The lens required is biconvex and to be shaped truly enough to focus its pencil of light at the particular distance of the sheet of photosensitive cells at the back, the retina. The biconvex lens is made of cells, like those of the skin but modified to be glass-clear. It is delicately slung with accurate centring across the path of the light which *will* in due time some months later enter the eye. In front of it a circular screen controls, like the iris-stop of a camera or microscope, the width of the beam and is adjustable, so that in a poor light more is taken for the image. In microscope, or photographic camera, this adjustment is made by the observer working the instrument. In the eye this adjustment is automatic, worked by the image itself!

The lens and screen cut the chamber of the eye into a front half and a back half, both filled with clear humour, practically water, kept under a certain pressure maintaining the eye-ball's right shape. The front chamber is completed by a layer of skin specialized to be glass-clear, and free from blood-vessels which if present would with their blood throw shadows within the eye. This living glass-clear sheet is covered with a layer of tear-water constantly renewed. This tear-water has the special chemical power of killing germs which might inflame the eye. This glass-clear bit of skin has only one of the fourfold set of the skin-senses; its touch is always 'pain', for it should *not* be touched. The skin above and below this window grows into movable flaps, dry outside like ordinary skin, but moist inside so as to wipe the window clean every minute or so from any specks of dust, by painting over it fresh tear-water.

The light-sensitive screen at the back is the key-structure. It registers a continually changing picture. It receives, takes and records a moving picture life-long without change of 'plate', through every waking day. It signals its shifting exposures to the brain.

This camera also focuses itself automatically, according to the distance of the picture interesting it. It makes its lens 'stronger' or 'weaker' as required. This camera also turns itself in the direction of the view required. It is moreover contrived as though with forethought of self-preservation. Should danger threaten, in a moment its skin shutters close protecting its transparent window. And the whole structure, with its prescience and all its efficiency, is produced by and out of specks of granular slime arranging themselves as of their own accord in sheets and layers and acting seemingly on an agreed plan. That done, and their organ complete, they abide by what they have accomplished. They lapse into relative quietude and change no more. It all sounds an unskilful overstated tale which challenges belief. But to faithful observation so it is. There is more yet.

The little hollow bladder of the embryo-brain, narrowing itself at two points so as to be triple, thrusts from its foremost chamber to either side a hollow bud. This bud pushes toward the overlying skin. That skin, as though it knew and sympathized, then dips down forming a cuplike hollow to meet the hollow brain-stalk growing outward. They meet. The round end of the hollow brain-bud dimples inward and becomes a cup. Concurrently, the ingrowth from the skin nips itself free from its original skin. It rounds itself into a hollow ball, lying in the mouth of the brain-cup. Of this stalked cup, the optic cup, the stalk becomes in a few weeks a cable of a million nerve-fibres connecting the nerve-cells within the eye-ball itself with the brain. The optic cup, at first just a two-deep layer of somewhat simple-looking cells, multiplies its layers at the bottom of the cup where, when light enters the eye – which will not be for some weeks yet – the photo-image will in due course lie. There the layer becomes a fourfold layer of great complexity. It is strictly speaking a piece of the brain lying within the eye-ball. Indeed the whole brain itself, traced back to its embryonic beginning, is found to be all of a piece with the primordial skin – a primordial gesture as if to inculcate Aristotle's maxim about sense and mind.

The deepest cells at the bottom of the cup become a photo-sensitive layer – the sensitive film of the camera. If light is to act on the retina – and it is from the retina that light's visual effect is known to start – it must be absorbed there. In the retina a delicate purplish pigment absorbs incident light and is bleached by it, giving a light-picture. The photo-chemical effect generates nerve-currents running to the brain.

The nerve-lines connecting the photo-sensitive layer with the brain are not simple. They are in series of relays. It is the primitive cells of the optic cup, they and their progeny, which become in a few weeks these relays resembling a little brain, and each and all so shaped and connected as to transmit duly to the right points of the brain itself each light-picture momentarily formed and 'taken'. On the sense-cell layer the 'image' has, picture-like, two dimensions. These space-relations 'reappear' in the mind; hence we may think their data in the picture are in some way preserved in the electrical patterning of the resultant disturbance in the brain. But reminding us that the step from electrical disturbance in the brain to the mental experience is the mystery it is, the mind adds the third dimension when interpreting the two-dimensional picture! Also it adds colour; in short it makes a three-dimensional visual scene out of an electrical disturbance.

All this the cells lining the primitive optic cup have, so to say, to bear in mind, when laying these lines down. They lay them down by becoming them themselves.

Cajal [see p. 252], the gifted Spanish neurologist, gave special study to the retina and its nerve-lines to the brain. He turned to the insect-eye thinking the nerve-lines there 'in relative simplicity' might display schematically, and therefore more readably, some general plan which Nature adopts when furnishing animal kind with sight. After studying it for two years this is what he wrote:

> The complexity of the nerve-structures for vision is even in the insect something incredibly stupendous. From the insect's faceted eye proceeds an inextricable criss-cross of excessively slender nerve-fibres. These then plunge into a cell-labyrinth which doubtless serves to integrate what comes from the retinal layers. Next follow a countless host of amacrine cells and with them again numberless centrifugal fibres. All these elements are more-over so small the highest powers of the modern microscope hardly avail for following them. The intricacy of the connexions defies description. Before it the mind halts, abased. *In tenuis labor.* Peering through the microscope into this Lilliputian life one wonders whether what we disdainfully term 'instinct' (Bergson's 'intuition') is not, as Jules Fabre claims, life's crowning mental gift. Mind with instant and decisive action, the mind which in these tiny and ancient beings reached its blossom ages ago and earliest of all.

. . . The human eye has about 137 million separate 'seeing' elements spread out in the sheet of the retina. The number of nerve-lines leading from them to the brain gradually condenses down to little over a million. Each of these has in the brain, we must think, to find its right nerve-exchanges. Those nerve-exchanges lie far apart, and are but stations on the way to further stations. The whole crust of the brain is one thick tangled jungle of exchanges and of branching lines going thither and coming thence. As the eye's cup develops into the nervous retina all this intricate orientation to locality is provided for by corresponding growth in the brain. To compass what is needed adjacent cells, although sister and sister, have to shape themselves quite differently the one from the other. Most become patterned filaments, set lengthwise in the general direction of the current of travel. But some thrust out arms laterally as if to embrace together whole cables of the conducting system.

Nervous 'conduction' is transmission of nervous signals, in this case to the brain. There is also another nervous process, which physiology was slower to discover. Activity at this or that point in the conducting system, where relays are introduced, can be decreased even to suppression. This lessening is called inhibition; it occurs in the retina as elsewhere. All this is arranged for by the developing eye-cup when preparing and carrying out its million-fold connections with the brain for the making of a seeing eye. Obviously there are almost illimitable opportunities for a false step. Such a false step need not count at the time because all that we have been considering is done months or weeks before the eye can be used. Time after time so perfectly is all performed that the infant eye is a good and fitting eye, and the mind soon is instructing itself and gathering knowledge through it. And the child's eye is not only an eye true to the human type, but an eye with personal likeness to its individual parent's. The many cells which made it have executed correctly a multitudinous dance engaging millions of performers in hundreds of sequences of particular different steps, differing for each performer according to his part. To picture the complexity and the precision beggars any imagery I have. But it may help us to think further.

There is too that other layer of those embryonic cells at the back of the eye. They act as the dead black lining of the camera; they with their black pigment kill any stray light which would blur the optical image. They can shift their pigment. In full daylight they screen, and at night they unscreen, as wanted, the special seeing elements which serve for seeing in dim light. These are the cells which manufacture the purple pigment, 'visual purple', which sensitizes the eye for seeing in low light.

Then there is that little ball of cells which migrated from the skin and thrust itself into the mouth of the eye-stalk from the brain. It makes a lens there; it changes into glass-clear fibres, grouped with geometrical truth, locking together by toothed edges. The pencil of light let through must come to a point at the right distance for the length of the eye-ball which is to be. Not only must the lens be glass-clear but its shape must be optically right, and its substance must have the right optical refractive index. That index is higher than that of anything else which transmits light in the body. Its two curved surfaces back and front must be truly centred on one and the right axis, and each of the sub-spherical curvatures must be curved to the right degree, so that, the refractive index being right, light is brought to a focus on the retina and gives there a shaped image. The optician obtains glass of the desired refractive index and skilfully grinds its curvatures in accordance with the mathematical formulae required.

With the lens of the eye, a batch of granular skin-cells are told off to travel from the skin to which they strictly belong, to settle down in the mouth of the optic cup, to arrange themselves in a compact and suitable ball, to turn into transparent fibres, to assume the right refractive index, and to make themselves into a subsphere with two correct curvatures truly centred on a certain axis. Thus it is they make a lens of the right size, set in the right place, that is, at the right distance behind the transparent window of the eye in front and the sensitive seeing screen of the retina behind. In short they behave as if fairly possessed.

I would not give a wrong impression. The optical apparatus of the eye is not all turned out with a precision equal to that of a first-rate optical workshop. It has defects which disarm the envy of the optician. It is rather as though the planet, producing all this as it does, worked under limitations. Regarded as a planet which 'would', we yet find it no less a planet whose products lie open to criticism. On the other hand, in this very matter of the eye the process of its construction seems to seize opportunities offered by the peculiarity in some ways adverse of the material it is condemned to use. It extracts from the untoward situation practical advantages for its instrument which human craftsmanship could never in that way provide. Thus the cells composing the core of this living lens are denser than those at the edge. This corrects a focussing defect inherent in ordinary glass-lenses. Again, the lens of the eye, compassing what no glass-lens can, changes its curvature to focus near objects as well as distant when wanted, for instance, when we read. An elastic capsule is spun over it and is arranged to be eased by a special muscle. Further, the pupil – the camera stop – is self-adjusting. All this without our having even to wish it; without even our knowing anything about it, beyond that we are seeing satisfactorily.

The making of this eye out of self-actuated specks, which draw together and multiply and move as if obsessed with one desire, namely, to make the eye-ball. In a few weeks they have done so. Then, their madness over, they sit down and rest, satisfied to be life-long what they have made themselves, and, so to say, wait for death.

The chief wonder of all we have not touched on yet. Wonder of wonders, though familiar even to boredom. So much with us that we forget it all our time. The eye sends, as we saw, in to the cell-and-fibre forest of the brain throughout the waking day continual rhythmic streams of tiny, individually evanescent, electrical potentials. This throbbing streaming crowd of electrified shifting points in the sponge-work of the brain bears no obvious semblance in space-pattern, and even in temporal relation resembles but a little remotely the tiny two-dimensional upside-down picture of the outside world which the eye-ball paints on the beginnings of its nerve-fibres to the brain. But that little picture sets up an electrical storm. And that electrical storm so set up is one which affects a whole population of brain-cells. Electrical charges having in themselves not the faintest elements of the visual – having, for instance, nothing of 'distance', 'right-side-upness', nor 'vertical', nor 'horizontal', nor 'colour', nor 'brightness', nor 'shadow', nor 'roundness', nor 'squareness', nor contour', nor 'transparency', nor 'opacity', nor 'near', nor 'far', nor visual anything – conjure up all these. A shower of little electrical leaks conjures up for me, when I look, the landscape; the castle on the height, or, when I look at him, my friend's face, and how distant he is from me they tell me. Taking their word for it, I go forward and my other senses confirm that he is there.

It is a case of 'the world is too much with us'; too banal to wonder at. Those other things we paused over, the building and shaping of the eye-ball, and the establishing of its nerve connections with the right points of the brain, all those other things and the rest pertaining to them we called in chemistry and physics and final causes to explain to us. And they did so, with promise of more help to come.

But this last, not the eye, but the 'seeing' by the brain behind the eye? Physics and chemistry there are silent to our every question. All they say to us is that the brain is theirs, that without the brain which is theirs the seeing is not. But as to how? They vouchsafe us not a word.

Source: Sir Charles Sherrington, *Man on His Nature*, 2nd edition, Cambridge, Cambridge University Press, 1951.

Green Mould in the Wind

Alexander Fleming (1881–1955), a Scottish farmer's son, began his working life as a clerk in a London shipping company. But in 1902 he won a scholarship to medical school and, after graduating, became a research bacteriologist. Service in France with the Medical Corps in the First World War developed his interest in preventing infection in wounds. After the war, at St Mary's Hospital, London, he discovered an enzyme, contained in tears and other bodily fluids, and in egg white, that dissolved bacteria. He called it *lysozyme* (meaning 'dissolving enzyme'). As it dissolved only relatively harmless bacteria, however, it was of no practical use. His next discovery was far more momentous. It is described here by Sarah Riedman and Elton Gustafson in their book on Nobel Prize-winners in Medicine.

In a laboratory at St Mary's Hospital in London there sat a row of petri dishes – those little plates on which bacteriologists grow microbes on a layer of solidified gelatin or agar. It was the workshop of Dr Alexander Fleming, a modest little man with shaggy brows and wire-rimmed spectacles. In the back of his mind he had long held the idea that he would like to find a better way to treat wounds than with harsh antiseptics. Even if they killed some bacteria, these chemicals did greater damage to the white blood cells. One morning in September 1928, he was doing some routine experiments, examining milky-white cultures. They contained colonies of staphylococcus (spherical organisms growing in groups of four), the germ that causes ugly boils. His eye fell on one petri dish that had become contaminated with a green mould.

Contamination is the curse of bacteriological work. Researchers are as familiar with these microbic 'weeds' as the farmer is with unwanted invaders among his planted crops. Bacteriologists just throw the cultures down the drain, and are glad when not too much work has gone into nurturing the particular spoiled crop of colonies. In Fleming's dish, a casual wanderer had drifted in through the window in a blast of dusty London air, and had settled itself among the staph colonies he was studying.

Fleming was just about to discard the contaminated contents, but before doing so he took another look. This was different, and he made a note of it: 'It was astonishing that for some considerable distance around the mould growth the staphylococcal colonies were undergoing lysis. What had formerly been a well-grown colony was now a faint shadow of its former self.' It suddenly struck him about this 'lysis' – something was *dissolving* his microbes! 'I was sufficiently interested to pursue the subject,' he later reported.

With a loop at the end of a platinum wire, the most used tool of microbe searchers, he fished up a speck of the mould from the colony and placed it in peptone broth, the food moulds feed on. As it grew there, he saw first a fuzzy white, then a tufted green mass. When he examined it under his microscope he decided that the 'weed' belonged to a large family of moulds – *Penicillium* (from the Latin, meaning *little brush*) – which streak roquefort cheese green and spoil apples and oranges. Someone else might have given it no more thought, but in Fleming's brain the wheels were set in motion – especially those that had been turning around the idea of benign antiseptics.

He called to his helpers, for his vague hunch would have to be worked on. If this mould was brewing a juice that killed his bug colonies, he would have to look for it in the broth in which it was flourishing. So he filtered off some of the mould filtrate and dropped a bit of it on the glass plate in which his healthy staph colonies were growing. Sure enough, after several hours his bacteria died, disappearing under his very eyes as he examined a speck of the culture through his lens.

Together with his assistants he began to dilute the mould-containing broth, finding that in one-hundredth of its strength it decimated the bacteria. Further and further dilutions still retained their killing power, several times as potent as pure carbolic acid, which, while killing bacteria, also burns the tissues. They repeated the procedure with other organisms, the deadly germs pneumococci and streptococci. These, too, were killed as surely as his staphs. But what would it do to animal tissues?

To find out, they next brought in the usual inhabitants of bacteriology laboratories: mice, rabbits – the living test-tubes. Into the bellies of the mice, and into the blood vessels that stand out plainly through the transparent tissues of rabbits' ears, he slipped his hypodermic needle, injecting a thimbleful of the broth from the syringe. If this mould filtrate was poisonous, his animals would soon show its effects. But the mice and rabbits showed no more effect than if he had given them salt water.

This was indeed a find worth recording: 'It has been demonstrated that a species of Penicillium produces in culture a very powerful antibacterial substance,' Fleming put down in his notes. 'It is a more powerful inhibitory agent than carbolic acid and it can be applied to an infected surface undiluted as it is non-irritating and non-toxic.' He christened his remarkable substance *penicillin.*

Not being a chemist, he did not try to separate the microbe killer from the broth. But he did go on to other experiments, finding that the substance did not slay other bacteria – those that caused typhoid fever, dysentery and other infections in intestinal disease. On a fresh plate of jelly-like agar he dropped some of his own saliva, which contains the myriad different bacteria in the mouth. When, in the warmth of the incubator, the assorted colonies grew in profusion, he added penicillin. Some colonies were wiped out; others continued to thrive. Ergo! Those that were destroyed were sensitive to penicillin; the others were helpless against this mould antiseptic. In this way, one could detect the bacillus in influenza when sputum smeared on an agar plate was treated with penicillin. The influenza bacillus wasn't touched by penicillin, and here was a convenient way to detect it in sputum!

When he published his results in 1929, he suggested that penicillin could be used as a helpful laboratory tool to separate the 'goats from the sheep', as it were, in a mixed culture.

This accidental discovery in Fleming's laboratory was close to being miraculous. How easily it could have been missed without the world ever being the wiser! Luckily it happened to a man to whom Pasteur would have applied his now famous saying: 'Chance only favours the mind prepared for it.' Fleming himself once said: 'Do not wait for fortune to smile on you; prepare yourself with knowledge.'

Fleming maintained the culture, and sent it on request to laboratories around the world. However, he and his co-workers were unable to extract pure penicillin from the penicillium broth. This task was taken up in 1935 by the Australian doctor Howard Florey (1898–1968), working in Oxford with the chemist Dr Ernst Boris Chain, a refugee from Hitler's terror.

The division of labour was between the microbe growers, whose job it was to find the right food, temperature, and habitat for the mould, and the chemist, who was to try to coax the pure substance out of the broth. His chore was particularly difficult because the fragile active substance presented many knotty problems in separation and purification.

During several months of trial with various solvents, Chain came closer and closer to the one that would pull out the killing substance from the filtrate. Then it was necessary to separate it from the solvent. What was left was a little bit of brown powder. The bacteriologists took up once more from here. They dropped tiny flecks of the powder onto the culture plates to test their action on the bacteria. To gain some idea of its power is to realize that one part in *two million* dilution checked the growth of the colonies of the disease-producing bacteria.

But was the killer safe to use in animals? Florey, the doctor, directed the next step – to see what penicillin would do inside the tissues of mice, rats, and rabbits. It passed this test too: the animals were none the worse for having shots of penicillin. One more question had to be answered: what would it do in animals infected with lethal bacteria?

Florey called for dozens of mice to which he gave killing doses of staphylococcus – an organism that causes clean wounds to become infected. These test animals he divided into two groups: one received injections of penicillin every three hours, the other group did not. Then he watched his mute 'patients'. Later, those that had not received penicillin began to sicken and die and within twenty-four hours they were all dead. During the same period those that were getting the medicine also sickened; they were bedraggled, miserable little mice almost overwhelmed by the deadly germs multiplying in their blood. But then came a turning point: they began to get better, some faster than others, until by the end of a week of regular injections of penicillin, all but one were alive and frisky.

Penicillin had passed the animal test too. It killed streptococci and staphylococci without harm to the animals' tissues. The team was convinced that penicillin was a good performer and would be safe to try in a human being. But now they were thwarted by a mechanical problem – how to get enough of the germ-killer to treat a man who required several thousand times the dose that would do for an infected mouse. Not only was penicillin hard to obtain in any quantity, it did not remain long enough in the blood. The material had to be injected every few hours to replace the amount that spilled over into the urine of the animals. 'You might just as well try to fill a bathtub when the plug is out,' Florey remarked.

It was chiefly a problem of making enough of the active material. The work of brewing broth filtrate and separating the brown medicine was intensified. Finally, working day and night, the laboratory workers extracted enough material from the many batches to supply a human. About a teaspoon of the powder was turned over to Dr Mary Florey, the chief researcher's wife, to try on a desperately sick patient.

A policeman was lying gravely ill in Radcliffe Hospital at Oxford University. He had nicked himself while shaving and the tiny wound had become infected with staphylococci which had entered the blood. His face was covered with discharging abscesses and he was burning with a fever of 105 degrees. The abscesses were teeming with pus-forming staph and strep. The sulfa drug he had been given was helpless against this rapidly spreading infection. Dr Florey went to work on 12 February 1941.

The brown powder was dissolved in salt water, and from a glass bottle suspended high above the bed the drug dripped down a rubber tube and through a large hypodermic needle inserted into the vein of the patient's arm. For three days it trickled into the policeman's blood, as the doctors watched the fever go down and the abscesses slowly getting better. Not to waste the limited supply, the patient's urine was collected so that Dr Chain could reclaim the material discarded by the kidneys. But after the fifth day, just as the patient was definitely getting better, the penicillin gave out and couldn't be replaced rapidly enough. As a result it was not possible to save the patient. More of the drug was acquired by careful hoarding and tried on a second patient who also died because again the small supply became exhausted. Despite these tragedies, due to scarcity of the drug, penicillin was definitely established as a microbe-killer. Then a third patient, a fifteen-year-old boy with an infected wound from an operation, was saved with penicillin recovered from the urine of the first two. The fourth, fifth and sixth patients also were cured.

Clearly, the problem now was one of taking the production of penicillin out of the laboratory, and going into large-scale manufacture. It was 1941, the darkest time of the war in Europe. In England, fighting for its very existence, mass production was out of the question. Yet the need for this miracle drug was most acute – more soldiers were dying on the battlefields from infected wounds than from direct hits. There was only one place to which the London research team could turn for help – the United States.

Florey and one of his bacteriologist-assistants were brought to America by a grant from the Rockefeller Foundation. The purpose of the mission was to find a way to increase the yield of penicillin from the mould and to manufacture it on a commercial scale. In this project the researchers secured the support of both the U.S. Department of Agriculture and several American drug companies. This was a giant cooperative effort involving research team, government, and industry. Starting with a tube of mould from Fleming's original culture, brought over by Florey, penicillin was made to grow in gargantuan fermentation tanks, each with a capacity of 12,000 gallons.

The story of how high-yielding mould strains were found, and the methods for purifying penicillin developed (absolutely pure penicillin is white instead of brown) requires a book in itself. The mould culture was taken from small flasks and milk bottles to huge fermentation vats; the fodder used – corn-steep and milk sugar – brought a greater harvest; sterile air under pressure and ultra-violet light prevented contamination. By the end of 1943 production was five billion units a month, and at the end of the following year, 300 billion units – enough to treat a half million human beings each month.

Thus, human ingenuity, efficiency, and industry solved the problem of the limited supply of penicillin, and no one had to die for lack of the drug. Penicillin could now be tested – and successfully – in patients not only with infected wounds, but with crippling heart disease, syphilis, gonorrhoea, discharging ears, certain types of pneumonia, bacterial invasion of bone, and in the treatment of burns as well as boils and carbuncles. It became abundantly clear that the discovery of the first antibiotic was perhaps the greatest single victory ever achieved by science over infectious disease.

For this epoch-making discovery the three principals – Fleming, Florey and Chain – received the Nobel Prize in 1945 with the citation: 'For the discovery of penicillin and its therapeutic effect for the cure of different infectious maladies.'

Source: Sarah R. Riedman and Elton T. Gustafson, *Portraits of Nobel Laureates in Medicine and Physiology*, London, New York, Toronto, Abelard-Schuman, 1963.

In the Black Squash Court: The First Atomic Pile

The director of the team that built the world's first atomic pile was the Italian physicist Enrico Fermi, who had quit Fascist Italy in 1938. In the same year he had been awarded the Nobel Prize for developing the technique of bombarding uranium atoms with neutrons. Following on this work, Otto Hahn and Fritz Strassman in Berlin had found that among the fragments obtained when uranium atoms were bombarded were atoms of barium, which has an atomic weight approximately half that of uranium. It was their colleague the Austrian Lise Meitner and her nephew Otto Frisch who realized that what had taken place was atomic fission – the uranium atom splitting to produce two of barium. Since a uranium atom undergoing fission would also emit neutrons, it occurred to Fermi and others that a chain reaction might occur – the emitted neutrons hitting other uranium atoms and splitting them, thus emitting other neutrons which would hit and split other atoms, and so on, a process that would release a huge amount of energy. The function of an atomic pile is to induce and control such a reaction. Fermi's team on the project included Herbert Anderson, the Hungarian Leo Szilard, and the Canadian Walter H. Zinn. This account is from his wife Laura Fermi's book, *Atoms in the Family*, 1955.

The operation of the atomic pile was the result of almost four years of sustained work, which started when discovery of uranium fission became known, arousing enormous interest among physicists.

Experiments at Columbia University and at other universities in the United States had confirmed Enrico's hypothesis that neutrons would be emitted in the process of fission. Consequently, a chain reaction appeared possible in theory. To achieve it in practice seemed a vague and distant possibility. The odds against it were so great that only the small group of stubborn physicists at Columbia pursued work in that direction. At once they were faced with two sets of difficulties.

The first lay in the fact that neutrons emitted in the process of uranium fission were too fast to be effective atomic bullets and to cause fission in uranium. The second difficulty was due to loss of neutrons: under normal circumstances most of the neutrons produced in fission escaped into the air or were absorbed by matter before they had a chance of acting as uranium splitters. Too few produced fission to cause a chain reaction.

Neutrons would have to be slowed down and their losses reduced by a large factor, if a chain reaction was to be achieved. Was this feasible?

To slow down neutrons was an old trick for Enrico, from the time when he and his friends in Rome had recognized the extraordinary action of paraffin and water on neutrons. So the group at Columbia – Szilard, Zinn, Anderson, and Enrico – undertook the investigation of fission of uranium under water. Water, in the physicists' language, was being used as a moderator.

After many months of research they came to the conclusion that neither water nor any other hydrogenated substance is a suitable moderator. Hydrogen absorbs too many neutrons and makes a chain reaction impossible.

Leo Szilard and Fermi suggested trying carbon for a moderator. They thought that carbon would slow down neutrons sufficiently and absorb fewer of them than water, provided it was of a high degree of purity. Impurities have an astounding capacity for swallowing neutrons.

Szilard and Fermi conceived a contrivance that they thought might produce a chain reaction. It would be made of uranium and very pure graphite disposed in layers: layers exclusively of graphite would alternate with layers in which uranium chunks would be embedded in graphite. In other words, it would be a 'pile'.

An atomic pile is, of necessity, a bulky object. If it were too small, neutrons would escape into the surrounding air before they had a chance to hit a uranium atom, and they would be lost to fission and chain reaction. How large the pile ought to be, nobody knew.

Did it matter whether the scientists did not know the size of the pile? All they had to do, one might think, was to put blocks of graphite over blocks of graphite, alternating them with lumps of uranium, and keep on at it until they had reached the critical size, at which a chain reaction would occur. They could also give the pile different shapes – cubical, pyramidal, oval, spherical – and determine which worked best.

It was not so simple. Only a few grams of metallic uranium were available in the United States, and no commercial graphite came close to the requirements of purity.

The 1951 edition of Webster's *New Collegiate Dictionary* states that graphite is 'soft, black, native carbon of metallic lustre; often called *plumbago* or *black lead*. It is used for lead pencils, crucibles, lubricants, etc . . .' The atomic pile built in 1942, clearly included in the 'etc.,' was to use as much graphite as would go into making a pencil for each inhabitant of the earth, man, woman, and child. Moreover, graphite for a pile must be of a state of purity absolutely inconceivable for any other purpose. Scientists would have to be patient.

Procurement became a big and important task, one for which Fermi was not suited and which he would rather leave to others. Luckily for him, Leo Szilard did not share his aversion to interrupting research and shopping around.

Szilard was a man with an astounding number of ideas, several of which turned out to be good. He had no fewer acquaintances than ideas, a not negligible percentage of whom were important persons in high positions. These two sets of circumstances made of Szilard a powerful and useful spokesman for the small group of researchers, one who could confront the difficulties of politics with sufficient impetus to overcome them successfully. Willingly and with determination he undertook the not easy task of turning grams into tons, both of metallic uranium and of highly pure graphite.

The first question one asks when undertaking a task of that kind is: 'Who is going to finance my enterprise and give me the cash that is needed?' Szilard hoped he knew the answer. During the summer of 1939, with Wigner, Teller, Einstein and Sachs, he had succeeded in arousing President Roosevelt's interest in uranium work. Now, at the very beginning of 1940, he scored his second victory and obtained the first tangible, if small, proof of that professed interest, when Columbia University received the first grant of $6,000 from the Army and Navy to purchase materials.

Thus by early spring 1940 a few tons of pure graphite started to arrive at the physics building of Columbia University. Fermi and Anderson turned into bricklayers and began to stack graphite bricks in one of their laboratories.

They were well aware that for many months, perhaps for years, there would not be uranium and graphite of good enough quality and in sufficient quantity to attempt a pile. That did not matter for the time being: they knew so very little about the properties of the substances they were to work with – of metallic uranium not even the melting point had been determined – that much study of these properties ought to be pursued and completed before they could in good conscience recommend that the Uranium Committee undertake the tremendous effort and expense that would go in the project.

So they stacked graphite bricks into a stocky column, placed a neutron source under it, observed what happened to the neutrons in the graphite and began to collect data.

This work, dull as it sounds, was considered very important; and when the Advisory Committee on Uranium met on April 28, 1940, it decided to wait for more results at Columbia University before making formal recommendations for the project. The committee made this decision despite the report that the Nazis had set aside a large section of the Kaiser Wilhelm Institute in Berlin for research on uranium.

After the study on graphite, came that on uranium: How does it absorb and re-emit neutrons? Under what conditions will it undergo fission? How many neutrons will be produced altogether?

The experiments proceeded slowly for lack of materials and Fermi would have liked to speed up his work. Besides, he was convinced that from the behaviour of a small pile he would obtain much more information pertinent to building a larger pile. Fermi and his group were able to start work on the 'small pile' by the spring of 1941. They demolished their column of graphite bricks and laid them down again, placing lumps of uranium among them. Slowly, as more graphite arrived at Columbia, a black wall grew up. The black wall reached the ceiling; but it was still far from being a chain-reacting pile: too many neutrons escaped from it or were absorbed inside it, and too few remained to produce fission.

It became evident that the experiment could not be pursued to find success in that same laboratory. A larger room, with higher ceiling, was needed. No such room was available at Columbia, and somebody would have to look for one elsewhere. Fermi was absorbed in his research. His work was too important to be interrupted. So Herbert Anderson took off his overalls, put on a suit, coat and a hat, and went scouting in New York City and its suburbs in search of a loft that could house a pile. He spotted several possibilities and began some bargaining aimed at the best deal.

Before Herbert could make a final choice, Enrico learned that he, his group, his equipment, and the materials he had gathered would have to move to Chicago. It was the very end of 1941 . . .

The best place Compton [Professor Arthur H. Compton of the University of Chicago, who had been appointed head of research into chain reaction] had been able to find for work on the pile was a squash court under the West Stands of Stagg Field, the University of Chicago stadium. President Hutchins had banned football from the Chicago campus, and Stagg Field was used for odd purposes. To the west, on Ellis Avenue, the stadium is closed by a tall grey stone structure in the guise of a medieval castle. Through a heavy portal is the entrance to the space beneath the West Stands. The Squash Court was part of this space. It was 30 feet wide, twice as long, and over 26 feet high.

The physicists would have liked more space, but places better suited for the pile, which Professor Compton had hoped he could have, had been requisitioned by the expanding armed forces stationed in Chicago. The physicists were to be contented with the Squash Court, and there Herbert Anderson had started assembling piles. They were still 'small piles,' because material flowed to the West Stands at a very slow, if steady, pace. As each new shipment of crates arrived, Herbert's spirits rose. He loved working and was of impatient temperament. His slender, almost delicate, body had unsuspected resilience and endurance. He could work at all hours and drive his associates to work along with his same intensity and enthusiasm.

A shipment of crates arrived at the West Stands on a Saturday afternoon, when the hired men who would normally unpack them were not working. A university professor, older by several years than Herbert, gave a look at the crates and said lightly: 'Those fellows will unpack them Monday morning.'

'Those fellows, Hell! We'll do them now,' flared up Herbert, who had never felt inhibited in the presence of older men, higher up in the academic hierarchy. The professor took off his coat and the two of them started wrenching at the crates.

Profanity was freely used at the Met. Lab. It relieved the tension built up by having to work against time. Would Germany get atomic weapons before the United States developed them? Would these weapons come in time to help win the war? These unanswered questions constantly present in the minds of the leaders in the project pressed them to work faster and faster, to be tense, and to swear.

Success was assured by the spring. A small pile assembled in the Squash Court showed that all conditions – purity of materials, distribution of uranium in the graphite lattice – were such that a pile of critical size would chain-react. . . .

While waiting for more materials, Herbert Anderson went to the Goodyear Tyre and Rubber Company to place an order for a square balloon. The Goodyear people had never heard of square balloons, they did not think they could fly. At first they threw suspicious glances at Herbert. The young man, however, seemed to be in full possession of his wits. He talked earnestly, had figured out precise specifications, and knew exactly what he wanted. The Goodyear people promised to make a square balloon of rubberized cloth. They delivered it a couple of months later to the Squash Court. It came neatly folded but, once unfolded, it was a huge thing that reached from floor to ceiling.

The Squash Court ceiling could not be pushed up as the physicists would have liked. They had calculated that their final pile ought to chain-react somewhat before it reached the ceiling. But not much margin was left, and calculations are never to be trusted entirely. Some impurities might go unnoticed, some unforeseen factor might upset theory. The critical size of the pile might not be reached at the ceiling. Since the physicists were compelled to stay within that very concrete limit, they thought of improving the performance of the pile by means other than size.

The experiment at Columbia with a canned pile had indicated that such an aim might be attained by removing the air from the pores of the graphite. To can as large a pile as they were to build now would be impracticable, but they could assemble it inside a square balloon and pump the air from it if necessary.

The Squash Court was not large. When the scientists opened the balloon and tried to haul it into place, they could not see its top from the floor. There was a movable elevator in the room, some sort of scaffolding on wheels that could raise a platform. Fermi climbed onto it, let himself be hoisted to a height that gave him a good view of the entire balloon, and from there he gave orders:

'All hands stand by!'

'Now haul the rope and heave her!'

'More to the right!'

'Brace the tackles to the left!'

To the people below he seemed an admiral on his bridge, and 'Admiral' they called him for a while.

When the balloon was secured on five sides, with the flap that formed the sixth left down, the group began to assemble the pile inside it. Not all the material had arrived, but they trusted that it would come in time.

From the numerous experiments they had performed so far, they had an idea of what the pile should be, but they had not worked out the details, there were no drawings nor blueprints and no time to spare to make them. They planned their pile even as they built it. They were to give it the shape of a sphere of about 26 feet in diameter, supported by a square frame, hence the square balloon.

The pile supports consisted of blocks of wood. As a block was put in place inside the balloon, the size and shape of the next were figured. Between the Squash Court and the near-by carpenter's shop there was a steady flow of boys, who fetched finished blocks and brought specifications for more on bits of paper.

When the physicists started handling graphite bricks, everything became black. The walls of the Squash Court were black to start with. Now a huge black wall of graphite was going up fast. Graphite powder covered the floor and made it black and as slippery as a dance floor. Black figures skidded on it, figures in overalls and goggles under a layer of graphite dust. There was one woman among them, Leona Woods; she could not be distinguished from the men, and she got her share of cussing from the bosses.

The carpenters and the machinists who executed orders with no knowledge of their purpose and the high-school boys who helped lay bricks for the pile must have wondered at the black scene. Had they been aware that the ultimate result would be an atomic bomb, they might have renamed the court Pluto's Workshop or Hell's Kitchen.

To solve difficulties as one meets them is much faster than to try to foresee them all in detail. As the pile grew, measurements were taken and further construction adapted to results.

The pile never reached the ceiling. It was planned as a sphere 26 feet in diameter, but the last layers were never put into place. The sphere remained flattened at the top. To make a vacuum proved unnecessary, and the balloon was never sealed. The critical size of the pile was attained sooner than was anticipated.

Only six weeks had passed from the laying of the first graphite brick, and it was the morning of December 2 [1942].

Herbert Anderson was sleepy and grouchy. He had been up until two in the morning to give the pile its finishing touches. Had he pulled a control rod during the night, he could have operated the pile and have been the first man to achieve a chain reaction, at least in a material, mechanical sense. He had a moral duty not to pull that rod, despite the strong temptation. It would not be fair to Fermi. Fermi was the leader. He had directed research and worked out theories. His were the basic ideas. His were the privilege and the responsibility of conducting the final experiment and controlling the chain reaction.

'So the show was all Enrico's, and he had gone to bed early the night before,' Herbert told me years later, and a bit of regret still lingered in his voice.

Walter Zinn also could have produced a chain reaction during the night. He, too, had been up and at work. But he did not care whether he operated the pile or not; he did not care in the least. It was not his job.

His task had been to smooth out difficulties during the pile construction. He had been some sort of general contractor: he had placed orders for material and made sure that they were delivered in time; he had supervised the machine shops where graphite was milled; he had spurred others to work faster, longer, more efficiently. He had become angry, had shouted, and had reached his goal. In six weeks the pile was assembled, and now he viewed it with relaxed nerves and with that vague feeling of emptiness, of slight disorientation, which never fails to follow completion of a purposeful task.

There is no record of what were the feelings of the three young men who crouched on top of the pile, under the ceiling of the square balloon. They were called the 'suicide squad.' It was a joke, but perhaps they were asking themselves whether the joke held some truth. They were like firemen alerted to the possibility of a fire, ready to extinguish it. If something unexpected were to happen, if the pile should get out of control, they would 'extinguish' it by flooding it with a cadmium solution. Cadmium absorbs neutrons and prevents a chain reaction.

A sense of apprehension was in the air. Everyone felt it but outwardly, at least, they were all calm and composed.

Among the persons who gathered in the Squash Court on that morning, one was not connected with the Met. Lab. – Mr Crawford H. Greenewalt of E. I duPont de Nemours, who later became the president of the company. Arthur Compton had led him there out of a nearby room where, on that day, he and other men from his company happened to be holding talks with top Army officers.

Mr Greenewalt and the duPont people were in a difficult position, and they did not know how to reach a decision. The Army had taken over the Uranium Project on the previous August and renamed it Manhattan District. In September General Leslie R. Groves was placed in charge of it. General Groves must have been of a trusting nature: before a chain reaction was achieved, he was already urging the duPont de Nemours Company to build and operate piles on a production scale.

In a pile, Mr Greenewalt was told, a new element, plutonium, is created during uranium fission. Plutonium would probably be suited for making atomic bombs. So Greenewalt and his group had been taken to Berkeley to see the work done on plutonium, and then flown to Chicago for more negotiations with the Army.

Mr Greenewalt was hesitant. Of course his company would like to help win the war! But piles and plutonium!

With the Army's insistent voice in his ears, Compton, who had attended the conference, decided to break the rules and take Mr Greenewalt to witness the first operation of a pile.

They all climbed onto the balcony at the north end of the Squash Court; all, except the three boys perched on top of the pile and except a young physicist, George Weil, who stood alone on the floor by a cadmium rod that he was to pull out of the pile when so instructed.

And so the show began.

There was utter silence in the audience, and only Fermi spoke. His grey eyes betrayed his intense thinking, and his hands moved along with his thoughts.

'The pile is not performing now because inside it there are rods of cadmium which absorb neutrons. One single rod is sufficient to prevent a chain reaction. So our first step will be to pull out of the pile all control rods but the one that George Weil will man.' As he spoke others acted. Each chore had been assigned in advance and rehearsed. So Fermi went on speaking, and his hands pointed out the things he mentioned.

'This rod, that we have pulled out with the others, is automatically controlled. Should the intensity of the reaction become greater than a pre-set limit, this rod would go back inside the pile by itself.

'This pen will trace a line indicating the intensity of the radiation. When the pile chain-reacts, the pen will trace a line that will go up and up and that will not tend to level off. In other words, it will be an exponential line.

'Presently we shall begin our experiment. George will pull out his rod a little at a time. We shall take measurements and verify that the pile will keep on acting as we have calculated.

'Weil will first set the rod at thirteen feet. This means that thirteen feet of the rod will still be inside the pile. The counters will click faster and the pen will move up to this point, and then its trace will level off. Go ahead, George!'

Eyes turned to the graph pen. Breathing was suspended. Fermi grinned with confidence. The counters stepped up their clicking; the pen went up and then stopped where Fermi had said it would. Greenewalt gasped audibly. Fermi continued to grin.

He gave more orders. Each time Weil pulled out some more, the counters increased the rate of their clicking, the pen raised to the point that Fermi predicted, then it levelled off.

The morning went by. Fermi was conscious that a new experiment of this kind, carried out in the heart of a big city, might become a potential hazard unless all precautions were taken to make sure that at all times the operation of the pile conformed closely with the results of the calculations. In his mind he was sure that if George Weil's rod had been pulled out all at once, the pile would have started reacting at a leisurely rate and could have been stopped at will by reinserting one of the rods. He chose, however, to take his time and be certain that no unforeseen phenomenon would disturb the experiment.

It is impossible to say how great a danger this unforeseen element constituted or what consequences it might have brought about. According to the theory, an explosion was out of the question. The release of lethal amounts of radiation through an uncontrolled reaction was improbable. Yet the men in the Squash Court were working with the unknown. They could not claim to know the answers to all the questions that were in their minds. Caution was welcome. Caution was essential. It would have been reckless to dispense with caution.

So it was lunch time, and, although nobody else had given signs of being hungry, Fermi, who is a man of habits, pronounced the now historical sentence:

'Let's go to lunch.'

After lunch they all resumed their places, and now Mr Greenewalt was decidedly excited, almost impatient.

But again the experiment proceeded by small steps, until it was 3.20.

Once more Fermi said to Weil:

'Pull it out another foot'; but this time he added, turning to the anxious group in the balcony: 'This will do it. Now the pile will chain-react.'

The counters stepped up; the pen started its upward rise. It showed no tendency to level off. A chain reaction was taking place in the pile.

In the back of everyone's mind was one unavoidable question.

'When do we become scared?'

Under the ceiling of the balloon the suicide squad was alert, ready with their liquid cadmium: this was the moment. But nothing much happened. The group watched the recording instruments for 28 minutes. The pile behaved as it should, as they all had hoped it would, as they had feared it would not.

The rest of the story is well known. Eugene Wigner, the Hungarian-born physicist who in 1939 with Szilard and Einstein had alerted President Roosevelt to the importance of uranium fission, presented Fermi with a bottle of Chianti. According to an improbable legend, Wigner had concealed the bottle behind his back during the entire experiment.

All those present drank. From paper cups, in silence, with no toast. Then all signed the straw cover on the bottle of Chianti. It is the only record of the persons in the Squash Court on that day.

Source: Laura Fermi, *Atoms in the Family: My Life with Enrico Fermi, Designer of the First Atomic Pile*, London, Allen Unwin, 1955.

A Death and the Bomb

Arch-enemy of gobbledegook and obscurity, Nobel Prize-winner Richard Feynman (1918–88) excelled at making science clear to the unscientific. The two books of memoirs and conversations compiled by Ralph Leighton, *Surely You're Joking, Mr Feynman* and *What Do You Care What Other People Think?* reveal a defiantly individual, iconoclastic personality, distrustful of 'intellectuals'. He once said that he would be just as happy if his children turned out to be truck-drivers or guitar-players, rather than scientists.

His main scientific work was to remake the theory of quantum electrodynamics (QED, for short) which explains the interaction of light (photons) and matter (electrons). Almost all natural phenomena, including all chemistry and biology, are covered by this theory. Explaining it to the general reader (in *QED: The Strange Theory of Light and Matter*), he begins, typically, with a familiar experience. Everyone knows that light is partially reflected from some surfaces – glass, for example. If you have a lamp in your room in daytime, and look out of the window, you can see things outside plus a dim reflection of your lamp. The fact that the lamp is partially reflected means that some photons (light particles) are bounced back by the electrons in the glass, while others pass through. Experiment shows that for every 100 photons an average of 4 bounce back, 96 go through. No one knows why. No one knows how a photon 'makes up its mind' which course to follow. No one can predict which course a given photon will opt for. Science can only work out the percentage probability.

The simple, graphic quality of this example is persistently evident in all Feynman's writing – about life or science. In the Second World War he worked at Los Alamos on the atom bomb project. His wife Arlene was dying of TB of the lymphatic gland. They had known she was fatally ill when they married. He took leave from the project, drove to the hospital, and was there when she died. Afterwards, he kissed her:

I was very surprised to discover that her hair smelled exactly the same. Of course, after I stopped and thought about it, there was no reason why her hair should smell different in such a short time. But to me it was a kind of shock, because in my mind, something enormous had just happened – and yet nothing had happened.

Arlene's death was followed by the successful testing of the first atomic bomb in the Nevada desert on 16 July 1945, recalled by Feynman in *Surely You're Joking*:

After we'd made the calculations, the next thing that happened, of course, was the test. I was actually at home on a short vacation at that time, after my wife died, and so I got a message that said, 'The baby is expected on such and such a day.'

I flew back, and I arrived *just* when the buses were leaving, so I went straight out to the site and we waited out there, twenty miles away. We had a radio, and they were supposed to tell us when the thing was going to go off and so forth, but the radio wouldn't work, so we never knew what was happening. But just a few minutes before it was supposed to go off the radio started to work, and they told us there was twenty seconds or something to go, for people who were far away like we were. Others were closer, six miles away.

They gave out dark glasses that you could watch it with. Dark glasses! Twenty miles away, you couldn't see a damn thing through dark glasses. So I figured the only thing that could really hurt your eyes (bright light can never hurt your eyes) is ultraviolet light. I got behind a truck windshield, because the ultraviolet can't go through glass, so that would be safe, and so I could *see* the damn thing.

Time comes, and this *tremendous* flash out there is so bright that I duck, and I see this purple splotch on the floor of the truck. I said, 'That's not it. That's an after-image.' So I look back up, and I see this white light changing into yellow and then into orange. Clouds form and disappear again – from the compression and expansion of the shock wave.

Finally, a big ball of orange, the center that was so bright, becomes a ball of orange that starts to rise and billow a little bit and get a little black around the edges, and then you see it's a big ball of smoke with flashes on the inside of the fire going out, the heat.

All this took about one minute. It was a series from bright to dark, and I had *seen* it. I am about the only guy who actually looked at the damn thing – the first Trinity test. Everybody else had dark glasses, and the people at six miles couldn't see it because they were all told to lie on the floor. I'm probably the only guy who saw it with the human eye.

Finally, after about a minute and a half, there's suddenly a tremendous noise – BANG, and then a rumble, like thunder – and that's what convinced me. Nobody had said a word during this whole thing. We were all just watching quietly. But this sound released everybody – released me particularly because the solidity of the sound at that distance meant that it had really worked.

The man standing next to me said, 'What's that?'

I said, 'That was the Bomb.'

The man was William Laurence [author of *Dawn Over Zero*]. He was there to write an article describing the whole situation. I had been the one who was supposed to have taken him around. Then it was found that it was too technical for him, and so later H. D. Smyth came and I showed him around. One thing we did, we went into a room and there on the end of a narrow pedestal was a small silver-plated ball. You could put your hand on it. It was warm. It was radioactive. It was plutonium. And we stood at the door of this room, talking about it. This was a new element that was made by man, that had never existed on the earth before, except for a very short period possibly at the very beginning. And here it was all isolated and radioactive and had these properties. And we had made it. And so it was *tremendously* valuable.

Meanwhile, you know how people do when they talk – you kind of jiggle around and so forth. He was kicking the doorstop, you see, and I said, 'Yes, the doorstop certainly is appropriate for this door.' The doorstop was a ten-inch hemisphere of yellowish metal – gold, as a matter of fact.

What had happened was that we needed to do an experiment to see how many neutrons were reflected by different materials, in order to save the neutrons so we didn't use so much material. We had tested many different materials. We had tested platinum, we had tested zinc, we had tested brass, we had tested gold. So, in making the tests with the gold, we had these pieces of gold and somebody had the clever idea of using that great ball of gold for a doorstop for the door of the room that contained the plutonium.

Source: *Surely You're Joking, Mr Feynman. Adventures of a Curious Character, Richard P. Feynman. As told by Ralph Leighton*, ed. Edward Hutchings, London, Unwin Hyman, 1985.

The Story of a Carbon Atom

Born in Turin in 1919, Primo Levi graduated in chemistry shortly before the Fascist race laws prohibited Jews like himself from taking university degrees. In 1943 he joined a partisan group in northern Italy, was arrested and deported to Auschwitz. His expertise as a chemist saved him from the gas chambers, however. He was set to work in a factory, and liberated in 1945.

His memoir *The Periodic Table* takes its title from the table of elements, arranged according to their atomic mass, which was originally devised by Dmitri Mendeleyev in 1869 (see p. 148). Levi links each episode of his life to a certain element. But in the book's final section, printed below, he sets himself to imagine the life of a carbon atom. This was, he says, his first 'literary dream', and came to him in Auschwitz.

Our character lies for hundreds of millions of years, bound to three atoms of oxygen and one of calcium, in the form of limestone: it already has a very long cosmic history behind it, but we shall ignore it. For it time does not exist, or exists only in the form of sluggish variations in temperature, daily or seasonal, if, for the good fortune of this tale, its position is not too far from the earth's surface. Its existence, whose monotony cannot be thought of without horror, is a pitiless alternation of hots and colds, that is, of oscillations (always of equal frequency) a trifle more restricted and a trifle more ample: an imprisonment, for this potentially living personage, worthy of the Catholic Hell. To it, until this moment, the present tense is suited, which is that of description, rather than the past tense, which is that of narration – it is congealed in an eternal present, barely scratched by the moderate quivers of thermal agitation.

But, precisely for the good fortune of the narrator, whose story could otherwise have come to an end, the limestone rock ledge of which the atom forms a part lies on the surface. It lies within reach of man and his pickax (all honor to the pickax and its modern equivalents; they are still the most important intermediaries in the millennial dialogue between the elements and man): at any moment – which I, the narrator, decide out of pure caprice to be the year 1840 – a blow of the pickax detached it and sent it on its way to the lime kiln, plunging it into the world of things that change. It was roasted until it separated from the calcium, which remained so to speak with its feet on the ground and went to meet a less brilliant destiny, which we shall not narrate. Still firmly clinging to two of its three former oxygen companions, it issued from the chimney and took the path of the air. Its story, which once was immobile, now turned tumultuous.

It was caught by the wind, flung down on the earth, lifted ten kilometers high. It was breathed in by a falcon, descending into its precipitous lungs, but did not penetrate its rich blood and was expelled. It dissolved three times in the water of the sea, once in the water of a cascading torrent, and again was expelled. It traveled with the wind for eight years: now high, now low, on the sea and among the clouds, over forests, deserts, and limitless expanses of ice; then it stumbled into capture and the organic adventure.

Carbon, in fact, is a singular element: it is the only element that can bind itself in long stable chains without a great expense of energy, and for life on earth (the only one we know so far) precisely long chains are required. Therefore carbon is the key element of living substance: but its promotion, its entry into the living world, is not easy and must follow an obligatory, intricate path, which has been clarified (and not yet definitively) only in recent years. If the elaboration of carbon were not a common daily occurrence, on the scale of billions of tons a week, wherever the green of a leaf appears, it would by full right deserve to be called a miracle.

The atom we are speaking of, accompanied by its two satellites which maintained it in a gaseous state, was therefore borne by the wind along a row of vines in the year 1848. It had the good fortune to brush against a leaf, penetrate it, and be nailed there by a ray of the sun. If my language here becomes imprecise and allusive, it is not only because of my ignorance: this decisive event, this instantaneous work *a tre* – of the carbon dioxide, the light, and the vegetal greenery – has not yet been described in definitive terms, and perhaps it will not be for a long time to come, so different is it from the other 'organic' chemistry which is the cumbersome, slow, and ponderous work of man: and yet this refined, minute, and quick-witted chemistry was 'invented' two or three billion years ago by our silent sisters, the plants, which do not experiment and do not discuss, and whose temperature is identical to that of the environment in which they live. If to comprehend is the same as forming an image, we will never form an image of a happening whose scale is a millionth of a millimeter, whose rhythm is a millionth of a second, and whose protagonists are in their essence invisible. Every verbal description must be inadequate, and one will be as good as the next, so let us settle for the following description.

Our atom of carbon enters the leaf, colliding with other innumerable (but here useless) molecules of nitrogen and oxygen. It adheres to a large and complicated molecule that activates it, and simultaneously receives the decisive message from the sky, in the flashing form of a packet of solar light: in an instant, like an insect caught by a spider, it is separated from its oxygen, combined with hydrogen and (one thinks) phosphorus, and finally inserted in a chain, whether long or short does not matter, but it is the chain of life. All this happens swiftly, in silence, at the temperature and pressure of the atmosphere, and gratis: dear colleagues, when we learn to do likewise we will be *sicut Deus* [like God], and we will have also solved the problem of hunger in the world.

But there is more and worse, to our shame and that of our art. Carbon dioxide, that is, the aerial form of the carbon of which we have up till now spoken: this gas which constitutes the raw material of life, the permanent store upon which all that grows draws, and the ultimate destiny of all flesh, is not one of the principal components of air but rather a ridiculous remnant, an 'impurity,' thirty times less abundant than argon, which nobody even notices. The air contains 0.03 percent; if Italy was air, the only Italians fit to build life would be, for example, the fifteen thousand inhabitants of Milazzo in the province of Messina. This, on the human scale, is ironic acrobatics, a juggler's trick, an incomprehensible display of omnipotence-arrogance, since from this ever renewed impurity of the air we come, we animals and we plants, and we the human species, with our four billion discordant opinions, our milleniums of history, our wars and shames, nobility and pride. In any event, our very presence on the planet becomes laughable in geometric terms: if all of humanity, about 250 million tons, were distributed in a layer of homogeneous thickness on all the emergent lands, the 'stature of man' would not be visible to the naked eye; the thickness one would obtain would be around sixteen thousandths of a millimeter.

Now our atom is inserted: it is part of a structure, in an architectural sense; it has become related and tied to five companions so identical with it that only the fiction of the story permits me to distinguish them. It is a beautiful ring-shaped structure, an almost regular hexagon, which however is subjected to complicated exchanges and balances with the water in which it is dissolved; because by now it is dissolved in water, indeed in the sap of the vine, and this, to remain dissolved, is both the obligation and the privilege of all substances that are destined (I was about to say 'wish') to change. And if then anyone really wanted to find out why a ring, and why a hexagon, and why soluble in water, well, he need not worry; these are among the not many questions to which our doctrine can reply with a persuasive discourse, accessible to everyone, but out of place here.

It has entered to form part of a molecule of glucose, just to speak plainly: a fate that is neither fish, flesh, nor fowl, which is intermediary, which prepares it for its first contact with the animal world but does not authorize it to take on a higher responsibility: that of becoming part of a proteic edifice. Hence it travels, at the slow pace of vegetal juices, from the leaf through the pedicel and by the shoot to the trunk, and from here descends to the almost ripe bunch of grapes. What then follows is the province of the winemakers: we are only interested in pinpointing the fact that it escaped (to our advantage, since we would not know how to put it in words) the alcoholic fermentation, and reached the wine without changing its nature.

It is the destiny of wine to be drunk, and it is the destiny of glucose to be oxidized. But it was not oxidized immediately: its drinker kept it in his liver for more than a week, well curled up and tranquil, as a reserve aliment for a sudden effort; an effort that he was forced to make the following Sunday, pursuing a bolting horse. Farewell to the hexagonal structure: in the space of a few instants the skein was unwound and became glucose again, and this was dragged by the bloodstream all the way to a minute muscle fiber in the thigh, and here brutally split into two molecules of lactic acid, the grim harbinger of fatigue: only later, some minutes after, the panting of the lungs was able to supply the oxygen necessary to quietly oxidize the latter. So a new molecule of carbon dioxide returned to the atmosphere, and a parcel of the energy that the sun had handed to the vine-shoot passed from the state of chemical energy to that of mechanical energy, and thereafter settled down in the slothful condition of heat, warming up imperceptibly the air moved by the running and the blood of the runner. 'Such is life,' although rarely is it described in this manner: an inserting itself, a drawing off to its advantage, a parasitizing of the downward course of energy, from its noble solar form to the degraded one of low-temperature heat. In this downward course, which leads to equilibrium and thus death, life draws a bend and nests in it.

Our atom is again carbon dioxide, for which we apologize: this too is an obligatory passage; one can imagine and invent others, but on earth that's the way it is. Once again the wind, which this time travels far; sails over the Apennines and the Adriatic, Greece, the Aegean, and Cyprus: we are over Lebanon, and the dance is repeated. The atom we are concerned with is now trapped in a structure that promises to last for a long time: it is the venerable trunk of a cedar, one of the last; it is passed again through the stages we have already described, and the glucose of which it is a part belongs, like the bead of a rosary, to a long chain of cellulose. This is no longer the hallucinatory and geological fixity of rock, this is no longer millions of years, but we can easily speak of centuries because the cedar is a tree of great longevity. It is our whim to abandon it for a year or five hundred years: let us say that after twenty years (we are in 1868) a wood worm has taken an interest in it. It has dug its tunnel between the trunk and the bark, with the obstinate and blind voracity of its race; as it drills it grows, and its tunnel grows with it. There it has swallowed and provided a setting for the subject of this story; then it has formed a pupa, and in the spring it has come out in the shape of an ugly gray moth which is now drying in the sun, confused and dazzled by the splendor of the day. Our atom is in one of the insect's thousand eyes, contributing to the summary and crude vision with which it orients itself in space. The insect is fecundated, lays its eggs, and dies: the small cadaver lies in the undergrowth of the woods, it is emptied of its fluids, but the chitin carapace resists for a long time, almost indestructible. The snow and sun return above it without injuring it: it is buried by the dead leaves and the loam, it has become a slough, a 'thing,' but the death of atoms, unlike ours, is never irrevocable. Here are at work the omnipresent, untiring, and invisible gravediggers of the undergrowth, the micro-organisms of the humus. The carapace, with its eyes by now blind, has slowly disintegrated, and the ex-drinker, ex-cedar, ex-wood worm has once again taken wing.

We will let it fly three times around the world, until 1960, and in justification of so long an interval in respect to the human measure we will point out that it is, however, much shorter than the average: which, we understand, is two hundred years. Every two hundred years, every atom of carbon that is not congealed in materials by now stable (such as, precisely, limestone, or coal, or diamond, or certain plastics) enters and reenters the cycle of life, through the narrow door of photosynthesis. Do other doors exist? Yes, some syntheses created by man; they are a title of nobility for man-the-maker, but until now their quantitative importance is negligible. They are doors still much narrower than that of the vegetable greenery; knowingly or not, man has not tried until now to compete with nature on this terrain, that is, he has not striven to draw from the carbon dioxide in the air the carbon that is necessary to nourish him, clothe him, warm him, and for the hundred other more sophisticated needs of modern life. He has not done it because he has not needed to: he has found, and is still finding (but for how many more decades?) gigantic reserves of carbon already organicized, or at least reduced. Besides the vegetable and animal worlds, these reserves are constituted by deposits of coal and petroleum: but these too are the inheritance of photosynthetic activity carried out in distant epochs, so that one can well affirm that photosynthesis is not only the sole path by which carbon becomes living matter, but also the sole path by which the sun's energy becomes chemically usable.

It is possible to demonstrate that this completely arbitrary story is nevertheless true. I could tell innumerable other stories, and they would all be true: all literally true, in the nature of the transitions, in their order and data. The number of atoms is so great that one could always be found whose story coincides with any capriciously invented story. I could recount an endless number of stories about carbon atoms that become colors or perfumes in flowers; of others which, from tiny algae to small crustaceans to fish, gradually return as carbon dioxide to the waters of the sea, in a perpetual, frightening round-dance of life and death, in which every devourer is immediately devoured; of others which instead attain a decorous semi-eternity in the yellowed pages of some archival document, or the canvas of a famous painter; or those to which fell the privilege of forming part of a grain of pollen and left their fossil imprint in the rocks for our curiosity; of others still that descended to become part of the mysterious shape-messengers of the human seed, and participated in the subtle process of division, duplication, and fusion from which each of us is born. Instead, I will tell just one more story, the most secret, and I will tell it with the humility and restraint of him who knows from the start that his theme is desperate, his means feeble, and the trade of clothing facts in words is bound by its very nature to fail.

It is again among us, in a glass of milk. It is inserted in a very complex, long chain, yet such that almost all of its links are acceptable to the human body. It is swallowed; and since every living structure harbors a savage distrust toward every contribution of any material of living origin, the chain is meticulously broken apart and the fragments, one by one, are accepted or rejected. One, the one that concerns us, crosses the intestinal threshold and enters the bloodstream: it migrates, knocks at the door of a nerve cell, enters, and supplants the carbon which was part of it. This cell belongs to a brain, and it is my brain, the brain of the *me* who is writing; and the cell in question, and within it the atom in question, is in charge of my writing, in a gigantic minuscule game which nobody has yet described. It is that which at this instant, issuing out of a labyrinthine tangle of yeses and nos, makes my hand run along a certain path on the paper, mark it with these volutes that are signs: a double snap, up and down, between two levels of energy, guides this hand of mine to impress on the paper this dot, here, this one.

Source: Primo Levi, *The Periodic Table*, translated from the Italian by Raymond Rosenthal, London, Abacus, 1986 (Sphere Books).

Tides

Arguably the most important book published this century was *Silent Spring* (1962), which prompted governments in many countries to restrict the use of pesticides. Its author, Rachel Carson (1907–64) was born in Springdale, Pennsylvania, and studied biology at Pennsylvania College for Women. In 1936 she joined the staff of the United States Bureau of Fisheries as a marine biologist. She had always wanted to be a writer, and she won international fame with *The Sea Around Us* (1961) which was translated into 30 languages. These extracts are from it.

The tides are a response of the mobile waters of the ocean to the pull of the moon and the more distant sun. In theory, there is a gravitational attraction between every drop of sea water and even the outermost star of the universe. In practice, however, the pull of the remote stars is so slight as to be obliterated in the vaster movements by which the ocean yields to the moon and the sun. Anyone who has lived near tide water knows that the moon, far more than the sun, controls the tides. He has noticed that, just as the moon rises later each day by fifty minutes, on the average, than the day before, so, in most places, the time of high tide is correspondingly later each day. And as the moon waxes and wanes in its monthly cycle, so the height of the tide varies. Twice each month, when the moon is a mere thread of silver in the sky, and again when it is full, we have the strongest tidal movements – the highest flood tides and the lowest ebb tides of the lunar month. These are called the spring tides. At these times sun, moon, and earth are directly in line and the pull of the two heavenly bodies is added together to bring the water high on the beaches, and send its surf leaping upward against the sea cliffs, and draw a brimming tide into the harbours so that the boats float high beside their wharfs. And twice each month, at the quarters of the moon, when sun, moon, and earth lie at the apexes of a triangle, and the pull of sun and moon are opposed, we have the moderate tidal movements called the neap tides.

Then the difference between high and low water is less than at any other time during the month.

That the sun, with a mass 27 million times that of the moon, should have less influence over the tides than a small satellite of the earth is at first surprising. But in the mechanics of the universe, nearness counts for more than distant mass, and when all the mathematical calculations have been made we find that the moon's power over the tides is more than twice that of the sun . . .

If the history of the earth's tides should one day be written by some observer of the universe, it would no doubt be said that they reached their greatest grandeur and power in the younger days of Earth, and that they slowly grew feebler and less imposing until one day they ceased to be. For the tides were not always as they are today, and as with all that is earthly, their days are numbered.

In the days when the earth was young, the coming in of the tide must have been a stupendous event. If the moon was formed by the tearing away of a part of the outer crust of the earth, it must have remained for a time very close to its parent. Its present position is the consequence of being pushed farther and farther away from the earth for some 2 billion years. When it was half its present distance from the earth, its power over the ocean tides was eight times as great as now, and the tidal range may even have been several hundred feet on certain shores. But when the earth was only a few million years old, assuming that the deep ocean basins were then formed, the sweep of the tides must have been beyond all comprehension. Twice each day, the fury of the incoming waters would inundate all the margins of the continents. The range of the surf must have been enormously extended by the reach of the tides, so that the waves would batter the crests of high cliffs and sweep inland to erode the continents. The fury of such tides would contribute not a little to the general bleakness and grimness and uninhabitability of the young earth.

Under such conditions, no living thing could exist on the shores or pass beyond them, and, had conditions not changed, it is reasonable to suppose that life would have evolved no further than the fishes. But over the millions of years the moon has receded, driven away by the friction of the tides it creates. The very movement of the water over the bed of the ocean, over the shallow edges of the continents, and over the inland seas carries within itself the power that is slowly destroying the tides, for tidal friction is gradually slowing down the rotation of the earth. In those early days we have spoken of, it took the earth a much shorter time – perhaps only about four hours – to make a complete rotation on its axis. Since then, the spinning of the globe has been so greatly slowed that a rotation now requires, as everyone knows, about 24 hours. This retarding will continue according to mathematicians, until the day is about 50 times as long as it is now.

And all the while the tidal friction will be exerting a second effect, pushing the moon farther away, just as it has already pushed it out more than 200,000 miles. (According to the laws of mechanics, as the rotation of the earth is retarded, that of the moon must be accelerated, and centrifugal force will carry it farther away.) As the moon recedes, it will, of course, have less power over the tides and they will grow weaker. It will also take the moon longer to complete its orbit around the earth. When finally the length of the day and of the month coincide, the moon will no longer rotate relatively to the earth, and there will be no lunar tides.

All this, of course, will require time on a scale the mind finds it difficult to conceive, and before it happens it is quite probable that the human race will have vanished from the earth. This may seem, then, like a Wellsian fantasy of a world so remote that we may dismiss it from our thoughts. But already, even in our allotted fraction of earthly time, we can see some of the effects of these cosmic processes. Our day is believed to be several seconds longer than that of Babylonian times. Britain's Astronomer Royal recently called the attention of the American Philosophical Society to the fact that the world will soon have to choose between two kinds of time. The tide-induced lengthening of the day has already complicated the problems of human systems of keeping time. Conventional clocks, geared to the earth's rotation, do not show the effect of the lengthening days. New atomic clocks now being constructed will show actual times and will differ from other clocks . . .

The influence of the tide over the affairs of sea creatures as well as men may be seen all over the world. The billions upon billions of sessile animals, like oysters, mussels, and barnacles, owe their very existence to the sweep of the tides, which brings them the food which they are unable to go in search of. By marvellous adaptations of form and structure, the inhabitants of the world between the tide lines are enabled to live in a zone where the danger of being dried up is matched against the danger of being washed away, where for every enemy that comes by sea there is another that comes by land, and where the most delicate of living tissues must somehow withstand the assault of storm waves that have the power to shift tons of rock or to crack the hardest granite.

The most curious and incredibly delicate adaptations, however, are the ones by which the breeding rhythm of certain marine animals is timed to coincide with the phases of the moon and the stages of the tide. In Europe it has been well-established that the spawning activities of oysters reach their peak on the spring tides, which are about two days after the full or the new moon. In the waters of northern Africa there is a sea urchin that, on the nights when the moon is full and apparently only then, releases its reproductive cells into the sea. And in tropical waters in many parts of the world there are small marine worms whose spawning behaviour is so precisely adjusted to the tidal calendar that, merely from observing them, one could tell the month, the day, and often the time of day as well.

Near Samoa in the Pacific, the palolo worm lives out its life on the bottom of the shallow sea, in holes in the rocks and among the masses of corals. Twice each year, during the neap tides of the moon's last quarter in October and November, the worms forsake their burrows and rise to the surface in swarms that cover the water. For this purpose, each worm has literally broken its body in two, half to remain in its rocky tunnel, half to carry the reproductive products to the surface and there to liberate the cells. This happens at dawn on the day before the moon reaches its last quarter, and again on the following day; on the second day of the spawning the quantity of eggs liberated is so great that the sea is discoloured.

The Fijians, whose waters have a similar worm, call them 'Mbalolo' and have designated the periods of their spawning 'Mbalola lailai' (little) for October and 'Mbalolo levu' (large) for November. Similar worms near the Gilbert Islands respond to certain phases of the moon in June and July; in the Malay Archipelago a related worm swarms at the surface on the second and third nights after the full moon of March and April, when the tides are running highest. A Japanese palolo swarms after the new moon and again after the full moon in October and November.

Concerning each of these, the question recurs but remains unanswered: is it the state of the tides that in some unknown way supplies the impulse from which springs this behaviour, or is it, even more mysteriously, some other influence of the moon? It is easier to imagine that it is the press and the rhythmic movement of the water that in some way brings about this response. But why is it only certain tides of the year, and why for some species is it the fullest tides of the month and for others the least movement of the waters that are related to the perpetuation of the race? At present, no one can answer.

No other creature displays so exquisite an adaptation to the tidal rhythm as the grunion – a small, shimmering fish about as long as a man's hand. Through no one can say what processes of adaptation, extending over no one knows how many millennia, the grunion has come to know not only the daily rhythm of the tides, but the monthly cycle by which certain tides sweep higher on the beaches than others. It has so adapted its spawning habits to the tidal cycle that the very existence of the race depends on the precision of this adjustment.

Shortly after the full moon of the months from March to August the grunion appear in the surf on the beaches of California. The tide reaches flood stage, slackens, hesitates, and begins to ebb. Now on these waves of the ebbing tide the fish begin to come in. Their bodies shimmer in the light of the moon as they are borne up the beach on the crest of a wave, they lie glittering on the wet sand for a perceptible moment of time, then fling themselves into the wash of the next wave and are carried back to sea. For about an hour after the turn of the tide this continues, thousands upon thousands of grunion coming up onto the beach, leaving the water, returning to it. This is the spawning act of the species.

During the brief interval between successive waves, the male and female have come together in the wet sand, the one to shed her eggs, the other to fertilize them. When the parent fish return to the water, they have left behind a mass of eggs buried in the sand. Succeeding waves on that night do not wash out the eggs because the tide is already ebbing. The waves of the next high tide will not reach them, because for a time after the full of the moon each tide will halt its advance a little lower on the beach than the preceding one. The eggs, then, will be undisturbed for at least a fortnight. In the warm, damp, incubating sand they undergo their development. Within two weeks the magic change from fertilized egg to larval fishlet is completed, the perfectly formed little grunion still confined within the membranes of the egg, still buried in the sand waiting for release. With the tides of the new moon it comes. Their waves wash over the places where the little masses of the grunion eggs were buried, the swirl and rush of the surf stirring the sand deeply. As the sand is washed away, and the eggs feel the touch of the cool sea water, the membranes rupture, the fishlets hatch, and the waves that released them bear them away to the sea.

But the link between tide and living creature I like best to remember is that of a very small worm, flat of body, with no distinction of appearance, but with one unforgettable quality. The name of this worm is *Convoluta roscoffensis*, and it lives on the sandy beaches of northern Brittany and the Channel Islands. Convoluta has entered into a remarkable partnership with a green alga, whose cells inhabit the body of the worm and lend to its tissues their own green colour. The worm lives entirely on the starchy products manufactured by its plant guest, having become so completely dependent upon this means of nutrition that its digestive organs have degenerated. In order that the algal cells may carry on their function of photosynthesis (which is dependent upon sunlight) Convoluta rises from the damp sands of the intertidal zone as soon as the tide has ebbed, the sand becoming spotted with large green patches composed of thousands of the worms. For several hours while the tide is out, the worms lie thus in the sun, and the plants manufacture their starches and sugars; but when the tide returns, the worms must again sink into the sand to avoid being washed away, out into deep water. So the whole lifetime of the worm is a succession of movements conditioned by the stages of the tide – upward into sunshine on the ebb, downward on the flood.

What I find most unforgettable about Convoluta is this: sometimes it happens that a marine biologist, wishing to study some related problem, will transfer a whole colony of the worms into the laboratory, there to establish them in an aquarium, where there are no tides. But twice each day Convoluta rises out of the sand on the bottom of the aquarium, into the light of the sun. And twice each day it sinks again into the sand. Without a brain, or what we would call a memory, or even any very clear perception, Convoluta continues to live out its life in this alien place, remembering, in every fibre of its small green body, the tidal rhythm of the distant sea.

Source: Rachel Carson, *The Sea Around Us*, 1951, reprinted in Rachel Carson, *The Sea*, with an Introduction by Brian Vesey-Fitzgerald, London, Toronto, Sydney, New York, Hart Davis, MacGibbon, Granada Publishing, 1976.

The Hot, Mobile Earth

Since the 1960s the study of plate tectonics has transformed the way scientists think about the earth. Charles Officer, Research Professor in Earth Sciences at Dartmouth College, and Jake Page, former editor of *Natural History* and *Smithsonian* magazines, give a vivid account of this development in their book *Tales of the Earth: Paroxysms and Perturbations of the Blue Planet* (1993).

Only in the last thirty years have we discarded the prevailing wisdom that the continents and ocean basins have always been in the same geographical locations. This was a comforting model, in a sense; it just happened to be entirely wrong. The competing hypothesis of 'continental drift' – or 'plate tectonics', as it is now known – has permitted us a far greater understanding of the nature of such things as volcanoes. Understanding them does nothing to tame them, of course, but it does serve to make the Earth appear a bit less whimsical in its outbursts.

In plate tectonic theory, the Earth's outer surface, or crust, is considered to be divided into a number of plates, which move horizontally at rates of a fraction of an inch to a few inches per year. New plate material is formed at their originating ends and old plate material is subducted back into the Earth at their trailing ends. The new plate material consists of molten magma which has been brought up from depth, particularly along the mid-ocean volcanic ridge system. The old plate material is carried back down into the mantle of the Earth, principally along the major earthquake zones surrounding the Pacific Ocean. The plates themselves move as rigid slabs over the viscous and underlying mantle and are considered to be driven by thermal convection currents in the mantle.

That the continents may have drifted about on the Earth's surface is an idea often attributed to Francis Bacon, essayist, lord chancellor to James I, candidate for those who don't believe Shakespeare wrote his own plays, and later subject of a bribery conviction. While computer analyses and other studies have shown that the author of Shakespeare's plays was almost certainly a man named William Shakespeare, a consultation of Bacon's own writings shows that while he did note in 1620 the obvious similarities of the continental outlines on either side of the Atlantic Ocean, he did not suggest that at one time they might have formed a unified land mass. That possibility was espoused in *The Origin of Continents and Oceans*, an elegant book by German meteorologist Alfred Wegener, first published in German in 1912 and translated into English in 1915. The idea was dismissed by most Earth scientists as inept and unscientific. After all, it challenged the very fundament of geologic thinking and, if accepted, would have called for a wholesale rethinking of how the Earth works.

Wegener was present at a meeting in 1926 of the American Association of Petroleum Geologists during which a symposium was held on the subject. Or perhaps it might be more aptly called an ambush. A professor from the University of Chicago commented that geologists might well ask if theirs could still be regarded as a science when it is 'possible for such a theory as this to run wild'. Another from Johns Hopkins University commented on Wegener's methodology: 'It is not scientific but takes the familiar course of an initial idea, a selective search through the literature for corroborative evidence, ignoring most of the facts that are opposed to the idea, and ending in a state of auto-intoxication in which the subjective idea comes to be considered as an objective fact.' In these words, not only Wegener's hypothesis but Wegener himself was under attack. Contrary to a common perception of scientists as dignified, objective investigators, they often play hardball with subjective zeal – especially when their basic premises are challenged.

Widely considered a pseudoscientific notion, the matter of continental drift rested until the 1960s. A sudden change in attitude toward the matter is generally attributed to the publication in 1963 of a scientific article by Fred Vine and Drum Matthews of Cambridge University. It was well known by then that the Earth acts like a great magnet that switches its magnetic polarity through geologic time. Sometimes, in essence, the North Pole becomes the South Pole, and vice versa. And as molten rock cools and hardens, magnetic particles in the lava are 'frozen' like little compass needles in the hard rock, their direction depending on the state of the Earth's magnetic polarity at the time. Vine and Matthews took note of the alternating positive and negative magnetic 'stripes' in the rock, which parallel the great ridges that occur on the mid-Atlantic sea floor, and suggested that they could best be explained if the sea floor itself were spreading out – moving away from the ridges. The particular magnetic signature would be picked up as the molten lava cooled, the signatures alternating in stripes as the Earth's polarity switched back and forth through the geologic ages. Thus, however dimly perceived, there was now a mechanism by which the continents might indeed have spread. Evidence in favor of this hypothesis began to cascade, and, in spite of a few serious contrarians and the waggish carping of a group that called itself the Stop Continental Drift Society, continental drift is now accepted as the explanation for the present configuration of the continents and is considered a major feature of the Earth's continuing metabolism . . .

Volcanism is directly associated with the mid-ocean ridges, where molten material fills the gaps that occur as the sea floor spreads. Such spreading occurs as two 'plates' move in opposite directions from the ridge. The ridges are thought to be fed, either directly or indirectly, with molten material that comes up as giant plumes from a great depth. Called *mantle plumes*, they are presumed to originate near the boundary between the lower mantle and the liquid core of the Earth, about halfway to the center of the Earth. The rising molten material in these plumes is basic in composition (as opposed to acidic), dominated by heavy minerals, and enriched in sulfur dioxide, which, along with the contained carbon dioxide, chlorine, and water, is vaporized when the molten magma erupts at the Earth's surface. These latter components of the magma are called its volatile constituents.

It stands to reason that if two plates are moving away from each other in one place, they will be crashing against something else at the other, leading end, and this is what happens. In many cases, where two plates collide, there is a subduction zone, where one or both of them descend back into the mantle. When the edge of a plate is subducted to a sufficient depth, its material reaches temperatures high enough to bring about at least partial melting, which in turn produces chambers of magma that tend to rise up. Volcanism then occurs at the Earth's surface. Volcanoes that occur in subduction zones (such as Indonesia) typically spew out more acid debris composed of lighter materials from the subsumed and overlying crustal materials. One such group of volcanoes circumscribing the Pacific Ocean is known as the Ring of Fire.

One of the mantle plumes that feeds the mid-Atlantic ridge rises in the North Atlantic directly under the island of Iceland, which can be thought of as the child of this plume, and is one of the several visible parts of the ridge, others being the Azores, Saint Peter and Saint Paul Rocks, and Tristan da Cunha. On Iceland, a volcano called Laki erupted with a gigantic lava flow in June 1783, and the eruption continued for eight months, a dramatic example of volcanic pollution. An enormous amount of sulfur dioxide was ejected into the atmosphere, returning to Earth as an estimated 100 million tons of sulfuric acid rain. This is about the same amount of acid rain attributable to human causes that today falls on the Earth in an entire year.

Happily for modern scientists, there is a record of this and other such events, a record as precise as the annual growth rings in trees but going further back than any living tree. In the more northerly latitudes of Greenland, snow and ice deposition increases in layers year by year, and the effects of a great variety of unusual atmospheric events become trapped in these layers, accessible by means of ice cores. Peaks of high acidity from the sulfur dioxide aerosols that have settled back to Earth in the ice cores have now been correlated with all known volcanic eruptions, and the Laki eruption in 1783 created higher-acidity peaks than any other volcanic eruption in the past thousand years. . . .

While it has been determined that so far as earthquake (and tornado) damage are concerned, the safest place to live in the United States is near a tiny town called Crossroads in south-eastern New Mexico, earthquakes can occur virtually anywhere. Their geographic distribution is generally categorized in the terms of plate tectonics. Thus we have either *interplate* or *intraplate* earthquakes.

Most quakes are of the *interplate* variety, occurring along the boundaries of the Earth's great plates where they grind against each other. The mechanisms of such quakes are fairly well understood in a general way. As one plate moves slowly past its neighbor, enormous strain builds up, not unlike the way in which strain builds up when you try to open a firmly closed jar. Eventually, the strain placed on the lid is enough, and it opens with a pop. Similarly, the strain built up by the plates eventually results in its quick release in the form of earthquake movement, and the plates return to a relatively unstrained state.

The process is more complicated than that, of course, depending on the type of boundary between plates, of which there are three kinds. At mid-ocean ridges, the plates form with the up-welling of magmatic material, and their lateral movement is away from the ridge in opposite directions. This is what is happening at the mid-Atlantic ridge, for example. On the other hand, there are places where two adjoining plates move horizontally relative to each other, usually at different velocities, along what are called *transform* faults (as is the case with the San Andreas fault and others in California). The third boundary type is when one plate is subducted under its neighbor and back into the deep interior of the Earth. Quakes along the mid-ocean ridges are relatively few in number and are usually small. Transform-fault quakes can be either small or large in magnitude. The subduction-zone quakes are the most numerous and are often among the largest; their focal zones (that is, where the energy is released) can extend to depths greater than 400 miles. In contrast, the focal zones at the other two boundary types are typically shallow.

The least-understood earthquakes are those that occur within a plate – the intraplate quakes. Though far less common than quakes that occur along the boundaries of plates, the intraplate quakes account for about half of the high-magnitude, shallow-depth earthquakes that rattle into human consciousness and wreak havoc on humankind's works. The Haicheng quake, the only one to have been so successfully predicted that the community could be evacuated, was an intraplate earthquake.

That the first such prediction took place in China seems fitting, since seismology as a science had its origins there with the development in AD 132 of the first instrument to record ground motion from earthquakes. Chinese concern with earthquakes is understandable, even at so early a time in history. These violent upheavals are common there and have taken a tremendous toll on the Middle Kingdom's huge population. The State Seismological Bureau of the People's Republic reported that in the thirty-seven years from 1949 to 1976, some 27 million people died and 76 million more were injured as a result of 100 earthquakes. If these figures are correct, the toll is nearly unimaginable. By comparison, the total number of Americans killed in the American Civil War has been estimated at 364,000; those in World War II, at 407,000. The greatest hazard to life in the United States, it is generally agreed, is the automobile, which accounts for 50,000 deaths each year (or in a 37-year period, by way of comparison 1,850,000) – not even a tenth of China's earthquake toll.

Source: Charles Officer and Jake Page, *Tales of the Earth: Paroxysms and Perturbations of the Blue Planet*, Oxford University Press, New York, 1993.

The Poet and the Surgeon

This was the title-poem of James Kirkup's first collection (1952). He has since become widely known as a poet, translator and travel-writer. In 1977 his poem about the love of a Roman centurion for Christ had the distinction of being the subject of the first prosecution for blasphemous libel for over 50 years. Philip Allison was, at the time of his death in 1972, Nuffield Professor of Surgery at Oxford. The operation Kirkup describes was a new development in the early 1950s. Designed to remedy the narrowing of the mitral valve of the heart, it entailed the surgeon putting his finger into the valve in order to split it.

A Correct Compassion

To Mr Philip Allison, after watching him perform a Mitral Stenosis Valvulotomy in the General Infirmary at Leeds.

Cleanly, sir, you went to the core of the matter,
using the purest kind of wit, a balance of belief and art,
You with a curious nervous elegance laid bare
The root of life, and put your finger on its beating heart.

The glistening theatre swarms with eyes, and hands, and eyes.
On green-clothed tables, ranks of instruments transmit a sterile gleam.
The masks are on, and no unnecessary smile betrays
A certain tension, true concomitant of calm.

Here we communicate by looks, though words,
Too, are used, as in continuous historic present
You describe our observations and your deeds.
All gesture is reduced to its result, an instrument.

She who does not know she is a patient lies
Within a tent of green, and sleeps without a sound
Beneath the lamps, and the reflectors that devise
Illuminations probing the profoundest wound.

A calligraphic master, improvising, you invent
The first incision, and no poet's hesitation
Before his snow-blank page mars your intent:
The flowing stroke is drawn like an uncalculated inspiration.

A garland of flowers unfurls across the painted flesh.
With quick precision the arterial forceps click.
Yellow threads are knotted with a simple flourish.
Transfused, the blood preserves its rose, though it is sick.

Meters record the blood, measure heart-beats, control the breath.
Hieratic gesture: scalpel bares a creamy rib; with pincer knives
The bone quietly is clipped, and lifted out. Beneath,
The pink, black-mottled lung like a revolted creature heaves,

Collapses; as if by extra fingers is neatly held aside
By two ordinary egg-beaters, kitchen tools that curve
Like extraordinary hands. Heart, laid bare, silently beats. It can hide
No longer yet is not revealed. – 'A local anaesthetic in the cardiac
nerve.'

Now, in firm hands that quiver with a careful strength,
Your knife feels through the heart's transparent skin; at first,
Inside the pericardium, slit down half its length,
The heart, black-veined, swells like a fruit about to burst,

But goes on beating, love's poignant image bleeding at the dart
Of a more grievous passion, as a bird, dreaming of flight, sleeps on
Within its leafy cage. – 'It generally upsets the heart
A bit, though not unduly, when I make the first injection'.

Still, still the patient sleeps, and still the speaking heart is dumb.
The watchers breathe an air far sweeter, rarer than the room's.
The cold walls listen. Each in his own blood hears the drum
She hears, tented in green, unfathomable calms.

'I make a purse-string suture here, with a reserve
Suture, which I must make first, and deeper,
As a safeguard, should the other burst. In the cardiac nerve
I inject again a local anaesthetic. Could we have fresh towels to cover

All these adventitious ones. Now can you all see.
When I put my finger inside the valve, there may be a lot

Of blood, and it may come with quite a bang. But I let it flow,
In case there are any clots, to give the heart a good clean-out.

Now can you give me every bit of light you've got.'
We stand on the benches, peering over his shoulder.
The lamp's intensest rays are concentrated on an inmost heart.
Someone coughs, 'If you have to cough, you will do it outside
 this theatre.' – 'Yes, sir.'

'How's she breathing, Doug.? Do you feel quite happy?' – 'Yes, fairly
Happy.' – 'Now. I am putting my finger in the opening of the valve.
I can only get the tip of my finger in. – It's gradually
Giving way. – I'm inside. – No clots. – I can feel the valve

Breathing freely now around my finger, and the heart working.
Not too much blood. It opened very nicely.
I should say that anatomically speaking
This is a perfect case. – Anatomically.

For of course, anatomy is not physiology.'
We find we breathe again, and hear the surgeon hum.
Outside, in the street, a car starts up. The heart regularly
Thunders. – 'I do not stitch up the pericardium.

It is not necessary.' – For this is imagination's other place,
Where only necessary things are done, with the supreme and grave
Dexterity that ignores technique; with proper grace
Informing a correct compassion, that performs its love, and makes it
 live.

A less reassuring poem, about an earlier stage in the history of surgery, is by the doctor-poet Dannie Abse:

In the Theatre

(A true incident)

Only a local anaesthetic was given because of the blood pressure problem. The patient, thus, was fully awake throughout the operation. But in those days – in 1918, in Cardiff, when I was Lambert Rogers' dresser – they could not locate a brain tumour with precision. Too much normal brain tissue was destroyed as the surgeon crudely searched for it, before he felt the resistance of it . . . all somewhat hit and miss. One operation I shall never forget . . .

Dr Wilfred Abse

Sister saying – 'Soon you'll be back in the ward,'
sister thinking – 'Only two more on the list,'
the patient saying – 'Thank you, I feel fine';
small voices, small lies, nothing untoward,
though, soon, he would blink again and again
because of the fingers of Lambert Rogers,
rash as a blind man's, inside his soft brain.

If items of horror can make a man laugh
then laugh at this: one hour later, the growth
still undiscovered, ticking its own wild time;
more brain mashed because of the probe's braille path;

Lambert Rogers desperate, fingering still;
his dresser thinking, 'Christ! Two more on the list,
a cisternal puncture and a neural cyst.'

Then, suddenly, the cracked record in the brain,
a ventriloquist voice that cried, 'You sod,
leave my soul alone, leave my soul alone,' –
the patient's dummy lips moving to that refrain,
the patient's eyes too wide. And, shocked,
Lambert Rogers drawing out the probe
with nurses, students, sister, petrified.

'Leave my soul alone, leave my soul alone,'
that voice so arctic and that cry so odd
had nowhere else to go – till the antique
gramophone wound down and the words began
to blur and slow, '. . . leave . . . my . . . soul . . . alone . . .'
to cease at last when something other died.
And silence matched the silence under snow.

Sources: James Kirkup, *A Correct Compassion and Other Poems*, London, Oxford University Press, 1952. Dannie Abse, *Collected Poems, 1948–76*, London, Hutchinson, 1977.

Enter Love and Enter Death

Joseph Wood Krutch (1893–1970) was both a distinguished literary scholar (he held Chairs of English and Drama at Columbia University) and the author of award-winning natural history books such as *The Desert Year* (1952) and *The Best of Two Worlds* (1953). The Volvox, the subject of this extract from *The Great Chain of Life* (1957), is a freshwater organism, visible to the naked eye as a green speck about the size of a pinhead. It is indeterminately both plant and animal: zoologists class it among protozoa, and botanists among algae. Its reproductive cells are differentiated from its other cells, a fact considered significant in tracing the evolution of higher animals from protozoa.

On the second of January 1700 Anthony van Leeuwenhoek [see p. 28], draper of Delft and self-taught Columbus of the littlest world, was writing to the Royal Society of London one of the many letters in which he described his voyages of discovery within a drop of water.

To William Dampier and other such rovers he left the exploration of the terrestrial globe. To another contemporary he left those equally adventurous voyages 'through strange seas of thought, alone' which took Newton across abysses of space to spheres much larger but not so little known as those to which Leeuwenhoek devoted his long life.

These worlds of his were not lifeless but teeming with life; and his discoveries, unlike those of Columbus, were discoveries in an absolute sense. He saw what no man, not merely no European man, had ever seen before. He had every right to be – he probably was – more amazed than Balboa.

The draper of Delft was already just short of seventy when he wrote:

> I had got the aforesaid water taken out of the ditches and runnels on the 30th of August: and on coming home, while I was busy looking at the multifarious very little animalcules a-swimming in this water, I saw floating in it, and seeming to move of themselves, a great many green round particles, of the bigness of sand-grains.

When I brought these little bodies before the microscope [actually a single very small lens which he had ground himself and fixed between two perforated metal plates] I saw that they were not simply round, but that their outermost membrane was everywhere beset with many little projecting particles, which seemed to me to be triangular, with the end tapering to a point: and it looked to me as if, in the whole circumference of that little ball, eight such particles were set, all orderly arranged and at equal distances from one another: so that upon so small a body there did stand a full two thousand of the said projecting particles.

This was for me a pleasant sight, because the little bodies aforesaid, how oft soever I looked upon them, never lay still; and because too their progression was brought about by a rolling motion . . .

Each of these little bodies had enclosed within it 5, 6, 7, nay some even 12, very little round globules, in structure like to the body itself wherein they were contained.

There is no mistaking the fact that what had just swum into Leeuwenhoek's ken was that very original and inventive organism, Volvox. Through hundreds of millions of years it had waited in countless places for man to become aware of its existence and, ultimately, to guess how important a step it had taken in the direction of both consciousness and that curiosity which was leading the Dutch draper to seek out in the ditches of Delft. The 'little projecting particles' are the peripheral cells which enclose the watery jelly in Volvox's interior. The '5, 6, 7, nay some even 12, very little round globules, in structure like to the body itself' are the vegetative 'daughter cells' produced by a sort of virgin birth between the sexual generations, much as in some of the higher plants 'offsets' as well as seeds are produced. Nor did Leeuwenhoek's observation stop there:

While I was keeping watch, for a good time on one of the biggest round bodies . . . I noticed that in its outermost part an opening appeared, out of which one of the enclosed round globules, having a fine green color, dropped out; and so one after another till they were all out, and each took on the same motion in the water as the body out of which it came. Afterwards, the first round body remained lying without any motion: and soon after a

> second globule, and presently a third, dropped out of it; and so one after another till they were all out and each took its proper motion.
>
> After the lapse of several days the first round body became as it were, again mingled with the water; for I could perceive no sign of it.

In other words Leeuwenhoek saw both the liberation of the daughter colonies and also, though he did not realize its importance, something even more remarkable. He saw Volvox yielding to one of its two remarkable inventions – natural (or inevitable) Death. The other half of the strange story eluded him completely. He did not know that after a few generations have been vegetatively reproduced by the process he observed there comes a generation that will produce eggs which must be fertilized by sperm before they can develop.

Nearly three hundred years later and more than five thousand miles from Delft, I, in my late turn, have also been looking at Volvox still rolling along in his gracefully expert way. Like most of the free-living protozoa, he has established himself pretty well over the whole earth outside the regions of eternal ice and there is no use speculating how the cosmopolitan distribution was achieved or how he got to America. During the vast stretches of time which have been his, routes were open at one time or another from every part of the earth to every other part.

Historical plant geographers come almost to blows over the question how, for instance, the sweet potato got to the South Sea islands. But Volvox's history goes back too far for even speculation or contention to reach. If only the fit survive and if the fitter they are the longer they survive, then Volvox must have demonstrated its superb fitness more conclusively than any higher animal ever has.

My equipment is as much superior to Leeuwenhoek's as his originality, ingenuity, and persistence were superior to mine. Instead of a single blob of glass fixed in front of a tiny tube holding water and held up to sun or candlelight, I use the compound microscope, which was not brought to its present state until the second half of the nineteenth century. Light passes from a mirror through a complicated set of lenses designed to place it at exactly the right spot. An image is formed by another series of lenses, cunningly designed to correct one another's faults and then form an image in a black tube, this image again magnified by another set of lenses. I can confine Volvox to a hanging drop of water; I can light him from below, from the side, or even from the top. I can slow him down with sticky substances introduced into the drop; and I keep a supply of his species thriving in an artificial culture medium prepared for me by a biological supply house dealing in all sorts of improbable things. But though the beauty of Volvox must be even clearer to me than it was to Leeuwenhoek I have seen only what he saw and described in unmistakable terms.

Under a magnification of no more than a hunded diameters – called by microscopists 'very low power' – Volvox looks about the size of a marble and when motionless less like a plant-animal than like some sort of jeweler's work intended, perhaps, as an earring. The surface of the crystal sphere is set with hundreds of tiny emeralds; its interior contains five or six larger emeralds disposed with careless effectiveness.

But Volvox is seldom motionless when alive and in good health. Bright as a jewel, intricate as a watch, and mobile as a butterfly, his revolutions bring one emerald after another into a position where they sparkle in the light. Though he is called the Roller [the English meaning of Latin, *Volvox*], he actually *revolves* rather than *rolls*, because he seems to turn on an invisible axis, much as the planets do, and he moves forward with this axis pointing in the direction of his motion. His speed varies and he frequently changes direction but I once counted the seconds it took him to cross the field of the microscope and calculated that his speed, in proportion to his size, is comparable to that of a man moving at a fast trot. Volvox, however, suggests nothing so undignified as a trot. There is something majestic and, one might almost imagine, irresistible about his revolutions – again like those of a planet. One half expects to hear some music of the spheres.

Because the microscope has temporarily abolished the barrier of size which separates the universe of Volvox from my own, I enter temporarily into a dreamlike relationship with him, though he is unaware of my world and perhaps equally unaware of his own. Nevertheless there is an easy purposefulness in his movements and from what I have learned from the careful research of others I know that it is all much more complex and astonishing than I would ever have guessed or than Leeuwenhoek did guess.

When I lift my head from the microscope the dream vanishes. But it is man and his consciousness which is really the fleeting dream. Volvox, or something very much like him, was leading his surprisingly complex life millions of years before man's dream began and may well continue to do so for millions of years after the dream ends. Harder to realize is the fact that the enterprise and adventures of Volvox typify certain of the innovations and inventions which are casually summed up in the word 'evolution' and hence constituted some of the earliest and most essential steps toward making possible our dream.

Such acquaintance with Volvox as I have gained from my own casual observations is as superficial as what one gets from a moment's examination of a flower or from peering with mild curiosity at some strange animal behind the bars of a zoo. Yet even so casual an examination will lead one to guess at some of the significant facts.

The predominant color of Volvox is green; the green looks like chlorophyl, and so it is. Moreover Volvox can be 'cultured' in a purely chemical solution. All of this suggests a plant rather than an animal, but biologists have decided that the classification is meaningless at this level. Both textbooks of zoology and textbooks of botany usually claim Volvox, and there are no hard feelings because zoologists and botanists agree that Volvox represents a stage of evolution at which plants have not diverged from animals. He is either the one or the other. Or, more properly, he is neither. He is not a plant or an animal; he is simply something which is alive.

Concerning another ambiguity the observer is likely to make an even better guess. The walls separating one cell from another are clearly marked out and each of the cells which lie upon the periphery of the sphere is exactly like the other peripheral cells. Every one of them looks like a complete one-celled creature, lashing its two flagella precisely as many such one-celled creatures do – although, as was mentioned a few pages back, it is easy to see that Volvox's cells do not crack their whips independently because they are co-ordinated in such a way that the organism as a whole moves purposefully forward in a given direction as though the individual flagella were controlled by a central intelligence.

Yet the individual cells not only look like separate animals but are in fact morphologically all but indistinguishable from a certain very common specific free-living organism that occurs by the millions in stagnant rain water and is one of the usual causes why such water takes on a green color. Like this creature each of Volvox's peripheral cells has a nucleus, two flagella, and an 'eye spot.' Any observer might easily suppose that a number of such tiny creatures had recently got together and formed a ball; or, that the one-celled creatures were the result of the dissolution of Volvoxes. But if the one-celled creatures did form aggregates which then became cooperative, all that happened millions of years ago and for a very long time Volvox has been much more than a mere casual grouping. One of his peripheral cells cannot live without the others. He is at least as much one creature as many and he must live or die as a whole. He can no longer dissolve into the myriad of individuals of which, long ago, his ancestors were no doubt composed.

If you want to call him a multicelled animal you have to admit that the peripheral cells have retained a good deal of their original equipment for independent life. On the other hand if you want to call him a mere aggregation you have to concede that the individuals composing the colony have been co-ordinated, disciplined, and socialized to a degree no human dictator has yet even hoped to achieve in his 'monolithic' state. But in any event the guess that Volvox represents a stage in the development of the 'higher' multicellular animals and suggests how the transition was made is more immediately persuasive than many of the other guesses biology finds itself compelled to make.

Without comment I pass over the suggestion made sometimes with horror and sometimes with approval that our present-day society is in the process of taking a step analogous to that once taken by Volvox; that just as the one-celled animal cooperated until he was no longer an individual but part of a multicelled body, so perhaps the highest of the multicelled animals is now in the process of uniting to make a society in which he will count for as much and as little as an individual cell counts for in the human body.

Now comes the most powerful argument of all for calling Volvox a unified individual rather than even a tight social group and it has to do with three different sorts of cells sometimes found within his central jelly. The least remarkable of the groups of special cells are those composing the 'daughter colonies' which Leeuwenhoek saw and which in time will break out of the parent colony to start life on their own. About them there is nothing so very surprising, since 'budding' of one kind or another is not uncommon among microscopic organisms. The other two groups of specialized cells are much more interesting because they seem to represent the first appearance of sexual differentiation.

One group is of spindle-shaped cells very much like the sperm of the higher animals. The other is a sort of egg considerably larger than the sperm because, like most eggs, it contains a rich store of reserve food to nourish a growing 'embryo.' Sooner or later a sperm cell will seek out an egg, the two will fuse, and thus they will pool the hereditary characteristics carried by the sperm and by the egg. Thus Volvox introduces a method of procedure almost universal among the higher plants and animals. Moreover, there is even the beginning of the distinction between Male and Female *individuals* as well as a distinction in the sex cells themselves, because, in the species which I have been observing, a given individual usually produces either eggs only or sperm only. The cynic who said that the two great errors in creation were the inclination of the earth's axis and the differentiation of the sexes probably had no idea how long ago the second error was made. To eliminate it we would not have to wipe the slate quite clean but we would have to go back at least as far as Volvox.

Being the inventor of sex would seem to be a sufficient distinction for a creature just barely large enough to be seen by the naked eye. But as we have already said, Volvox brought Natural Death as well as Sex into the world. The amoeba and the paramecium are potentially immortal. From time to time each divides itself into two, but in the course of this sort of reproduction no new individual is ever produced – only fragments of the original individuals, whose life has thus been continuous back to the time when life itself was first created. Though individuals can be killed there is no apparent reason why amoeba should ever die.

Individuals have been kept alive in laboratories for years by carefully isolating one-half of the organism after each division. What memories an amoeba would have, if it had any memories at all! How fascinating would be its firsthand account of what things were like in Protozoic or Paleozoic times! But for Volvox, death seems to be as inevitable as it is in a mouse or in a man. Volvox must die as Leeuwenhoek saw it die because it has had children and is no longer needed. When its time comes it drops quietly to the bottom and joins its ancestors. As Hegner, the Johns Hopkins zoologist, once wrote: 'This is the first advent of inevitable natural death in the animal kingdom and all for the sake of sex.' And as he asked: 'Is it worth it?'

Nature's answer during all the years which have intervened between the first Volvox and quite recent times has been a pretty steady *Yes* . . .

No detectable difference in structure exists between two conjugating protozoa. There are no males and no females. Neither are there any specific sex cells – no eggs and no sperm. It is in respect to these two facts that Volvox takes a great step toward sexuality as it is commonly known.

Volvox never indulges in the kind of conjugation we have been describing. Neither does it ever divide into two halves. Instead it produces what may properly be called 'children' – sometimes by the vegetative process which Leeuwenhoek described and sometimes by a more surprising method which no free-living, single-celled animal ever practices.

Somewhere inside its sphere appear certain groups of small cells which might at first sight be mistaken for vegetative buds. But they develop quite differently in either one of two ways. Sometimes they become quite large. Sometimes on the other hand they split up into a great number of extremely small mobile cells. The first are eggs; the second sperm. Neither is good for anything by itself. But each is ready to do what the egg and the sperm of all the higher animals do. The egg waits. The sperm seeks it out. Then the two fuse and the fertilized egg is endowed with the hereditary traits contributed by the sperm as well as with those originally its own. Some species of Volvox are hermaphroditic as many lower organisms are. A single individual, that is to say, produces both male and female cells. But Volvox is also inventing the sexual differentiation of the whole organism. Certain species are commonly either male or female; the one producing only eggs, the other only sperm.

In what ways, one is bound to wonder, is this differentiation of sex cells and the further differentiation of male and female organisms superior to the simpler arrangement which the protozoa have managed to get along with during millions of years?

It has been, as satirists have so frequently pointed out, the cause of a lot of trouble in the world. Yet there must be compensating advantages, because as one moves upward along the evolutionary scale sexuality becomes universal and even hermaphroditism tends to disappear.

That sexual differentiation provides a richer emotional experience is a reason that few biologists are likely to admit as relevant and indeed it would be hard to prove that Volvox finds life more colorful than a paramecium does. Hence the biologist has to fall back upon such things as the superior viability of an egg, which can be heavy with reserve food resources because it does not have to be active when a small mobile sperm is there to seek it out. Possibly another fact is even more important. In all the higher animals sperm and egg cells differ from every other cell in the body of that organism in that they have only half the normal number of chromosomes and that the normal number is re-established when the sperm's half-number is added to the egg's half-number – which arrangement certainly shuffles hereditary characteristics more thoroughly than when the offspring has the whole inheritance from both sides of the family. In any event (and to repeat) there must be some advantage, since every animal above the protozoan level tends to adopt the novel arrangement first observable in Volvox.

Source: Joseph Wood Krutch, *The Great Chain of Life*, London, Eyre & Spottiswoode, 1957.

In the Primeval Swamp

In *A Land* (1951), from which this extract is taken, Jacquetta Hawkes, the English archaeologist, evokes the geological shaping of Britain. Hugh Miller (1802–56), author of *The Old Red Sandstone* (1841), was at various times poet, journalist, bank-clerk and stone-mason. His work in quarries aroused his interest in geology, which he combined with devout religious faith, believing that each great geological age was a separate creation by God.

In his account of his first discovery of Devonian fishes in the Old Red Sandstone, Hugh Miller describes how he split open a calcareous nodule and found inside 'finely enamelled' fish scales. 'I wrought on with the eagerness of a discoverer entering for the first time a *terra incognita* of wonders. Almost every fragment of clay, every splinter of sandstone, every limestone nodule contained its organism – scales, spines, plates, bones, entire fish . . . I wrought on until the advancing tide came splashing over the nodules, and a powerful August sun had risen towards the middle of the sky; and were I to sum up all my happier hours, the hour would not be forgotten in which I sat down on a rounded boulder of granite by the edge of the sea and spread out on the beach before me the spoils of the morning.' This August day was in 1830. The young man's hammer had discovered the remains of the earliest fishes, the *Ostracoderms* whose leathery skins were armoured with plates and spines, and who, lacking a jaw, fed through a slit set below the pointed snout. The Devonian seas were full of these creatures.

Occasionally, when an inland sea dried up, there must have been a horrible flapping and floundering, a dull rattling of horny armour before they suffocated and the bodies of untellable shoals were buried, later to form a dense mass of fossilized remains.

Such happenings, however, were no more than local catastrophes, for elsewhere these vertebrates and their successors, so crucial in the evolution of species, throve and multiplied to such an extent that the

Devonian is sometimes called the Age of Fishes. By the middle of the period as well as the *Ostracoderms* (many would wish to withhold the name of fish from an animal that could not open its mouth) there were more developed fishes of many kinds, some of them already wearing scales. A few species such as *Dinichthys* grew to as much as twenty feet and had heavily armoured jaws as ruthless as a mechanical excavator. It is true that before them the eighteen-inch trilobites, the six-foot arachnids, had their relative power to tyrannize, but it seems that these great predatory vertebrates must have brought the first keen fear into the sea. Something akin to human emotion ran along those newly evolved spines when *Dinichthys* hurled himself among the helpless shoals.

Among the scaled fish one Devonian group seems to have held the secret of the future. These were the varieties that had paired fins and lungs enabling them, if stranded by seasonal drying, to shuffle back to the water. From them, so far as we know, is descended the whole train of the land vertebrates.

Already before the close of the Devonian Age, the land had taken the place of the seas as the stage on which the great scenes of evolution were to be played. Algæ and seaweed had already breathed out the free oxygen that made life on land possible. With this invisible atmospheric envelope of the earth ready to receive it, life came up from the sea. The lunged fish had given rise to true amphibians; all manner of insects, not yet able to fly, had crawled on to the land, and there were millipedes, mites, and spiders. The land that had always been silent and undisturbed began not only to be minutely stirred by small burrowings and by the growth of plants, but was marked by the impress of feet, even though between the footsteps went the groove of a scaly tail.

The country which the eyes of these amphibians saw sharply if vacuously was already green. With a virgin environment to exploit, the new land plants flourished amazingly. They were of those smooth, spiny and militant kinds we have come to associate with tropical conservatories, but already they had much in common with modern plants; sap flowed in them and they breathed through open pores. Indeed, by the end of the age the vegetation had developed far towards the luxuriance of the Carboniferous forests. There were the fountain-like tree ferns, and seed ferns carrying little nuts below their fronds; the big horsetails had a tree growth and there were even forerunners of true conifers. All these forms are extinct, yet they were so near to what has become familiar that I doubt whether the ordinary, unobservant passer-by would notice them if they could spring up again in hedgerow or wood.

In no geological scheme is the Devonian accepted as a major turning-point; it does not mark either the beginning or the end of one of the great eras. To me, in this effort of recollection, it appears to be one. However broken up and unrecognizable, some of the land that was to be Britain was clear of the sea and green with vegetation. The main masses of our mountains had been formed, and the Old Red Sandstone was ready to support heavy cornfields and cider orchards. To watch the close of a Devonian day would not have been the unimaginable experience of a few hundred million years earlier. As the shadows of the trees lengthened there would have been a clapping and harsh rustling of the big leaves on the river bank as clumsy animals pushed among them; if there was no bird-song or even the humming of insects at least there was that most characteristic evening sound, the occasional splash of fish in quiet water.

Perhaps more than any other, the age that followed was to reach through time and effect the face and fortune of the British Isles. This it was to do by creating a substance – coal – which at a certain moment in their historical evolution men sought as eagerly as food, so eagerly that they were ready to leave their habitat and become pale-skinned burrowing creatures, coming to the surface only at night. To move away from the pleasanter places and huddle their dwellings round the grimy entrance to their tunnels.

At first, with some spread of warm and shallow seas, limestone formed, the Carboniferous or Mountain Limestone that was to be built into some of the most solid and respectable piles in England, buttresses of its pride and self-confidence. The work of silting up these Carboniferous seas was completed by deposits brought from the northern continent of Atlantis, then hot, mountainous, and swept by monsoons. A large river with tributaries drawn from territories stretching from the north of Scotland to Norway poured out its coarse sediments across north-eastern England. So were Norwegian pebbles brought to Yorkshire and held in the Millstone Grits that were laid down as the deltas of this northern river. Silting, combined with the elevation of expanses of low-lying land and the influence of the warm rains of the southern monsoons, led to the formation of marsh and brackish swamps where the Coal Measure forest grew in sombre luxuriance.

It is sombre in these swamps, for the foliage is dark green and there are nowhere any flowers. Yet there is scent in the air. Here already is the rich aromatic breath of resins, a presage of the smell of pinewoods on summer days when pine cones crack in the sun. In many places the trees grow straight from the tepid water that carries a dull film where clouds of pollen have blown across it. Ferns feather the mud-banks and there are thickets of horsetails with the radiate whorls and neatly socketed stems of their diminished and weedy descendants. When, as a very small child, I was playing with a horsetail that had been growing as a weed in one of our flower-beds, dismantling it section by section like a constructional toy, I remember how my father told me it was one of the oldest plants on earth, and I experienced a curious confusion of time. I was holding the oldest plant in my hand, and so I, too, was old. Now huge horsetails are growing in the Carboniferous swamp while above them the fern trees with their sprouting leaves cut off most of such sunlight as has succeeded in straying through the still loftier canopy of the scale trees – the lycopods whose slender trunks are chequered like snake skin. Across the hundred-foot verticals of the growing scale trees are the diagonals of many that have fallen and lodged against their fellows, while others lie horizontal, already half-digested by the swamp. Here decay is active among growth, trees and ferns thrusting towards the summit of their life, while others are slowly reverting to inorganic forms.

Among these imperceptible rhythms of growth and decay are the quicker movements of the swamp creatures. There are shoals of fish in the pools and slow streams of the forest; vast beds of molluscs line the edges of the lagoon. Dragging their wide bellies across the mudbanks, sagging heavily back into the water, go amphibious monsters like grosser crocodiles. Over the streams and pools, through the oppressive greenish light, with a clittering of glassy wings, twist gigantic dragonflies, the largest insects the earth will ever know.

There is still no spring in these forests, for all the foliage is evergreen, no seasonal rise and fall but only, continuously, life going on beside decay. The toll of decay mounts with the centuries, the swamp lives above a tremendous accumulation of its own past, tree-trunks, leaves, and fronds, and scattered among them the broken bodies of the animal population – bones, empty shells, the wings of dragonflies.

The swamp itself mounts slowly, but meanwhile the whole platform of land is sinking until somewhere far away the sea breaks in, sea water invades this stagnant world, fishes choke, the amphibians, if they can, move away and the insects go – as insects do. For a time forlorn, ragged trunks of dead scale trees stick through the water. But they sink, the whole scene sinks and the particles of sediment begin to fall again burying all the dead stuff of the swamps and forests in layers of forgetfulness. It is a drowsy scene to contemplate, and sleep muffles me. I see *Loxomma*, the amphibian, his flesh fallen away to reveal the long column of his spine and the little bones of his hands and feet. The spine is lengthening, vertebra after vertebra, without end, and running through the vista of their bony arches there is a mounting current, a sense of the passage of some energy and power. The vista of arches – I see now that it is a tunnel and that there are living creatures crawling along it, each with a single eye shining in its head. I am stupid, they are only lamps, and the roaring in my ears is nothing but a drill, one of those confounded drills. 'Christ, look at the old blighter,' some one says, and I notice that *Loxomma* is there again (perhaps he had never gone) and they have excavated him with their drill. 'Makes your spine creep a bit, don't it? Christ, look at that hand . . .'

Source: Jacquetta Hawkes, *A Land*, London, Cresset Press, 1951.

Krakatau: The Aftermath

The return of life to Krakatau has been of absorbing interest to biogeography, the branch of science pioneered by Alfred Russel Wallace (see p. 246). This account is by the distinguished writer and biologist Edward O. Wilson (b. 1929), whose study of the behaviour of competing animal populations, including human populations, effectively founded the subject of sociobiology. Wilson's books include *Sociobiology: the New Synthesis*, the Pulitzer Prizewinning *The Ants* (with Bert Holldobler), an autobiography, *Naturalist*, and *The Diversity of Life*, from which this extract is taken.

Krakatau, earlier misnamed Krakatoa, an island the size of Manhattan located midway in the Sunda Strait between Sumatra and Java, came to an end on Monday morning, August 27, 1883. It was dismembered by a series of powerful volcanic eruptions. The most violent occurred at 10:02 a.m., blowing upward like the shaped explosion of a large nuclear bomb, with an estimated force equivalent to 100–150 megatons of TNT. The airwave it created traveled at the speed of sound around the world, reaching the opposite end of the earth near Bogotá, Colombia, nineteen hours later, whereupon it bounced back to Krakatau and then back and forth for seven recorded passages over the earth's surface. The audible sounds, resembling the distant cannonade of a ship in distress, carried southward across Australia to Perth, northward to Singapore, and westward 4,600 kilometers to Rodriguez Island in the Indian Ocean, the longest distance traveled by any airborne sound in recorded history.

As the island collapsed into the subterranean chamber emptied by the eruption, the sea rushed in to fill the newly formed caldera. A column of magma, rock, and ash rose 5 kilometers, into the air, then fell earthward, thrusting the sea outward in a tsunami 40 meters in height. The great tidal waves, resembling black hills when first sighted on the horizon, fell upon the shores of Java and Sumatra, washing away entire towns and killing 40,000 people. The segments traversing the channels and reaching the open sea continued on as spreading waves around the world. The waves were still a meter high when they came ashore in Ceylon, now Sri Lanka, where they drowned one person, their last casualty. Thirty-two hours after the explosion, they rolled in to Le Havre, France, reduced at last to centimeter-high swells.

The eruptions lifted more than 18 cubic kilometers of rock and other material into the air. Most of this tephra, as it is called by geologists, quickly rained back down onto the surface, but a residue of sulfuric-acid aerosol and dust boiled upward as high as 50 kilometers and diffused through the stratosphere around the world, where for several years it created brilliant red sunsets and 'Bishop's rings,' opalescent coronas surrounding the sun . . .

In the following weeks, the Sunda Strait returned to outward normality, but with an altered geography. The center of Krakatau had been replaced by an undersea crater 7 kilometers long and 270 metres deep. Only a remnant at the southern end still rose from the sea. It was covered by a layer of obsidian-laced pumice 40 meters or more thick and heated to somewhere between 300° and 850°C, enough at the upper range to melt lead. All traces of life had, of course, been extinguished.

Rakata, the ash-covered mountain of old Krakatau, survived as a sterile island. But life quickly enveloped it again. In a sense, the spinning reel of biological history halted, then reversed, like a motion picture run backward, as living organisms began to return to Rakata. Biologists quickly grasped the unique opportunity that Rakata afforded: to watch the assembly of a tropical ecosystem from the very beginning. Would the organisms be different from those that had existed before? Would a rain forest eventually cover the island again?

The first search for life on Rakata was conducted by a French expedition in May 1884, nine months after the explosions. The main cliff was eroding rapidly, and rocks still rolled down the sides incessantly, stirring clouds of dust and emitting a continuous noise 'like the rattling of distant musketry.' Some of the stones whirled through the air, ricocheting down the sides of the ravines and splashing into the sea. What appeared to be mist in the distance turned close up into clouds of dust stirred by the falling debris. The crew and expedition members eventually found a safe landing site and fanned out to learn what they could. After searching for organisms in particular, the ship's naturalist wrote that 'notwithstanding all my researches, I was not able to observe any symptom of animal life. I only discovered one microscopic spider – only one; this strange pioneer of the renovation was busy spinning its web.'

A baby spider? How could a tiny wingless creature reach the empty island so quickly? Arachnologists know that a majority of species 'balloon' at some point in their life cycle. The spider stands on the edge of a leaf or some other exposed spot and lets out a thread of silk from the spinnerets at the posterior tip of its abdomen. As the strand grows it catches an air current and stretches downwind, like the string of a kite. The spider spins more and more of the silk until the thread exerts a strong pull on its body. Then it releases its grip on the surface and soars upward. Not just pinhead-sized babies but large spiders can occasionally reach thousands of meters of altitude and travel hundreds of kilometers before settling to the ground to start a new life. Either that or land on the water and die. The voyagers have no control over their own descent.

Ballooning spiders are members of what ecologists, with the accidental felicity that sometimes pops out of Greek and Latin sources, have delightfully called the aeolian plankton. In ordinary parlance, plankton is the vast swarm of algae and small animals carried passively by water currents; aeolian refers to the wind. The creatures composing the aeolian plankton are devoted almost entirely to long-distance dispersal. You can see some of it forming over lawns and bushes on a quiet summer afternoon, as aphids use their feeble wings to rise just high enough to catch the wind and be carried away. A rain of planktonic bacteria, fungus spores, small seeds, insects, spiders, and other small creatures falls continuously on most parts of the earth's land surface. It is sparse and hard to detect moment by moment, but it mounts up to large numbers over a period of weeks and months. This is how most of the species colonized the seared and smothered remnant of Krakatau.

The potential of the planktonic invasion has been documented by Ian Thornton and a team of Australian and Indonesian biologists who visited the Krakatau area in the 1980s. While studying Rakata they also visited Anak Krakatau ('Child of Krakatau'), a small island that emerged in 1930 from volcanic activity along the submerged northern rim of the old Krakatau caldera. On its ash-covered lava flows they placed traps made from white plastic containers filled with seawater. This part of the surface of Anak Krakatau dated from localized volcanic activity from 1960 to 1981 and was nearly sterile, resembling the condition on Rakata soon after the larger island's violent formation. During ten days the traps caught a surprising variety of windborne arthropods. The specimens collected, sorted, and identified included a total of 72 species of spiders, springtails, crickets, earwigs, barklice, hemipterous bugs, moths, flies, beetles, and wasps.

There are other ways to cross the water gaps separating Rakata from nearby islands and the Javan and Sumatran coasts. The large semiaquatic monitor lizard *Varanus salvator* probably swam over. It was present no later than 1899, feasting on the crabs that crawl along the shore. Another long-distance swimmer was the reticulated python, a giant snake reaching up to 8 meters in length. Probably all of the birds crossed over by powered flight. But only a small percentage of the species of Java and Sumatra were represented because it is a fact, curiously, that many forest species refuse to cross water gaps even when the nearest island is in full view. Bats, straying off course, made the Rakata landfall. Winged insects of larger size, especially butterflies and dragonflies, probably also traveled under their own power. Under similar conditions in the Florida Keys, I have watched such insects fly easily from one small island to another, as though they were moving about over meadows instead of salt water.

Rafting is a much less common but still important means of transport. Logs, branches, sometimes entire trees fall into rivers and bays and are carried out to sea, complete with microorganisms, insects, snakes, frogs, and occasional rodents and other small mammals living on them at the moment of departure. Blocks of pumice from old volcanic islands, riddled with enough closed air spaces to keep them afloat, also serve as rafts.

Once in a great while a violent storm turns larger animals such as lizards or frogs into aeolian debris, tearing them loose from their perches and propelling them to distant shores. Waterspouts pick up fish and transport them live to nearby lakes and streams.

Swelling the migration further, organisms carry other organisms with them. Most animals are miniature arks laden with parasites. They also transport accidental hitchhikers in soil clinging to the skin, including bacteria and protozoans of immense variety, fungal spores, nematode worms, tardigrades, mites, and feather lice. Seeds of some species of herbs and trees pass live through the guts of birds, to be deposited later in feces, which serves as instant fertilizer. A few arthropods practice what biologists call phoresy, deliberate hitch-hiking on larger animals. Pseudoscorpions, tiny replicas of true scorpions but lacking stings, use their lobster-like claws to seize the hairs of dragonflies and other large winged insects, then ride these magic carpets for long distances.

The colonists poured relentlessly into Rakata from all directions. A 100-meter-high electrified fence encircling the island could not have stopped them. Airborne organisms would still have tumbled in from above to spawn a rich ecosystem. But the largely happenstance nature of colonization means that flora and fauna did not return to Rakata in a smooth textbook manner, with plants growing to sylvan thickness, then herbivores proliferating, and finally carnivores prowling. The surveys made on Rakata and later on Anak Krakatau disclosed a far more haphazard buildup, with some species inexplicably going extinct and others flourishing when seemingly they should have quickly disappeared. Spiders and flightless carnivorous crickets persisted almost miraculously on bare pumice fields; they fed on a thin diet of insects landing in the aeolian debris. Large lizards and some of the birds lived on beach crabs, which subsisted in turn on dead marine plants and animals washed ashore by waves. (The original name of Krakatau was Karkata, or Sanskrit for 'crab'; Rakata also means crab in the old Javanese language.) Thus animal diversity was not wholly dependent on vegetation. And for its part vegetation grew up in patches, alternately spreading and retreating across the island to create an irregular mosaic.

If the fauna and flora came back chaotically, they also came back fast. In the fall of 1884, a little more than a year after the eruption, biologists encountered a few shoots of grass, probably *Imperata* and *Saccharum*. In 1886 there were fifteen species of grasses and shrubs, in 1897 forty-nine, and in 1928 nearly three hundred. Vegetation dominated by *Ipomoea* spread along the shores. At the same time grassland dotted with *Casuarina* pines gave way here and there to richer pioneer stands of trees and shrubs. In 1919 W. M. Docters van Leeuwen, from the Botanical Gardens at Buitenzorg, found forest patches surrounded by nearly continuous grassland. Ten years later he found the reverse: forest now clothed the entire island and was choking out the last of the grassland patches. Today Rakata is covered completely by tropical Asian rain forest typical in outward appearance. Yet the process of colonization is far from complete. Not a single tree species characterizing the deep, primary forests on Java and Sumatra has made it back. Another hundred years or more may be needed for investment by a forest fully comparable to that of old, undisturbed Indonesian islands of the same size.

Some insects, spiders, and vertebrates aside, the earliest colonists of most kinds of animals died on Rakata soon after arrival. But as the vegetation expanded and the forest matured, increasing numbers of species took hold. At the time of the Thornton expeditions of 1984–85, the inhabitants included thirty species of land birds, nine bats, the Indonesian field rat, the ubiquitous black rat, and nine reptiles, including two geckos and *Varanus salvator*, the monitor lizard. The reticulated python, recorded as recently as 1933, was not present in 1984–85. A large host of invertebrate species, more than six hundred in all, lived on the island. They included a terrestrial flatworm, nematode worms, snails, scorpions, spiders, pseudoscorpions, centipedes, cockroaches, termites, barklice, cicadas, ants, beetles, moths, and butterflies. Also present were microscopic rotifers and tardigrades and a rich medley of bacteria.

A first look at the reconstituted flora and fauna of Rakata, in other words Krakatau a century after the apocalypse, gives the impression of life on a typical small Indonesian island. But the community of species remains in a highly fluid state. The number of resident bird species may now be approaching an equilibrium, the rise having slowed markedly since 1919 to settle close to thirty. Thirty is also about the number on other islands of Indonesia of similar size. At the same time, the *composition* of the bird species is less stable. New species have been arriving, and earlier ones have been declining to extinction. Owls and flycatchers arrived after 1919, for example, while several old residents such as the bulbul (*Pycnonotus aurigaster*) and gray-backed shrike (*Lanius schach*) disappeared. Reptiles appear to be at or close to a similar dynamic equilibrium. So are cockroaches, nymphalid butterflies, and dragonflies. Flightless mammals, represented solely by the two kinds of rats, are clearly not. Nor are plants, ants, or snails. Most of the other invertebrates are still too poorly explored on Rakata over sufficiently long periods of time to judge their status, but in general the overall number of species appears to be still rising.

Rakata, along with Panjang and Sertung, and other islands of the Krakatau archipelago blasted and pumice-coated by the 1883 explosion, have within the span of a century rewoven a semblance of the communities that existed before, and the diversity of life has largely returned. The question remains as to whether endemic species, those found only on the archipelago prior to 1883, were destroyed by the explosion. We can never be sure because the islands were too poorly explored by naturalists before Krakatau came so dramatically to the world's attention in 1883. It seems unlikely that endemic species ever existed. The islands are so small that the natural turnover of species may have been too fast to allow evolution to attain the creation of new species, even without volcanic episodes.

In fact the archipelago has suffered turbulence that destroyed or at least badly damaged its fauna and flora every few centuries. According to Javanese legend, the volcano Kapi erupted violently in the Sunda Strait in 416 AD: 'At last the mountain Kapi with a tremendous roar burst into pieces and sunk into the deepest of the earth. The water of the sea rose and inundated the land.' A series of smaller eruptions, burning at least part of the forest, occurred during 1680 and 1681.

Today you can sail close by the islands without guessing their violent history, unless Anak Krakatau happens to be smoldering that day. The thick green forest offers testimony to the ingenuity and resilience of life. Ordinary volcanic eruptions are not enough, then, to break the crucible of life.

Source: Edward O. Wilson, *The Diversity of Life*, Cambridge, Mass., Harvard University Press, 1992.

Gorillas

Man's view of the gorilla illustrates in a dramatic way the change from slaughter to conservation that distinguishes modern attitudes to wildlife. To nineteenth-century explorers and naturalists gorillas were evil. Paul Du Chaillu of the Boston Society of Natural History, who explored West and Central Africa in the 1850s, described the 'ferocity and malignity' of the gorilla, and its resemblance to a 'spirit of the damned'. A more enlightened American, Carl Akeley, persuaded the authorities to create the first national park in Africa, the Albert National Park, in 1925, to protect the mountain gorilla. It was here that in 1959 George Schaller (b. 1932) carried out the first systematic field study of the species. Schaller's work, and his pioneering of new forms of scientific observation in the wild in the 1960s and 1970s, were landmarks in modern ecology. He concluded that gorillas were remarkably like people. They enjoyed 'close and affectionate' social interactions, much like a human family, and their mating system was polygamous – 'a type for which man certainly has a predilection'. This extract from his book *The Year of the Gorilla* (1964) records his first sighting.

It is an exhilarating experience to wander alone through unknown forests when everything is still new and mysterious. The senses sharpen, bringing into quick focus all that is seen and heard. I knew that there were leopards on these slopes, and the trails of the black buffalo crisscrossed the area. Both animals have the reputation of being unpredictable, and I was watchful. Each creature has its own distance at which it will take flight from an intruder, and each will allow itself to be approached only so far before defending itself. It behoves man to learn the responses of each species; until he has done so, and is familiar with the sights, sounds, and smells around him, there is an element of danger. Even so, danger that is understood does not detract from but rather enhances the pleasures of tracking.

The first signs of gorillas I came across were three nests on the ground. Branches from the undergrowth had been pulled toward the center creating crude platforms on which the animals had slept for the night. But the broken vegetation was wilted and the dung at the edge of the nests was covered with small red fungi, indicating that the site was old. I was heartened by my find, and, after examining the nests, I pushed on. Neither Doc [Dr J. T. Emlen, Professor of Zoology, University of Wisconsin] nor I had much luck that day. We found nothing but old sites.

The following morning, Rousseau wanted to check the Rukumi rest house, which lies another thousand feet higher, on the slopes of Mt Karisimbi. We joined him, and I took my sleeping bag in the event that I decided to remain there for the night. For an hour we climbed upward through the *Hagenia* woodland. Then these trees grew sparser and were replaced by *Hypericum lanceolatum*, a bushy tree with small, lanceolate leaves and striking yellow blossoms. Abruptly the terrain flattened, the forest ceased, and we stood at the edge of a huge meadow with the summit cone of Mt Karisimbi towering three thousand feet above us. The dark branches and trunks of the trees growing here and there in the meadow were in vivid contrast to the yellow of the grass. The small bell-shaped flowers on the branches of these trees I recognized as heather – not the small shrub most of us know but a heather tree over thirty feet high. Among the heather grew wild blackberries, large and delicious, on which we feasted until our fingers and mouths were purple-stained. The rest house stood against the base of the mountain with a magnificent view across the meadow, past the black silhouettes of the heather to the bleak summit of Mt Mikeno.

The slopes of Mt Karisimbi beckoned, and I knew that I must climb the peak. Doc came a short distance to examine the most unusual forest we had ever seen. Tree groundsels or giant senecios, weird and gnarled plants with an other-worldly look, were scattered over the open slopes to an altitude of 13,500 feet. In the temperate zones of the world, senecios are insignificant weeds, but here in the cold and humid mountains they are giants over twenty feet high, with thick stems and flowering heads a foot or more long. The large ovate leaves are clustered at the apex of the branches and they glisten as if polished. The only other tall plants in this curious forest were giant lobelias, consisting of a single stem, a cluster of narrow leaves, and pointing skyward like a candle a long flowering head covered with tiny purplish blossoms.

For the next two days we continued our search for gorillas. We found some fresh feeding sites, and Doc once thought he heard a gorilla in the distance. On the third day, far down along the Kanyamagufa Canyon, I heard a sound that electrified me – a rapid *pok-pok-pok*, the sound of a gorilla pounding its chest. I followed the edge of the canyon until I found a game trail that crossed it. Carefully I scouted along the slope where I expected the animal to be. But I had no luck. Only later did I learn that there is an almost ventriloqual quality in the sound of chest-beating that makes distance very difficult to judge.

When Doc and I returned to the canyon in the morning, we were greeted by the same noise. Evidently a gorilla had spotted us. I climbed into the crown of a tree to look over the shrubs that obscured our view, and Doc circled up the slope. Suddenly, as he told me later, the undergrowth swayed forty feet ahead, and Doc heard the soft grumbling sound of contented animals. Unaware of him, the gorillas approached to within thirty feet. Two black, shaggy heads peered for ten seconds from the vegetation. Uncertain of how to react, Doc raised his arms. The animals screamed and walked away. We both examined the swath of freshly trampled vegetation and the torn remnants of wild celery and nettles on which the gorillas had been feeding. While Doc took notes on the spoor, I followed the trail. The musty, somewhat sweet odor of gorilla hung in the air. Somewhere ahead and out of sight, a gorilla roared and roared again, *uuua-uuua!* an explosive, half-screaming sound that shattered the stillness of the forest and made the hairs on my neck rise. I took a few steps and stopped, listened, and moved again. The only sound was the buzzing of insects. Far below me white clouds crept up the slopes and fingered into the canyons. Then another roar, but farther away. I continued over a ridge, down, and up again. Finally I saw them, on the opposite slope about two hundred feet away, some sitting on the ground, others in trees.

An adult male, easily recognizable by his huge size and gray back, sat among the herbs and vines. He watched me intently and then roared. Beside him sat a juvenile, perhaps four years old. Three females, fat and placid, with sagging breasts and long nipples, squatted near the male, and up in the fork of a tree crouched a female with a small infant clinging to the hair on her shoulders. A few other animals moved around in the dense vegetation. Accustomed to the drab gorillas in zoos, with their pelage lusterless and scuffed by the cement floors of their cages, I was little prepared for the beauty of the beasts before me. Their hair was not merely black, but a shining blue-black, and their black faces shone as if polished.

We sat watching each other. The large male, more than the others, held my attention. He rose repeatedly on his short, bowed legs to his full height, about six feet, whipped his arms to beat a rapid tattoo on his bare chest, and sat down again. He was the most magnificent animal I had ever seen. His brow ridges overhung his eyes, and the crest on his crown resembled a hair miter; his mouth when he roared was cavernous, and the large canine teeth were covered with black tartar. He lay on the slope, propped on his huge shaggy arms, and the muscles of his broad shoulders and silver back rippled. He gave an impression of dignity and restrained power, of absolute certainty in his majestic appearance. I felt a desire to communicate with him, to let him know by some small gesture that I intended no harm, that I wished only to be near him. Never before had I had this feeling on meeting an animal. As we watched each other across the valley, I wondered if he recognized the kinship that bound us.

After a while the roars of the male became less frequent, and the other members of the group scattered slowly. Some climbed ponderously into shrubby trees and fed on the vines that draped from the branches; others reclined on the ground, either on the back or on the side, lazily reaching out every so often to pluck a leaf. They still kept their eyes on me, but I was amazed at their lack of excitement.

'George,' called Doc. 'George!'

At the sound all the gorillas rose and disappeared silently at a fast walk. Doc had become concerned for my safety after first hearing the many roars and finally the prolonged silence. We ate our lunch – crackers, cheese, and chocolate – before we checked the site where I had watched them. Their trail angled upward across a valley and up another ridge, where we found the group again two hours later. A female sat on a mound, her infant beside her. The male, ever alert, roared when he spotted us and stalked back and forth in the usual posture of gorillas – feet flat on the ground and the upper part of the body supported on the knuckles of the hands. When he approached the female on the mound, she moved rapidly to one side, and he claimed her place. As before, the group settled down and seemingly paid us scant attention. The male, who must have weighed over four hundred pounds, rested on the mound looking over the mountains and the plains, truly the master of his domain. A female holding an infant gently to her chest walked to his side.

'It must have been just born,' I whispered to Doc. 'It's still wet.' And he nodded in agreement.

The female leaned heavily against the side of the male. Her hairy arm almost obscured her spidery offspring, whose hairless arms and legs waved about in unoriented fashion. The male leaned over and with one hand fondled the infant. For two hours, enthralled, we watched this family scene. But the way home was long, and reluctantly we left the animals, but not before we spotted another gorilla far uphill and barely visible. Was it another group or a single animal? With buoyant steps we moved down the slope. We only hoped that in our presence the gorillas would always be as tranquil as today. Perhaps, we feared, they had merely been loath to move because of the imminent or recent birth of the infant.

But something else took our minds off the apes. Just ahead and close to the trail that crossed the Kanyamagufa Canyon an elephant wheezed, and then another. We stopped and strained our eyes, trying to locate the gray forms gliding softly through the bamboo and the brush around us. It was my first close contact with elephants, and I was nervous. I clambered up a tree with much noise as the dry branches snapped under my weight. It was embarrassing, for after all my scrambling I was still no higher than the elephant's back. Doc motioned me to come down, and we hurried in and out of the canyon, hoping that no one would bar our way.

On the following morning we returned to the ridge where we had seen the gorillas. The animals had descended into the valley and were now feeding leisurely. It struck us immediately that there were now more gorillas than yesterday. We counted, recounted, and agreed there were twenty-two: four adult or silver-backed males, one young or black-backed male, eight females, three juveniles, five infants, and one medium-sized gorilla of whose sex we were uncertain. Did two separate groups join, or had we seen only part of the group the day before? The big males roared and slapped their chests, but, as on the previous day, they seemed little concerned as they rested beneath the canopy of trees. As we watched, most of them lay down and went to sleep. One female sat with a large infant in her arms. Another small, woolly infant left its mother and bumbled over to the first female. Briefly she cuddled both youngsters to her chest. A male rose, casually ambled by a sitting female, suddenly grabbed her by the leg, yanked her two or three feet down the slope, then cantered off. It was a wonderful feeling to sit near these animals and to record their actions as no one had ever done before. We had the chance to observe significant and characteristic incidents, but we knew that to explain what we were seeing – and even to predict what might occur in a particular circumstance – would take many, many hours of observation.

After resting for some three hours, the animals spread over the hillside to feed. An infant ran along, put some leaves into its mouth, and spat them out. A juvenile, perhaps three or four years old, held to the end of a three-foot log with both hands. It bit off the rotten bark and appeared to lick a whitish fluid from the wood. Another juvenile came up and, having wiped the log with one finger, licked it. Soon all the gorillas were actively moving across the slope, feeding as they went. Their movements were restrained and rather phlegmatic; only the youngsters behaved in exuberant fashion. One infant dashed along, pounced on the back of another infant, and both disappeared rolling over and over into the undergrowth. After five hours with the gorillas we returned home.

A guard and I tried to find the group again the following day, but we had no luck. I realized how much I had to learn about the ways of the forest and about tracking animals over long distances. The park had no teachers, for the local bantus were all agriculturalists; they avoided the forest and the wild animals and evil spirits it contained. I knew that before I could study gorillas successfully I would have to teach myself to recognize the age of spoor, the number of animals involved, and the direction they had taken. The forest was vast, the animals few, and I could not depend on luck in finding gorillas as we had done so far.

Luck, however, was still with us, for on the way home, at a place where the Kanyamagufa Canyon veers sharply toward the upper slopes of Mt Mikeno, we came across a fresh trail. Early the following morning, when Doc and I had barely proceeded three hundred feet along this trail, the brush crackled ahead. A black arm reached from the undergrowth, pulled a strand of vine from a branch, and disappeared from view. We stepped behind the bole of a tree and peered at the feeding gorillas about one hundred feet away. Without warning, a female with infant walked toward our tree, a large male behind her. I nudged Doc and quietly climbed up on a branch without being seen by the animals. The female stopped some thirty feet from us and sat quietly with her large infant beside her. Once the infant glanced up at me, then stared intently for fifteen seconds without giving the alarm. But when the female inadvertently looked in my direction, her relaxed gaze hardened as she saw me. She grabbed her young with one arm, pulling it to her, and with the same motion rushed away, emitting a high-pitched scream. The male answered with a roar and looked around, and Doc, having failed to interpret the purpose of my nudge, was surprised to see me in the branches above him. The members of the group assembled around the male after a moment of tense alertness. The animals were still within about one hundred feet of us, and we wondered what would happen. To our relief, one face after another turned toward us in a quiet, quizzical stare as curiosity replaced alarm. They craned their necks, and two juveniles climbed into the surrounding trees to obtain a better view. One juvenile with a mischievous look on its face beat its chest, then quickly ducked into the vegetation, only to peer furtively through the screen of weeds as if to judge the effect of its commotion. Slowly the animals dispersed and went about their daily routine. I particularly remember one female who left the deep shade and settled herself at the base of a tree in a shaft of sunlight. She stretched her short legs in front of her and dangled her arms loosely at her sides. Her face was old and kind and creased by many wrinkles. She seemed utterly at peace and relaxed as she basked in the morning sun.

Source: George Schaller, *The Year of the Gorilla*, Chicago, University of Chicago Press, 1964.

Toads

George Orwell (1903–50) was not a trained naturalist, but he had the gift, essential to scientists, of noticing what other people did not. This piece was written for *Tribune* in April 1946.

Before the swallow, before the daffodil, and not much later than the snowdrop, the common toad salutes the coming of spring after his own fashion, which is to emerge from a hole in the ground, where he has lain buried since the previous autumn, and crawl as rapidly as possible towards the nearest suitable patch of water. Something – some kind of shudder in the earth, or perhaps merely a rise of a few degrees in the temperature – has told him that it is time to wake up: though a few toads appear to sleep the clock round and miss out a year from time to time – at any rate, I have more than once dug them up, alive and apparently well, in the middle of summer.

At this period, after his long fast, the toad has a very spiritual look, like a strict Anglo-Catholic towards the end of Lent. His movements are languid but purposeful, his body is shrunken, and by contrast his eyes look abnormally large. This allows one to notice, what one might not at another time, that a toad has about the most beautiful eye of any living creature. It is like gold, or more exactly it is like the golden-coloured semi-precious stone which one sometimes sees in signet-rings, and which I think is called a chrysoberyl.

For a few days after getting into the water the toad concentrates on building up his strength by eating small insects. Presently he has swollen to his normal size again, and then he goes through a phase of intense sexiness. All he knows, at least if he is a male toad, is that he wants to get his arms round something, and if you offer him a stick, or even your finger, he will cling to it with surprising strength and take a long time to discover that it is not a female toad. Frequently one comes upon shapeless masses of ten or twenty toads rolling over and over in the water, one clinging to another without distinction of sex. By degrees, however, they sort themselves out into couples, with the male duly sitting on the female's back. You can now distinguish males from females, because the male is smaller, darker and sits on top, with his arms tightly clasped round the female's neck. After a day or two the spawn is laid in long strings which wind themselves in and out of the reeds and soon become invisible. A few more weeks, and the water is alive with masses of tiny tadpoles which rapidly grow larger, sprout hind-legs, then forelegs, then shed their tails: and finally, about the middle of the summer, the new generation of toads, smaller than one's thumb-nail but perfect in every particular, crawl out of the water to begin the game anew.

I mention the spawning of the toads because it is one of the phenomena of spring which most deeply appeal to me, and because the toad, unlike the skylark and the primrose, has never had much of a boost from the poets.

Source: *The Collected Essays, Journalism and Letters of George Orwell, Volume 4: In Front of Your Nose, 1945–50*, edited by Sonia Orwell and Ian Angus, Harmondsworth, Penguin Books, 1970.

Russian Butterflies

The novelist Vladimir Nabokov (1899–1977) was a passionate lepidopterist. Several thousand specimens caught and preserved by him are now in the American Museum of Natural History and the Cornell University Museum of Entomology. These memories of his boyhood in pre-revolutionary Russia are from his autobiography *Speak, Memory.*

Near the intersection of two carriage roads (one, well-kept, running north-south in between our 'old' and 'new' parks, and the other, muddy and rutty, leading, if you turned west, to Batovo) at a spot where aspens crowded on both sides of a dip, I would be sure to find in the third week of June great blue-black nymphalids striped with pure white, gliding and wheeling low above the rich clay which matched the tint of their undersides when they settled and closed their wings. Those were the dung-loving males of what the old Aurelians used to call the Poplar Admirable, or, more exactly, they belonged to its Bucovinan subspecies. As a boy of nine, not knowing that race, I noticed how much our North Russian specimens differed from the Central European form figured in Hofmann, and rashly wrote to Kuznetsow, one of the greatest Russian, or indeed world, lepidopterists of all time, naming my new subspecies '*Limenitis populi rossica*.' A long month later he returned my description and aquarelle of '*rossica* Nabokov' with only two words scribbled on the back of my letter: '*bucovinensis* Hormuzaki.' How I hated Hormuzaki! And how hurt I was when in one of Kuznetsov's later papers I found a gruff reference to 'schoolboys who keep naming minute varieties of the Poplar Nymph!' Undaunted, however, by the *populi* flop, I 'discovered' the following year a 'new' moth. That summer I had been collecting assiduously on moonless nights, in a glade of the park, by spreading a bedsheet over the grass and its annoyed glow-worms, and casting upon it the light of an acytelene lamp (which, six years later, was to shine on Tamara [Nabokov's first love]). Into that arena of radiance, moths would come drifting out of the solid blackness around me, and it was in that manner, upon that magic sheet, that I took a beautiful *Plusia* (now *Phytometra*) which, as I saw at once, differed from its closest ally by its mauve-and-maroon (instead of golden-brown) forewings, and narrower bractea mark and was not recognizably figured in any of my books. I sent its description and picture to Richard South, for publication in *The Entomologist*. He did not know it either, but with the utmost kindness checked it in the British Museum collection – and found it had been described long ago as *Plusia excelsa* by Kretschmar. I received the sad news, which was most sympathetically worded ('. . . should be congratulated for obtaining . . . very rare Volgan thing . . . admirable figure . . .') with the utmost stoicism; but many years later, by a pretty fluke (I know I should not point out these plums to people), I got even with the first discoverer of *my* moth by giving his own name to a blind man in a novel.

Let me also evoke the hawkmoths, the jets of my boyhood! Colors would die a long death on June evenings. The lilac shrubs in full bloom before which I stood, net in hand, displayed clusters of a fluffy gray in the dusk – the ghost of purple. A moist young moon hung above the mist of a neighboring meadow. In many a garden have I stood thus in later years – in Athens, Antibes, Atlanta – but never have I waited with such a keen desire as before those darkening lilacs. And suddenly it would come, the low buzz passing from flower to flower, the vibrational halo around the streamlined body of an olive and pink Hummingbird moth poised in the air above the corolla into which it had dipped its long tongue. Its handsome black larva (resembling a diminutive cobra when it puffed out its ocellated front segments) could be found on dank willow herb two months later. Thus every hour and season had its delights. And, finally, on cold, or even frosty, autumn nights, one could sugar for moths by painting tree trunks with a mixture of molasses, beer, and rum. Through the gusty blackness, one's lantern would illumine the stickily glistening furrows of the bark and two or three large moths upon it imbibing the sweets, their nervous wings half open butterfly fashion, the lower ones exhibiting their incredible crimson silk from beneath the lichen-gray primaries. '*Catocala adultera!*' I would triumphantly shriek in the direction of the lighted windows of the house as I stumbled home to show my captures to my father . . .

There came a July day – around 1910, I suppose – when I felt the urge to explore the vast marshland beyond the Oredezh. After skirting the river for three or four miles, I found a rickety foot-bridge. While crossing over, I could see the huts of a hamlet on my left, apple trees, rows of tawny pine logs lying on a green bank, and the bright patches made on the turf by the scattered clothes of peasant girls, who, stark naked in shallow water, romped and yelled, heeding me as little as if I were the discarnate carrier of my present reminiscences.

On the other side of the river, a dense crowd of small, bright blue male butterflies that had been tippling on the rich, trampled mud and cow dung through which I trudged rose all together into the spangled air and settled again as soon as I had passed.

After making my way through some pine groves and alder scrub I came to the bog. No sooner had my ear caught the hum of diptera around me, the guttural cry of a snipe overhead, the gulping sound of the morass under my foot, than I knew I would find here quite special arctic butterflies, whose pictures, or, still better, nonillustrated descriptions I had worshiped for several seasons. And the next moment I was among them. Over the small shrubs of bog bilberry with fruit of a dim, dreamy blue, over the brown eye of stagnant water, over moss and mire, over the flower spikes of the fragrant bog orchid (the *nochnaya fialka* of Russian poets), a dusky little Fritillary bearing the name of a Norse goddess passed in low, skimming flight. Pretty Cordigera, a gem-like moth, buzzed all over its uliginose food plant. I pursued rose-margined Sulphurs, gray-marbled Satyrs. Unmindful of the mosquitoes that furred my forearms, I stooped with a grunt of delight to snuff out the life of some silver-studded lepidopteron throbbing in the folds of my net. Through the smells of the bog, I caught the subtle perfume of butterfly wings on my fingers, a perfume which varies with the species – vanilla, or lemon, or musk, or a musty, sweetish odor difficult to define. Still unsated, I pressed forward. At last I saw I had come to the end of the marsh. The rising ground beyond was a paradise of lupines, columbines, and pentstemons. Mariposa lilies bloomed under Ponderosa pines. In the distance, fleeting cloud shadows dappled the dull green of slopes above timber line, and the gray and white of Longs Peak.

Source: Vladimir Nabokov, *Speak, Memory*, Harmondsworth, Penguin Books, 1969.

Discovering a Medieval Louse

John Steinbeck's interest in science was stimulated by his long friendship with the marine biologist Ed Ricketts. Even more consuming was his interest in the Arthurian legends. This extract is from a letter written in September 1962.

The Morgan Library has a very fine 11th-century Launcelot in perfect condition. I was going over it one day and turned to the rubric of the first known owner dated 1221, the rubric a squiggle of very thick ink. I put a glass on it and there imbedded deep in the ink was the finest crab louse, *pfithira pulus*, I ever saw. He was perfectly preserved even to his little claws. I knew I would find him sooner or later because people of that period were deeply troubled with lice and other little beasties – hence the plagues. I called the curator over and showed him my find and he let out a cry of sorrow. 'I've looked at that rubric a thousand times,' he said. 'Why couldn't I have found him?'

Source: *Steinbeck: A Life in Letters*, ed. Elaine Steinbeck and Robert Wallsten, London, Heinemann, 1975.

The Gecko's Belly

Italo Calvino (1923–85) was born in Cuba and grew up in San Remo, Italy. He was an essayist and journalist as well as a novelist. The following is from *Mr Palomar* (1983).

On the terrace, the gecko has returned, as he does every summer. An exceptional observation point allows Mr Palomar to see him not from above, as we have always been accustomed to seeing geckos, treefrogs, and lizards, but from below. In the living room of the Palomar home there is a little show-case window and display case that opens on to the terrace; on the shelves of this case a collection of Art Nouveau vases is aligned; in the evening a 75-Watt bulb illuminates the objects; a plumbago plant trails its pale blue flowers from the wall against the outside glass; every evening, as soon as the light is turned on, the gecko, who lives under the leaves on that wall, moves onto the glass, to the spot where the bulb shines, and remains motionless, like a lizard in the sun. Gnats fly around, also attracted by the light; the reptile, when a gnat comes within range, swallows it.

Mr Palomar and Mrs Palomar every evening end up shifting their chairs from the television set to place them near the glass; from the interior of the room they contemplate the whitish form of the reptile against the dark background. The choice between television and gecko is not always made without some hesitation; each of the two spectacles has some information to offer that the other does not provide: the television ranges over continents gathering luminous impulses that describe the visible face of things; the gecko, on the other hand, represents immobile concentration and the hidden side, the obverse of what is displayed to the eye.

The most extraordinary thing are the claws, actual hands with soft fingers, all pad, which, pressed against the glass, adhere to it with their minuscule suckers: the five fingers stretch out like the petals of little flowers in a childish drawing, and when one claw moves, the fingers close like a flower, only to spread out again and flatten against the glass, making tiny streaks, like fingerprints. At once delicate and strong, these hands seem to contain a potential intelligence, so that if they could only be freed from their task of remaining stuck there to the vertical surface they could acquire the talents of human hands, which are said to have become skilled after they no longer had to cling to boughs or press on the ground.

Bent, the legs seem not so much all knee as all elbow, elastic in order to raise the body. The tail adheres to the glass only along a central strip, from which the rings begin that circle it from one side to the other and make of it a sturdy and well-protected implement; most of the time it is listless, idle, and seems to have no talent or ambition beyond subsidiary support (nothing like the calligraphic agility of lizards' tails); but when called upon, it proves well-articulated, ready to react, even expressive.

Of the head, the vibrant, capacious gullet is visible, and the protruding, lidless eyes at either side. The throat is a limp sack's surface extending from the tip of the chin, hard and all scales like that of an alligator, to the white belly that, where it presses against the glass, also reveals a grainy, perhaps adhesive, speckling.

When a gnat passes close to the gecko's throat, the tongue flicks and engulfs, rapid and supple and prehensile, without shape, capable of assuming whatever shape. In any case, Mr Palomar is never sure if he has seen it or not seen it: what he surely does see, now, is the gnat inside the reptile's gullet: the belly pressed against the illuminated glass is transparent as if under X-rays; you can follow the shadow of the prey in its course through the viscera that absorb it.

If all material were transparent – the ground that supports us, the envelope that sheathes our body – everything would be seen not as a fluttering of impalpable wings but as an inferno of grinding and ingesting. Perhaps at this moment a god of the nether world situated in the center of the earth with his eye that can pierce granite is watching us from below, following the cycle of living and dying, the lacerated victims dissolving in the bellies of their devourers until they, in their turn, are swallowed by another belly.

The gecko remains motionless for hours; with a snap of his tongue he gulps down a mosquito or a gnat every now and then; other insects, on the contrary, identical to the first, light unawares a few millimeters from his mouth and he seems not to perceive them. Is it the vertical pupil of his eyes, separated at the sides of his head, that does not notice? Or does he have criteria of choice and rejection that we do not know? Or are his actions prompted by chance or by whim?

The segmentation of legs and tail into rings, the speckling of tiny granulous plates on his head and belly give the gecko the appearance of a mechanical device; a highly elaborate machine, its every microscopic detail carefully studied, so that you begin to wonder if all that perfection is not squandered, in view of the limited operations it performs. Or is that perhaps the secret: content to be, does he reduce his doing to the minimum? Can this be his lesson, the opposite of the morality that, in his youth, Mr Palomar wanted to make his: to strive always to do something a bit beyond one's means?

Now a bewildered nocturnal butterfly comes within range. Will he overlook it? No, he catches this, too. His tongue is transformed into a butterfly net and he pulls it into his mouth. Will it all fit? Will he spit it out? Will he explode? No, the butterfly is there in his throat; it flutters, in a sorry state, but still itself, not touched by the insult of chewing teeth, now it passes the narrow limits of the neck, it is a shadow that begins its slow and troubled journey down along a swollen esophagus.

The gecko, emerging from its impassiveness, gasps, shakes its convulsed throat, staggers on legs and tail, twists its belly, subjected to a severe test. Will this be enough for him, for tonight? Will he go away? Was this the peak of every desire he yearned to satisfy? Was this the nearly impossible test in which he wanted to prove himself? No, he stays. Perhaps he has fallen asleep. What is sleep like for someone who has eyes without eyelids?

Mr Palomar is unable to move from there either. He sits and stares at the gecko. There is no truce on which he can count. Even if he turned the television back on, he would only be extending the contemplation of massacres. The butterfly, fragile Eurydice, sinks slowly into her Hades. A gnat flies, is about to light on the glass. And the gecko's tongue whips out.

Source: Italo Calvino, *Mr Palomar*, trans. William Weaver, London, Picador, Pan Books, 1986.

On The Moon

At 3.18 p.m. (Houston time) on 20 July 1969, astronaut Neil Armstrong sent back to the Manned Spacecraft Center at Houston, Texas, the message 'The Eagle has landed', indicating that the lunar module from the Apollo 11 Spacecraft had touched down on the moon, in the area known as the Sea of Tranquillity. Armstrong's heart rate at touchdown had risen to 156 beats a minute. The landing was the culmination of a 24 billion dollar space programme – money that critics protested should have been spent on reducing world poverty.

Recalling the approach to the moon, Armstrong said:

The most dramatic recollections I had were the sights themselves. Of all the spectacular views we had, the most impressive to me was on the way to the moon, when we flew through the shadow. We were still thousands of miles away, but close enough so that the moon almost filled our circular window. It was eclipsing the sun, from our position, and the corona of the sun was visible around the limb of the moon as a gigantic lens-shaped or saucer-shaped light, stretching out to several lunar diameters. It was magnificent, but the moon was even more so. We were in its shadow, so there was no part of it illuminated by the sun. It was illuminated only by earthshine. It made the moon appear blue-gray, and the entire scene looked decidedly three-dimensional.

I was really aware, visually aware, that the moon was in fact a sphere, not a disc. It seemed almost as if it were showing us its roundness, its similarity in shape to our earth, in a sort of welcome. I was sure that it would be a hospitable host. It had been awaiting its first visitors for a long time.

Before Armstrong stepped onto the moon, his fellow astronaut Buzz Aldrin celebratred Communion, using a wine chalice given him by the minister of his Presbyterian church. Armstrong found the view out of the lunar module's window comparable to a night-time scene lighted for photography:

The sky is black, you know. It's a very dark sky. But it still seemed more like daylight than darkness as we looked out the window. It's a peculiar thing, but the surface looked very warm and inviting. It looked as if it would be a nice place to take a sunbath. It was the sort of situation in which you felt like going out there in nothing but a swimming suit to get a little sun. From the cockpit, the surface seemed to be tan. It's hard to account for that, because later when I held this material in my hand, it wasn't tan at all. It was black, gray and so on. It's some kind of lighting effect, but out the window the surface looks much more like light desert sand than black sand.

Clambering down the ladder from Eagle's cabin Armstrong reported:

The L[unar] M[odule] footpads are only depressed in the surface about one or two inches. Although the surface appears to be very, very fine-grained, as you get close to it. It's almost like a powder. Now and then, it's very fine . . . I'm going to step off the LM now.

'I had thought about what I was going to say,' he later admitted, 'largely because so many people had asked me to think about it . . . It wasn't really decided until after we got onto the lunar surface.'

Stepping off the dish-shaped landing pad onto the moon at 9.56 p.m. (Houston time) he uttered the usually misquoted sentence: 'That's one small step for a man, one giant leap for mankind.'

On the moon the two astronauts collected rock samples, erected an American flag with a metal strip woven into its top edge so that it would appear to fly despite the windless conditions on the moon's surface, and talked to President Nixon on the phone. Re-entering the lunar module they left behind a plaque which read: 'Here Men from the planet Earth first set foot upon the Moon, July 1969, A.D. We came in peace for all Mankind.'

Back in the module, Armstrong thought it all over:

My impression was that we were taking a snapshot of a steady-state process, in which rocks were being worn down on the surface of the moon with time and other rocks were being thrown out on top as a result of new events somewhere near or far away. In other words, no matter when you had been to this spot before, a thousand years ago or a hundred thousand years ago, or if you came back to it a million years from now, you would see some different things each time, but the scene would generally be the same.

Buzz Aldrin found himself wondering how long his footsteps would linger on the moon's surface:

The moon was a very natural and very pleasant environment in which to work. It had many of the advantages of zero-gravity, but it was in a sense less *lonesome* than zero G, where you always have to pay attention to securing attachment points to give you some means of leverage. In one-sixth gravity, on the moon, you had a distinct feeling of being *somewhere*, and you had a constant, though at many times ill defined, sense of direction and force . . .

As we deployed our experiments on the surface we had to jettison things like lanyards, retaining fasteners, etc., and some of these we tossed away. The objects would go away with a slow, lazy motion. If anyone tried to throw a baseball back and forth in that atmosphere he would have difficulty, at first, acclimatizing himself to that slow, lazy trajectory; but I believe he could adapt to it quite readily.

Odor is very subjective, but to me there was a distinct smell to the lunar material – pungent, like gunpowder or spent cap-pistol caps. We carted a fair amount of lunar dust back inside the vehicle with us, either on our suits and boots or on the conveyor system we used to get boxes and equipment back inside. We did notice the odor right away.

It was a unique, almost mystical environment up there.

Source: *First on the Moon. A Voyage with Neil Armstrong, Michael Collins, Edwin E. Aldrin Jr.*, written with Grace Farmer and Dora Jane Hamblin, Epilogue by Arthur C. Clarke, Boston and Toronto, Little, Brown, 1970.

Gravity

Gravity is the weakest of the four forces in nature, so weak that only astronomical bodies exert it significantly. The others are the electromagnetic force, the strong nuclear force, which binds together atomic nuclei, and the weak nuclear force, which causes subatomic particles to scatter. For human life to exist on Earth it was essential that gravity should be strong enough to stop the atmosphere being ripped away into space, and weak enough to let us stand up and move around. These are the conditions celebrated in John Frederick Nims's poem. 'Klutz' (in the last stanza) is American slang for a clumsy person.

Mildest of all the powers of earth: no lightnings
For her – maniacal in the clouds. No need for
Signs with their skull and crossbones, chain-link gates:
Danger! Keep Out! High Gravity! she's friendlier.
Won't nurse – unlike the magnetic powers – repugnance;
Would reconcile, draw close: her passion's love.

No terrors lurking in her depths, like those
Bound in that buzzing strongbox of the atom,
Terrors that, loosened, turn the hills vesuvian,
Trace in cremation where the cities were.

No, she's our quiet mother, sensible.
But therefore down-to-earth, not suffering
Fools who play fast and loose among the mountains,
Who fly in her face, or, drunken, clown on cornices.

She taught our ways of walking. Her affection
Adjusted the morning grass, the sands of summer
Until our soles fit snug in each, walk easy.
Holding her hand, we're safe. Should that hand fail,
The atmosphere we breathe would turn hysterical,
Hiss with tornadoes, spinning us from earth
Into the cold unbreathable desolations.

Yet there – in fields of space – is where she shines,
Ring-mistress of the circus of the stars,
Their prancing carousels, their ferris wheels
Lit brilliant in celebration. Thanks to her
All's gala in the galaxy.

Down here she
Walks us just right, not like the jokey moon
Burlesquing our human stride to kangaroo hops;
Not like vast planets, whose unbearable mass
Would crush us in a bear hug to their surface
And into the surface, flattened. No: deals fairly.
Makes happy each with each: the willow bend
Just so, the acrobat land true, the keystone
Nestle in place for bridge and for cathedral.
Lets us pick up – or mostly – what we need:
Rake, bucket, stone to build with, logs for warmth,
The fallen fruit, the fallen child . . . ourselves.

Instructs us too in honesty: our jointed
Limbs move awry and crisscross, gawky, thwart;
She's all directness and makes that a grace,
All downright passion for the core of things,
For rectitude, the very ground of being:
Those eyes are leveled where the heart is set.

See, on the tennis court this August day:
How, beyond human error, she's the one
Whose will the bright balls cherish and obey
– As if in love. She's tireless in her courtesies
To even the klutz (knees, elbows all a-tangle),
Allowing his poky serve Euclidean whimsies,
The looniest lob its joy: serene parabolas.

Source: John Frederick Nims, *The Six-Cornered Snowflake and Other Poems*, New York, New Directions, 1990.

Otto Frisch Explains Atomic Particles

Otto Frisch (1904–79) was the physicist who, in collaboration with his aunt Lise Meitner, first realized that uranium atoms, bombarded with neutrons, split into atoms of lighter elements. Frisch and Meitner named the process 'fission'. At the time (1939) Frisch was working in Copenhagen with Niels Bohr, who passed on the news to Einstein and others in the USA. During the Second World War Frisch was a member of the Los Alamos atomic research project. He wrote this popular introduction to atomic particles in 1960.

For good reasons this has become known as the atomic age. Power from atomic nuclei is about to transform our world – and threatens to destroy it. Nearly every recent advance in engineering is based on what we know about the structure of atoms – high-strength alloys and plastics no less than fluorescent lamps, transistors and ferrites. Even in the study of the phenomena of life the stage has been reached when we examine effects of single atoms on living organisms. For these reasons alone it is clear that the study of atoms and their components is of great importance, but physicists have another strong incentive: curiosity. They just want to know what the world is made of, what are the smallest particles and how they behave. Therefore they are always ahead of practical applications; there always exists some knowledge about which people can ask 'what is it good for?' It is true that at the moment nobody can see any use for the 'newer' particles – mesons and hyperons – yet in the past any new discovery has invariably, within a few decades, found some practical use, or at least has become such an indispensable part of our knowledge that many practical advances would have been impossible without it. I do not foresee meson guns or hyperon boilers, but if applications for these particles are ever found, it is unlikely that even the most imaginative of present science-fiction writers will have envisaged them correctly.

Let me first recapitulate what is known about the structure of atoms. Since 1911 we have known that each atom consists of a heavy core or nucleus, surrounded by a number of much lighter particles called electrons. Different atoms have different kinds of nuclei, but electrons are all alike. Electrons all weigh the same and have the same negative electric charge, but an atom is not electrically charged, the negative charge of its electron being offset by an equal positive charge of the nucleus. To put it the other way round: the nucleus has a positive charge which is Z times the charge of an electron; hence Z electrons become arranged around the nucleus to form an atom, which is electrically neutral (i.e., uncharged).

The number (Z) of positive electronic charges on the nucleus, or of electrons around it, is called the atomic number. This determines the chemical properties of the atom, and many of its physical properties as well. Thus atoms with the same Z will stay together and not become separated when passed through chemical reactions; they belong to the same chemical element. Each element is characterized by its Z. Thus hydrogen has Z = 1; carbon, 6; copper, 29; and uranium, 92.

Though all the nuclei in any one element carry the same positive charge, they have not necessarily the same mass. Nuclei with the same charge but different mass are said to belong to different isotopes of the same element. The mass of a nucleus is, however, always very nearly an exact multiple of a certain unit mass which, in turn, is very nearly equal to the mass of the lightest nucleus of all, that of the lightest (and most common) hydrogen isotope, the proton. For instance, a gold nucleus weighs about 197 of these units; we say its mass number *A* is 197. But it cannot consist simply of 197 protons, for then its atomic number Z would also be 197 whereas it is really only 79. We now know that it consists of 197 nuclear particles of which, however, only 79 are protons; the other 118 have practically the same mass but no electric charge and are called neutrons. Protons and neutrons are known collectively as nucleons.

Free neutrons were first recognized in 1932 when it was found that they could be knocked out of certain light nuclei. For the knocking, nature had very kindly supplied fast-moving helium nuclei which are emitted from the nuclei of certain heavy elements such as uranium, radium, and others; when they were first observed their nature was not known and they were labelled 'alpha particles,' a name which has stuck. An alpha particle – like any other fast-moving charged particle – damages the atoms in the air through which it passes, leaving behind a trail of atoms or molecules which are electrically charged by having either lost an electron or gained an extra one, and which are called 'ions.' But a neutron slips through the air without making ions and is for that reason hard to detect; that is why it was discovered so late. A neutron can only be detected if it happens to strike a nucleus, which is then sent flying or is broken up; in that case, fast charged particles are formed which makes ions and can thereby be detected.

C. T. R. Wilson's cloud chamber provides a means of making trails of ions visible as fine tracks of water droplets, and so of detecting neutrons at second hand. It was a very convincing proof of the existence of neutrons to see the track of a nucleus, struck by a neutron somewhere in the gas of a cloud chamber, and sent flying by the collision. By placing the chamber between the poles of a strong electromagnet one can deflect the nuclei so that their tracks become curved; the degree of curvature indicates the speed of the nucleus, and hence the speed of the neutron that sent it flying. This kind of technique has been used again and again in the discovery and study of new particles.

When the neutron was first found, its study amounted to research of the purest kind. Nobody could have foreseen any practical use for it. This changed dramatically with the discovery (1939) that neutrons could cause the fission of uranium nuclei, with the liberation of more neutrons; in this way a chain reaction became possible, and this is now our chief source of neutrons and an increasingly important source of power. Today neutrons are an industrial commodity; several tons have been produced (and immediately consumed) in the brief history of atomic energy. Only ten years elapsed between the discovery of the neutron and the operation of the first atomic pile!

A nucleus always weighs a little less than the sum of the neutrons and protons – or briefly the nucleons – of which it consists. The reason is that the nucleons are bound together by strong forces; it would require a considerable energy, the so-called binding energy, to take a nucleus completely to pieces. Now it is one of the consequences of Einstein's theory of relativity that an energy E possesses a mass m, the exchange rate being given by the famous formula $E = mc^2$ where c is the speed of light. The energy contained in 700 domestic units of electricity corresponds to only one millionth of an ounce; so in ordinary life the mass equivalent of energy can be neglected. But if you could assemble a nucleus out of its nucleons the energy liberated would amount to about one per cent of the total mass. Even minor rearrangements inside nuclei cause changes in mass which can be accurately measured with the so-called mass spectrometer. The energies, too, can be measured, and in this way Einstein's equation has been checked and confirmed many times. Energies here are measured in MeV (the energy gained by an electron on being accelerated through one million volts). 1 MeV corresponds to one 940th of the mass of a proton, or to twice the mass of an electron.

Almost at the same time as the neutron, the positive electron or positron was discovered, and this opened entirely new vistas. Physicists had often asked themselves why protons were always positively and electrons negatively charged when the fundamental laws of electricity were quite symmetrical in respect of charge. Indeed the quantum theory of the electron, as developed in England by P. A. M. Dirac (1928), showed that positively charged electrons must be possible, and that one could produce a positron and a (negative) electron 'out of nothing,' provided the necessary energy was supplied. Indeed when gamma rays pass through matter, positrons are produced, and when they pass through a cloud chamber the production of pairs of electrons, one deflected to the right, the other to the left, by the electromagnet, can be clearly seen. The opposite process also occurs: when a positron meets an electron both disappear in a flash of gamma radiation – they are said to annihilate each other. This is the reason why positrons are so rare: they disappear within less than a millionth of a second in contact with matter although they can exist indefinitely in a perfect vacuum. The positron is said to be the 'antiparticle' of the electron, and vice versa.

As soon as the positron was discovered, it was realized that there ought to be an antiproton as well, a proton of negative charge, capable of annihilating itself with an ordinary proton. But since the proton is 1,836 times heavier than the electron, 1,836 times as much energy is needed to produce a proton-antiproton pair than to produce a positron-electron pair. For the latter process one needs about 1 MeV energy, an amount possessed by many ordinary gamma rays, but gamma rays of several thousands MeV are rare even in the cosmic radiation, the fine rain of very fast particles that comes to us from outer space. An accelerator was therefore built (in Berkeley, California) which could accelerate protons to about 6,000 MeV, and a determined and successful search was made for antiprotons. Particles were found which weighed as much as protons, but had the opposite (*i.e.*, negative) charge, and which suffered annihilation in the expected manner on passing through matter.

In the same experimental setup, antineutrons were also found. It may sound surprising that the neutron, which has no charge, should possess an 'electric mirror image' different from itself; but the neutron is a little magnet, which spins about its magnetic axis rather like the earth. In the antineutron, not only the electric but also the magnetic properties are reversed; so a neutron and an antineutron, spinning in the same direction, have their magnetic poles pointing in opposite ways, and this makes them different particles. Nobody has yet thought of a way of measuring the magnetic properties of antineutrons (though with neutrons it can be done); but there can be little doubt that the neutral particles, created under the same conditions as antiprotons and suffering annihilation in the same way, are indeed antineutrons.

Let us take stock. We have mentioned six particles so far: the electron, the proton, and the neutron, each with its respective antiparticle – protons and neutrons are jointly called nucleons. The positive electron (the 'antielectron') is usually called the positron. The word electron is sometimes used for both kinds and sometimes just for the negative kind, the meaning being usually clear from the context.

There is also the photon, the quantum of electromagnetic radiation, recognized by Einstein as early as 1905. Its existence, he realized, was a necessary consequence of the quantum theory of Planck (Germany) who had concluded – from a rather subtle argument about heat radiation – that radiation must be emitted and absorbed, not continuously as people had previously thought, but in packets whose energy-content was proportional to the frequency of oscillation in the radiation in question. For instance, the main part of the gamma radiation of radium has an oscillation frequency of $5{\cdot}3 \times 10^{20}$ per second, which corresponds with an energy content of 2·2 MeV for photons of this radiation. The frequency of visible light is a million times lower and hence its photons have a million times less energy; on the other hand, photons a million times more energetic are found in the cosmic radiation. But all these photons are basically the same thing, only endowed with different energies.

Next on my list is the neutrino, which has an extremely interesting story. Almost as soon as radioactivity was discovered at the turn of the century the various types of radiation were roughly classified and labelled with Greek letters. I have already mentioned alpha particles, the least penetrating of all, which were later identified as fast helium nuclei, and gamma rays, which are energetic photons rather like X-rays. Those with intermediate penetrating power were called beta rays and were soon recognized as fast electrons. Some of them were just atomic electrons which had been set in motion by energy from the nucleus, but others came out of the nucleus itself, created in the transformation of a neutron into a proton. It was possible to calculate with what energy they ought to come out (from the mass of the nucleus before and after), and the disturbing fact emerged that they never came out with this full energy, but with a distribution of lower energies. In each of these 'beta transformations' a random amount of energy was missing and could not be traced.

Where something is missing there must be a thief, so a young Austrian theoretician, W. Pauli (in Switzerland), suggested that the beta transformation consisted in the emission, not of one particle (the electron that is observed), but of two, which share the available energy, one of them escaping unobserved. Lead blocks had been set up by the experimental physicists in order to trap the missing energy, and Pauli had to assume that his unobserved particle was penetrating enough to go through all of them, and so had to be electrically neutral. It was therefore soon nicknamed *il neutrino* (the little neutral one) by the Italian physicist E. Fermi, who was the first to take it seriously. It also had to be very light in order to give good agreement with the observed distribution of the electrons. More and more indirect evidence accumulated that Pauli had been right, but the neutrino itself escaped all the traps set for it and proved to be millions of times more elusive than Pauli had assumed at first. Even so it could not be completely elusive: if atomic nuclei could send out neutrinos they must also be capable of stopping them. It was possible to calculate the minimum stopping-power and it turned out to be extremely small: the chance that out of a million neutrinos traversing the entire earth a single one could be stopped was about one in a million. Yet, in 1956, it was announced that neutrinos had been caught and their existence definitely confirmed. This was made possible by using an atomic pile as a source (some 10^{18} neutrinos every second) and by means of special counters capable of detecting any neutrino that got stopped in a cubic yard or so of scintillating fluid. Even so only a few neutrinos per hour were recorded, but this was enough to identify them, and a proud day it was for Pauli, whose bold guess of some twenty-five years ago had at last been fully vindicated.

There are also antineutrinos, and the difference between the two is that they spin in opposite directions about their direction of flight, like bullets fired respectively from barrels with right-handed and left-handed rifling. You may object that this is not a valid distinction, that to an observer overtaking a neutrino its direction of flight would appear the opposite, and hence it would look like an antineutrino to him. But a neutrino always travels with the speed of light and thus cannot be overtaken since no object can move faster. It looks like a neutrino to any observer, whichever way and however fast he moves.

You may feel that I have spent too much time on an almost unobservable particle, but in the first place I think the physicists can be proud of having traced and pinned down something so elusive, and secondly, the effects of neutrinos in the large are not negligible. The energy of the sun and stars comes from nuclear transformations in their interiors. Some of the transformations are of the beta kind, and it has been computed that no less than about one-sixth of the energy is taken away by neutrinos and lost in the depths of space.

How did the photon, the quantum of electromagnetic radiation, come to be discovered? Our eyes being sensitive to a particular kind of radiation which we call light, once the laws of the quantum theory were understood, the existence of light quanta or photons followed of necessity. But would a race of blind physicists have discovered the photon? I think they would. They could have discovered electric and magnetic phenomena and found out their laws, and from these laws they would have deduced the existence of electromagnetic waves (just as Maxwell did in 1864, though the fact that he knew light might have helped him!) and hence of photons. Once the electromagnetic forces were known the rest could be done by mathematics alone.

Do other forces also give rise to waves and hence to associated quanta? What about gravity? One can show that gravitational waves (if they exist; the mathematicians are still arguing) would have exceedingly small effects, and so the 'graviton' – the quantum of gravitational waves – is hardly even a matter of speculation. It is certainly not identical with the neutrino though some people suspect the neutrino may in some way be connected with gravity.

A force that is much more important on the subatomic scale is the so-called nuclear force which holds the protons and neutrons together in nuclei. Some twenty-five years ago a young Japanese mathematician, H. Yukawa, began to wonder whether the nuclear forces might be capable of wave motion, and what the associated quanta would be like. His task was difficult, for very little was known about the nuclear force, except that it fell off much more rapidly with distance than the electric force, *i.e.*, much more rapidly than the inverse square of the distance. But just how it fell off was not known. Yukawa did what mathematicians do in such cases: he made the simplest assumption that was compatible with the scanty experimental data, and went ahead. His result was startling: the quanta would be heavy, with a mass about 300 times that of the electron, and they might carry a positive or negative electric charge, equal to that of a proton. They would be very different from the photon which has no charge and no intrinsic mass.

Yukawa's heavy quanta were eventually (1947) discovered in cosmic rays and they are now usually called pions (brief for pi-mesons; the term meson denoting the fact that their mass is intermediate between that of the electron and the proton). They are rare in nature, at least at sea level, but high up in the stratosphere they are quite common because there they are constantly produced by the impact of the fast cosmic ray particles entering the atmosphere from outside. As Yukawa had guessed, they are unstable and live on an average for only one forty-millionth of a second. Yukawa expected that they would break up into an electron and a neutrino, and indeed occasionally they do so, but mostly they break up into a neutrino and a particle that was quite unexpected: the muon (brief for mu-meson). Muons are much tougher than the original pions. They live about 80 times longer and can go clean through atomic nuclei: they can travel right down through the atmosphere, and every square inch of ground is struck by several muons a minute. Muons are lighter than the pions – only 207 times as heavy as electrons, while pions are 273 times as heavy. Being much more common near sea level, they were the first to be discovered (1937) and at first were thought to be Yukawa's 'heavy quanta.' The war interfered with their study, and it was not until 1947 – when pions were discovered – that the confusion was straightened out.

In some ways the muon is the most mysterious of the new particles – precisely because it is so commonplace. Apart from its instability (it breaks up into an electron, a neutrino, and an antineutrino) it is just an overweight electron. It resembles the electron completely in every respect, that is, spin, magnetic properties, and indifference to nuclear forces. Why the electron should exist in two sizes we do not know. None of the other particles do.

A few years ago there was a brief flutter of excitement when, for a short time, it looked as if the muon might be tremendously important as the key that would unlock the energy of fusion reactions. By a fusion reaction is meant a collision between two light nuclei that results in the formation of a heavier nucleus from them. This process, on a grand scale, keeps the sun hot. Explosively it is the energy source in the hydrogen bomb. In each case, a temperature of millions of degrees is needed to make the process start, but the muon, it was found, could cause 'cold fusion.' In an ordinary hydrogen molecule, two hydrogen nuclei are held together by the electrons that circle around them; the distance – about one 400-millionth of an inch – is too large for fusion to happen. But when a muon is present it will tend to take the place of one of the electrons and will describe an orbit 200 times smaller; the two nuclei will be pulled 200 times closer together, and fusion will quickly occur. (One nucleus must be 'heavy hydrogen,' that is, the hydrogen isotope of mass number 2; two protons will not fuse.) The muon merely acts as a catalyst; when fusion occurs it finds itself loose again and immediately starts to round up another pair of nuclei. Unfortunately each fusion requires about a millionth of a second, and so the muon in its short life cannot do the trick more than once or twice, and the energy it liberates is much less than that needed to produce the muon in the first place. The excitement passed, but the episode showed again how the most academic type of research can perhaps lead to important industrial applications. If the muon had happened to have a lifetime a few thousand times longer, the process would have worked.

Both the pion and the muon exist with either a positive or a negative charge, one the antiparticle of the other. In addition, there is a neutral pion; this has an extremely short life, millions of times shorter even than that of the charged pion, and breaks up into two photons. A photon must be considered to be its own antiparticle, for if all the electromagnetic fields that make up a photon are reversed, the photon is still the same as it was before. The neutral pion must also be considered its own antiparticle since it breaks up into two photons almost at once.

We have now dealt with 14 particles: the electron, proton, neutron, muon, and neutrino, each with its antiparticle; and the photon and the three pions – positive, neutral, and negative.

There are 16 others which are called the 'strange particles.' Some are heavier than protons and are called hyperons; there are six of these – each with its antiparticle, not all of which have as yet been observed. The others are about half the weight of a proton and are called kaons (brief for *K*-mesons). They were called strange particles because when they were first discovered (from 1948 onward) their behavior was very puzzling. Even now, though we have a scheme that accounts for most of their properties, they are still pretty mysterious.

Consider, for example, the particle that comes next in mass after the proton. It is called the Lambda particle (written with a capital L because its symbol is Λ, Greek capital lambda). It is uncharged and after a short life of about one 10,000-millionth of a second it breaks up into a proton and a negative pion. It can also be made by what looks like the inverse process, namely by hitting a proton with a negative pion; but the pion needs much more energy than one would compute – by Einstein's formula – from the masses of the particles concerned. It turns out that together with a Lambda, a kaon is always produced; neither a kaon nor a Lambda can be produced alone. The present explanation is that the kaon and the Lambda are saddled with opposite amounts of something which has been given the slightly facetious name 'strangeness' and which cannot be created in the brief space of a collision. In this it resembles an electric charge, but it has no other physical effect, and it is not permanent. It takes some time to disappear and therefore delays the break-up of the Lambda which otherwise ought to occur much more rapidly. The same scheme applies also to the three Sigma particles (positive, neutral, and negative), which come next in mass. But beyond these are two particles – the Xi particles (negative and neutral) which have twice as much 'strangeness.' They are therefore produced with two kaons (not one) and they break up in two steps with the Lambda as an intermediate stage. The theory accounts for a great many complexities of behavior, but what 'strangeness' is we do not know.

There are two kaons (positive and neutral) and six hyperons. If we add the antiparticles, we have 16 strange particles. Together with the 14 mentioned earlier that makes 30. Are they all fundamental? We are fairly certain that none of them is 'simply' a compound of two or more. But perhaps some are more fundamental than others. There is still a great deal we do not know.

The many new particles discovered during the 1960s and 70s led physicists to consider the possibility that they were not all fundamental, but composed of smaller units of matter. In 1963 the American physicist Murray Gell-Mann introduced the concept of quarks (a term borrowed from James Joyce's *Finnegans Wake*) which he proposed as fundamental constituents of matter. Another American physicist, George Zweig, developed a similar theory independently, calling his fundamental particles 'aces'. According to Gell-Mann's model quarks come in three types or 'flavours' ('up', 'down' and 'strange'). 'Up' and 'down' quarks are the constituents of protons and neutrons. 'Strange' quarks occur in K-mesons and in combination with other quarks or antiquarks, and the attraction between them increases as they go further apart. The mass-less particles that bind quarks together are called gluons. They are emitted and absorbed by quarks and they have the ability to create other gluons as they move between quarks.

In *The Quark and the Jaguar* (1994), Gell-Mann recalls the naming of quarks:

For a long time it was thought that among the particles accompanying the electron on the list of fundamental fermions [a class of elementary particles] would be the neutron and proton, the constituents of atomic nuclei. However, that turned out to be false; the neutron and proton are not elementary. Physicists have learned on other occasions as well that objects originally thought to be fundamental turn out to be made of smaller things. Molecules are composed of atoms. Atoms, although named from the Greek for uncuttable, are made of nuclei with electrons around them. Nuclei in turn are composed of neutrons and protons, as physicists began to understand around 1932, when the neutron was discovered. Now we know that the neutron and proton are themselves composite: they are made of quarks. Theorists are now quite sure that it is the quarks that are analogues of the electron. (If, as seems unlikely today, the quarks should turn out to be composite, then the electron would have to be composite as well.)

In 1963, when I assigned the name 'quark' to the fundamental constituents of the nucleon, I had the sound first, without the spelling, which could have been 'kwork'. Then, in one of my occasional perusals of *Finnegans Wake*, by James Joyce, I came across the word 'quark' in the phrase 'Three quarks for Muster Mark.' Since 'quark' (meaning, for one thing, the cry of a gull) was clearly intended to rhyme with 'Mark', as well as 'bark' and other such words, I had to find an excuse to pronounce it as 'kwork'. But the book represents the dream of a publican named Humphrey Chimpden Earwicker. Words in the text are typically drawn from several sources at once, like the 'portmanteau words' in *Through the Looking Glass*. From time to time, phrases occur in the book that are partially determined by calls for drinks at the bar. I argued, therefore, that perhaps one of the multiple sources of the cry 'Three quarks for Muster Mark' might be 'Three quarts for Mister Mark,' in which case the pronunciation 'kwork' would not be totally unjustified. In any case, the number three fitted perfectly the way quarks occur in nature.

The recipe for making a neutron or proton out of quarks is, roughly speaking, 'Take three quarks.' The proton is composed of two '*u* quarks' and one '*d* quark', while the neutron contains two '*d* quarks' and one '*u* quark'. The *u* and *d* quarks have different values of the electric charge. In the same units in which the electron has an electric charge of -1, the proton has a charge of $+1$, while the neutron has charge 0. The charge of the *u* quark in those same units is $\frac{2}{3}$ and that of the *d* quark $-\frac{1}{3}$. Sure enough, if we add $\frac{2}{3}$, $\frac{2}{3}$, and $-\frac{1}{3}$, we get 1 for the charge of the proton; and if we add $-\frac{1}{3}$, $-\frac{1}{3}$, and $\frac{2}{3}$, we get 0 for the charge of the neutron.

The *u* and *d* are said to be different 'flavors' of quark. Besides flavor, the quarks have another, even more important property that is called 'color', although it has no more to do with real color than flavor in this context has to do with the flavors of frozen yoghurt. While the name color is mostly a joke, it also serves as a kind of metaphor. There are three colors, labeled red, green, and blue after the three basic colors of light in a simple theory of human color vision. (In the case of paints, the three primary colors are often taken to be red, yellow, and blue, but for mixing lights instead of paints for their effect on human observers, yellow is replaced by green.) The recipe for a neutron or proton is to take one quark of each color, that is, a red quark, a green quark, and a blue quark, in such a way that color averages out. Since, in vision, white can be regarded as a mixture of red, green, and blue, we can use the metaphor to say that the neutron and proton are white.

Not many poets have written about atomic particles. An exception is John Updike:

Cosmic Gall

Every second, hundreds of billions of these neutrinos pass through each square inch of our bodies, coming from above during the day and from below at night, when the sun is shining on the other side of the earth!

From 'An Explanatory Statement of Elementary Particle Physics', by M. A. Ruderman and A. H. Rosenfeld, in American Scientist

Neutrinos, they are very small.
  They have no charge and have no mass
And do not interact at all.
The earth is just a silly ball
  To them, through which they simply pass,
Like dustmaids down a drafty hall
  Or photons through a sheet of glass.
  They snub the most exquisite gas,
Ignore the most substantial wall
  Cold-shoulder steel and sounding brass,
Insult the stallion in his stall
  And, scorning barriers of class,
Infiltrate you and me! Like tall
And painless guillotines, they fall
  Down through our heads into the grass.
At night, they enter at Nepal
  And pierce the lover and his lass
From underneath the bed – you call
  It wonderful; I call it crass.

Sources: *Science Survey I*, ed. A. W. Haslett and John St John, New York, Vista Books, 1960; Murray Gell-Mann, *The Quark and the Jaguar: Adventures in the Simple and the Complex*, Little, Brown and Company (UK) Ltd, London, 1994; John Updike, *Telephone Poles and Other Poems*, London, André Deutsch, 1964.

From Stardust to Flesh

First broadcast on BBC2 on 27 January 1977, Nigel Calder's TV programme *The Key to the Universe* communicated to a mass audience the revolutionary advances made during the 1970s in astronomy and subatomic physics. This excerpt is from the book version.

In a sense human flesh is made of stardust.

Every atom in the human body, excluding only the primordial hydrogen atoms, was fashioned in stars that formed, grew old and exploded most violently before the Sun and the Earth came into being. The explosions scattered the heavier elements as a fine dust through space. By the time it made the Sun, the primordial gas of the Milky Way was sufficiently enriched with heavier elements for rocky planets like the Earth to form. And from the rocks atoms escaped for eventual incorporation in living things: carbon, nitrogen, oxygen, phosphorus and sulphur for all living tissue; calcium for bones and teeth; sodium and potassium, indispensable for the workings of nerves and brains; the iron colouring blood red . . . and so on.

No other conclusion of modern research testified more clearly to mankind's intimate connections with the universe at large and with the cosmic forces at work among the stars. An American nuclear physicist, William Fowler, and three British astonomers, Geoffrey Burbidge, Margaret Burbidge, and Fred Hoyle, carried out a classic study (1957–64) on how the stars made the elements. One motive for it was a wish to show that the elements had not been made in the Big Bang, at the birth of the universe. Fred Hoyle in particular was a spirited opponent of the Big Bang theory, as one of the authors of the rival Steady State theory. While Steady State's main assertion of an unchanging universe perished, the particular argument that the stars made all but the lightest elements prevailed.

A medium-sized star like the Sun was known to burn steadily in the nuclear fashion for billions of years. When it eventually began to run out of hydrogen fuel it would swell and puff away some of its contents into the surrounding space, before collapsing into a white dwarf star. Stars substantially bigger than the Sun burned much more fiercely and quickly: they were 'blue-hot' instead of white-hot. Because of their greater mass the force of gravity, acting like a pressure cooker, kept a big star hot and dense and so allowed more thorough stewing of the material of the stars.

And the big stars eventually exploded. In our galaxy, the Milky Way, such events were clearly seen only five times in a thousand years. But remains of stars that had exploded were quite plentiful. Arc-shaped clouds of dispersing debris glowed faintly among the other stars. More strident were the pulsars, the immensely compressed cores of exploding stars. They stood flashing like police beacons, each marking the scene of a cosmic accident.

Stellar explosions did remarkable things to the nuclei of atoms. The medieval alchemists had tried to change one chemical element into another, especially hoping to make gold. Their successors in the twentieth century could say why their efforts were in vain. The essential character of an element was fixed by the number of protons (positively charged particles) in the nucleus of each of its atoms. You could transmute an element only by reaching into the nucleus itself, which the alchemists had no means of doing. But stars were playing the alchemist all the time.

Stars in the normal state, whether big or small, burned the lightest element, hydrogen, and formed from it helium, the next heaviest element. The process gave off copious energy. In very massive stars, or in less massive stars going through a phase of internal collapse, the temperature might climb high enough for the helium to burn. It changed into carbon and oxygen, with a further release of energy. Then the carbon and oxygen could burn, too, to form still heavier elements.

The escalation through the table of elements became progressively more difficult. The heavier the element, the more protons it had in each nucleus, and the more powerful was the electric repulsion between two nuclei, preventing them from fusing together. By the time you wanted oxygen to burn to make sulphur and silicon, or silicon to burn to make iron, you needed temperatures of billions of degrees so that the nuclei were colliding with sufficient frenzy to crash through the electric barrier. Iron-making marked the limit to nuclear burning in stars, and there was known to be a great deal of iron about. The Earth inherited a huge core of molten iron and meteorites often contained iron, too, all of it forged in stars. If the nuclear forces had their way, the whole universe would consist of iron.

After iron, the making of heavier elements in stars began to consume energy rather than releasing it. No star could earn a steady living that way. But in the explosion of a big star some of the enormous energy released went into building up dozens of chemical elements heavier than iron: gold, lead . . . all the way through the table of elements to uranium and beyond. Even so, heavy elements remained far less abundant in the cosmos than the lighter elements.

Many of the atoms so formed, and later incorporated into the Earth, were radioactive. Their nuclei were overcharged with energy and unable to survive indefinitely. But 'not indefinitely' could mean billions of years. From uranium, thorium, potassium and other radioactive elements, energy stored during the explosions of the ancestral stars slowly trickled out into the rocks of the Earth. It generated the heat that fired volcanoes, shifted the continents and built mountains. The great creakings called earthquakes, which accompanied these processes, were thus direct consequences – albeit greatly delayed and translocated – of those stellar explosions that made the stuff of the Earth available.

The idea that all living things – humans, gnats, slugs, trees – have their origin in stardust, and the theory, referred to by Calder, that the universe is united in a constant process of creation and destruction, provide the key to Ted Hughes's difficult poem 'Fire-Eater':

> Those stars are the fleshed forebears
> Of these dark hills, bowed like labourers,
>
> And of my blood.
>
> The death of a gnat is a star's mouth: its skin,
> Like Mary's or Semele's, thin
>
> As the skin of fire:
> A star fell on her, a sun devoured her.
>
> My appetite is good
> Now to manage both Orion and Dog

With a mouthful of earth, my staple.
Worm-sort, root-sort, going where it is profitable.

A star pierces the slug,

The tree is caught up in the constellations.
My skull burrows among antennae and fronds.

Sources: Nigel Calder, *The Key to the Universe: A Report on the New Physics*, London, British Broadcasting Corporation, 1977. Ted Hughes, *Lupercal*, London, Faber and Faber, 1960.

Black Holes

The son of a Russian Jewish immigrant candy-store owner in New York City, Isaac Asimov (1920–92) began writing science fiction in his teens, and became the twentieth century's most prolific as well as its most masterly, lucid and imaginative explainer of science to the common reader. This matchlessly clear and compact account of black holes was first printed in the *Daily Telegraph* in 1979.

Of all the odd creatures in the astronomical zoo, the 'black hole' is the oddest. To understand it, concentrate on gravity.

Every piece of matter produces a gravitational field. The larger the piece, the larger the field. What's more, the field grows more intense the closer you move to its center. If a large object is squeezed into a smaller volume, its surface is nearer its center and the gravitational pull on that surface is stronger.

Anything on the surface of a large body is in the grip of its gravity, and in order to escape it must move rapidly. If it moves rapidly enough, then even though gravitational pull slows it down continually it can move sufficiently far away from the body so that the gravitational pull, weakened by distance, can never quite slow its motion to zero.

The minimum speed required for this is the 'escape velocity.' From the surface of the earth, the escape velocity is 7.0 miles per second. From Jupiter, which is larger, the escape velocity is 37.6 miles per second. From the sun, which is still larger, the escape velocity is 383.4 miles per second.

Imagine all the matter of the sun (which is a ball of hot gas 864,000 miles across) compressed tightly together. Imagine it compressed so tightly that its atoms smash and it becomes a ball of atomic nuclei and loose electrons, 30,000 miles across. The sun would then be a 'white dwarf.' Its surface would be nearer its center, the gravitational pull on that surface would be stronger, and escape velocity would now be 2,100 miles per second.

Compress the sun still more to the point where the electrons melt into the nuclei. There would then be nothing left but tiny neutrons, and they will move together till they touch. The sun would then be only 9 miles across, and it would be a 'neutron star.' Escape velocity would be 120,000 miles per second.

Few things material could get away from a neutron star, but light could, of course, since light moves at 186,282 miles per second.

Imagine the sun shrinking past the neutron-star stage, with the neutrons smashing and collapsing. By the time the sun is 3.6 miles across, escape velocity has passed the speed of light, and light can no longer escape. Since nothing can go faster than light, *nothing* can escape.

Into such a shrunken sun anything might fall, but nothing can come out. It would be like an endlessly deep hole in space. Since not even light can come out, it is utterly dark – it is a 'black hole.'

In 1939, J. Robert Oppenheimer first worked out the nature of black holes in the light of the laws of modern physics, and ever since astronomers have wondered if black holes exist in fact as well as in theory.

How would they form? Stars would collapse under their own enormous gravity were it not for the enormous heat they develop, which keeps them expanded. The heat is formed by the fusion of hydrogen nuclei, however, and when the hydrogen is used up the star collapses.

A star like our sun will eventually collapse fairly quietly to a white dwarf. A more massive star will explode before it collapses, losing some of its mass in the process. If the portion that survives the explosion and collapses is more than 1.4 times the mass of the sun, it will surely collapse into a neutron star. If it is more than 3.2 times the mass of the sun, it must collapse into a black hole.

Since there are indeed massive stars, some of them have collapsed by now and formed black holes. But how can we detect one? Black holes are only a few miles across after all, give off no radiation, and are trillions of miles away.

There's one way out. If matter falls into a black hole, it gives off X-rays in the process. If a black hole is collecting a great deal of matter, enough X-rays may be given off for us to detect them.

Suppose two massive stars are circling each other in close proximity. One explodes and collapses into a black hole. The two objects continue to circle each other, but as the second star approaches explosion it expands. As it expands, some of its matter spirals into the black hole, and there is an intense radiation of X-rays as a result.

In 1965, an X-ray source was discovered in the constellation Cygnus and was named 'Cygnus X-1.' Eventually, the source was pinpointed to the near neighborhood of a dim star, HD-226868, which is only dim because it is 10,000 light-years away. Actually, it is a huge star, 30 times the mass of our sun.

That star is one of a pair and the two are circling each other once every 5.6 days. The X-rays are coming from the other star, the companion of HD-226868. That companion is Cygnus X-1. From the motion of HD-226868, it is possible to calculate that Cygnus X-1 is 5 to 8 times the mass of our sun.

A star of that mass should be visible if it is an ordinary star, but no telescope can detect any star on the spot where X-rays are emerging. Cygnus X-1 must be a collapsed star that is too small to see. Since Cygnus X-1 is at least 5 times as massive as our sun, it is too massive to be a white dwarf; too massive, even, to be a neutron star.

It can be nothing other than a black hole; the first to be discovered.

Source: Isaac Asimov, *The Roving Mind: A Panoramic View of Fringe Science, Technology, and the Society of the Future*, London, Oxford University Press, 1987.

The Fall-Out Planet

The Gaia hypothesis was the brainchild of the scientist J. E. Lovelock, but its name was suggested by his friend, the novelist William Golding. Gaia was the Greek earth-goddess, also known as Ge, and the hypothesis states that the biosphere (i.e. the whole region of the earth's surface, the sea and the air that is inhabited by living organisms) is a self-regulating entity with the capacity to keep our planet healthy by controlling the chemical and physical environment. This extract is taken from Lovelock's *Gaia: A New Look at Life on Earth*.

It seems almost certain that close in time and space to the origin of our solar system, there was a supernova event. A supernova is the explosion of a large star. Astronomers speculate that this fate may overtake a star in the following manner: as a star burns, mostly by fusion of its hydrogen and, later, helium atoms, the ashes of its fire in the form of other heavier elements such as silicon and iron accumulate at the centre. If this core of dead elements, no longer generating heat and pressure, should much exceed the mass of our own sun, the inexorable force of its own weight will be enough to cause its collapse in a matter of seconds to a body no larger than a few thousand cubic miles in volume, although still as heavy as a star. The birth of this extraordinary object, a neutron star, is a catastrophe of cosmic dimensions. Although the details of this and other similar catastrophic processes are still obscure, it is obvious that we have here, in the death throes of a large star, all the ingredients for a vast nuclear explosion. The stupendous amount of light, heat, and hard radiation produced by a supernova event equals at its peak the total output of all the other stars in the galaxy.

Explosions are seldom one hundred per cent efficient. When a star ends as a supernova, the nuclear explosive material, which includes uranium and plutonium together with large amounts of iron and other burnt-out elements, is distributed around and scattered in space just as is the dust cloud from a hydrogen bomb test. Perhaps the strangest fact of all about our planet is that it consists largely of lumps of fall-out from a star-sized hydrogen bomb. Even today, aeons later, there is still enough of the unstable explosive material remaining in the Earth's crust to enable the reconstitution on a minute scale of the original event.

Binary, or double, star systems are quite common in our galaxy, and it may be that at one time our sun, that quiet and well-behaved body, had a large companion which rapidly consumed its store of hydrogen and ended as a supernova. Or it may be that the debris of a nearby supernova explosion mingled with the swirl of interstellar dust and gases from which the sun and its planets were condensing. In either case, our solar system must have been formed in close conjunction with a supernova event. There is no other credible explanation of the great quantity of exploding atoms still present on the Earth. The most primitive and old-fashioned Geiger counter will indicate that we stand on fall-out from a vast nuclear explosion. Within our bodies, no less than three million atoms rendered unstable in that event still erupt every minute, releasing a tiny fraction of the energy stored from that fierce fire of long ago.

The Earth's present stock of uranium contains only 0.72 per cent of the dangerous isotope U235. From this figure it is easy to calculate that about four aeons ago the uranium in the Earth's crust would have been nearly 15 per cent U235. Believe it or not, nuclear reactors have existed since long before man, and a fossil natural nuclear reactor was recently discovered in Gabon, in Africa. It was in action two aeons ago when U235 was only a few per cent. We can therefore be fairly certain that the geochemical concentration of uranium four aeons ago could have led to spectacular displays of natural nuclear reactions. In the current fashionable denigration of technology, it is easy to forget that nuclear fission is a natural process. If something as intricate as life can assemble by accident, we need not marvel at the fission reactor, a relatively simple contraption, doing likewise.

Thus life probably began under conditions of radioactivity far more intense than those which trouble the minds of certain present-day environmentalists. Moreover, there was neither free oxygen nor ozone in the air, so that the surface of the Earth would have been exposed to the fierce unfiltered ultra-violet radiation of the sun. The hazards of nuclear and of ultra-violet radiation are much in mind these days and some fear that they may destroy all life on Earth. Yet the very womb of life was flooded by the light of these fierce energies.

Source: J. E. Lovelock, *Gaia: A New Look at Life on Earth*, London, Oxford University Press, 1979.

Galactic Diary of an Edwardian Lady

The 'big bang' theory of the beginning of the universe was originally proposed by A. G. E. Lemaitre in 1927 and revised by George Gamow in 1946. According to the theory the universe began about 15 billion years ago in a hot, dense explosive phase. This would explain why, as the astronomer Edwin Hubble discovered in the late 1920s, the universe is expanding, with the galaxies receding from us and from one another. In 1965 two American radio engineers accidentally discovered microwave background radiation, seemingly coming from all directions in space, which is believed to be a remnant of the primordial fireball of the big bang. Edward Larrissy's poem commemorates the fact that two best-selling books of recent years have been *The Country Diary of An Edwardian Lady* and Stephen Hawking's *A Brief History of Time*, which is about big-bang theory.

In the beginning was a black bomb
That blew apart. A blinding smoke
Kept growing, growing

To a tropical fog, intolerably bright.
From this, white whorls of moonshine mist
Distilled, and then distilled

To petal-eddies on a dark pool.
And now they spin in clusters
Farther and farther apart

Like shining catkins, twisted into spools.
All forms, all time, all complexity,
From the first snowdrop to muffins and tea

Lay in that round black bomb
And will return there
When the hot afternoon is done.

Source: Edward Larrissy. Poem printed in the *Independent*, February, 1994.

The Light of Common Day

One of the greatest and most prolific writers of popular science and science fiction, Arthur C. Clarke was born in Minehead, Somerset, in 1917. As a child he made a map of the moon, using a home-made telescope, and started writing short stories while at Huish's Grammar School, Taunton, under the influence of the English master, Captain B. E. ('Mitty') Mitford, who would assemble his budding writers once a week after school round a table on which was a large bag of toffees. Unable to afford a university education, Clarke worked as an auditor and, at the outbreak of the Second World War, went into the RAF as a radar instructor. He began publishing science-fiction stories towards the end of the war. His fiction anticipates various developments in space technology, and in a 1945 article he correctly predicted the development of satellite radio and TV. In the 1960s he collaborated with Stanley Kubrick on the film *2001: A Space Odyssey*, based on a Clarke short story. The following essay, first published in 1963, was reprinted in his collection *By Space Possessed* (1993). Clarke says: 'I am particularly proud of the concluding paragraphs.'

No man has ever seen the Sun, or ever will. What we call 'sunlight' is only a narrow span of the entire solar spectrum – the immensely broad band of vibrations which the Sun, our nearest star, pours into space. All the colours visible to the eye, from warm red to deepest violet, lie within a single octave of this band – for the waves of violet light have twice the frequency, or 'pitch' if we think in musical terms, of red. On either side of this narrow zone are ranged octave after octave of radiations to which we are totally blind.

The musical analogy is a useful one. Think of one octave on the piano – less than the span of the average hand. Imagine that you were deaf to all notes outside this range; how much, then, could you appreciate of a full orchestral score when everything from contra-bassoon to piccolo is going full blast? Obviously you could get only the faintest idea of the composer's intentions. In the same way, by eye alone we can obtain only a grossly restricted conception of the true 'colour' of the world around us.

However, let us not exaggerate our visual handicap. Though visible light is merely a single octave of the Sun's radiation, this octave contains most of the power; the higher and lower frequencies are relatively feeble. It is, of course, no coincidence that our eyes are adapted to the most intense band of sunlight; if that band had been somewhere else in the spectrum, as is the case with other stars, evolution would have given us eyes appropriately tuned.

Nevertheless, the Sun's invisible rays are extremely important, and affect our lives in a manner undreamed of only a few years ago. Some of them, indeed, may control our destinies – and even, as we shall see in a moment, our very existence.

The visible spectrum is, quite arbitrarily, divided up into seven primary colours – the famous sequence, red, orange, yellow, green, blue, indigo, violet, if we start from the longest waves and work down to the shortest. Seven main colours in the one octave; but the complete band of solar radiations covers at least thirty octaves, or a total frequency range of ten thousand million to one. If we could see the whole of it, therefore, we might expect to discern more than two hundred colours as distinct from each other as orange is from yellow, or green is from blue.

Starting with the Sun's visible rays, let us explore outwards in each direction and see (though that word is hardly applicable) what we can discover. On the long-wave side we come first to the infra-red rays, which can be perceived by our skin but not by our eyes. Infra-red rays are heat radiation; go out of doors on a summer's day, and you can tell where the Sun is even though your eyes may be tightly closed.

Thanks to special photographic films, we have all had glimpses of the world of infra-red. It is an easily recognizable world, though tone values are strangely distorted. Sky and water are black, leaves and grass dazzling white as if covered with snow. It is a world of clear, far horizons, for infra-red rays slice through the normal haze of distance – hence their great value in aerial photography.

The further we go down into the infra-red, the stranger are the sights we encounter and the harder it becomes to relate them to the world of our normal senses. It is only very recently (partly under the spur of guided missile development) that we have invented sensing devices that can operate in the far infra-red. They see the world of heat; they can 'look' at a man wearing a brilliantly coloured shirt and smoking a cigarette – and see only the glowing tip. They can also look down on a landscape hidden in the darkness of night and see all the sources of heat from factories, cars, taxiing aircraft. Hours after a jet has taken off, they can still read its signature on the warm runway.

Some animals have developed an infra-red sense, to enable them to hunt at night. There is a snake which has two small pits near its nostrils, each holding a directional infra-red detector. These allow it to 'home' upon small, warm animals like mice, and to strike at them even in complete darkness. Only in the last decade have our guided missiles learned the same trick.

Below the infra-red, for several octaves, is a no man's land of radiation about which very little is known. It is hard to generate or to detect waves in this region, and until recently few scientists paid it much attention. But as we press on to more familiar territory, first we encounter the inch-long waves of radar, then the yard-long one of the shortwave bands, then the hundred-yard waves of the broadcast band.

The existence of all these radiations was quite unknown a century ago; today, of course, they are among the most important tools of our civilization. It is a bare twenty years since we discovered that the Sun also produces them, on a scale we cannot hope to match with our puny transmitters.

The Sun's radio output differs profoundly from its visible light, and the difference is not merely one of greater length. Visible sunlight is practically constant in intensity; if there are any fluctuations, they are too slight to be detected. Not only has the Sun shone with unvarying brightness throughout the whole span of human history, but we would probably notice no difference if we could see it through the eyes of one of the great reptiles.

But if you saw only the 'radio' Sun, you would never guess that it was the same object. Most of the time it is very dim – much dimmer, in fact, than many other celestial bodies. To the eye able to see only by radio waves, there would be little difference between day and night; the rising of the Sun would be a minor and inconspicuous event.

From time to time, however, the radio Sun explodes into nova brightness. It may, within *seconds*, flare up to a hundred, a thousand or even a million times its normal brilliance. These colossal outbursts of radio energy do not come from the Sun as a whole, but from small localized areas of the solar disc, often associated with sunspots.

This is one excellent reason why no animals have ever developed radio senses. Most of the time, such a sense would be useless, because the radio landscape would be completely dark – there would be no source of illumination.

In any event, 'radio eyes' would pose some major biological problems, because radio waves are millions of times larger than normal eyes, if they were to have the same definition. Even a radio eye which showed the world as fuzzily as a badly out-of-focus TV picture would have to be hundreds of yards in diameter; the gigantic antennas of our radar systems and radio telescopes dramatize the problem involved. If creatures with radio senses do exist anywhere in the Universe, they must be far larger than whales and can, therefore, only be inhabitants of gravity-free space.

Meanwhile, back on Earth, let us consider the other end of the spectrum – the rays shorter than visible light. As the blue deepens into indigo and then violet, the human eye soon fails to respond. But there is still 'light' present in solar radiation: the ultraviolet. As in the case of the infra-red, our skins can react to it, often painfully; for ultraviolet rays are the cause of sunburn.

And here is a very strange and little-known fact. Though I have just stated that our eyes do not respond to ultraviolet, the actual situation is a good deal more complicated. (In nature, it usually is.) The sensitive screen at the back of the eye – the retina, which is the precise equivalent of the film in a camera – does react strongly to ultraviolet. If it were the only factor involved, we could see by the invisible ultraviolet rays.

Then why don't we? For a purely technical reason. Though the eye is an evolutionary marvel, it is a very poor piece of optics. To enable it to work properly over the whole range of colours, a good camera has to have four, six or even more lenses, made of different types of glass and assembled with great care into a single unit. The eye has only one lens, and it already has trouble coping with the two-to-one range of wavelengths in the visible spectrum. You can prove this by looking at a bright red object on a bright blue background. They won't both be in perfect focus; when you look at one, the other will appear slightly fuzzy.

Objects would be even fuzzier if we could see by ultraviolet as well as by visible light, so the eye deals with this insoluble problem by eliminating it. There is a filter in the front of the eye which blocks the ultraviolet, preventing it from reaching the retina. The haze filter which photographers often employ when using colour film does exactly the same job, and for a somewhat similar reason.

The eye's filter is the lens itself – and here at last is the punch line of this rather long-winded narrative. If you are ever unlucky enough to lose your natural lenses (say through a cataract operation) and have them replaced by artificial lenses of clear glass, you will be able to see quite well in the ultraviolet. Indeed, with a source of ultraviolet illumination, like the so-called 'black light' lamps, you will be able to see perfectly in what is, to the normal person, complete darkness! I hereby donate this valuable information to the CIA, James Bond, or anyone else who is interested.

Normal sunlight, as you can discover during a day at the beach, contains plenty of ultraviolet. It all lies, however, in a narrow band – the single octave just above the visible spectrum in frequency. As we move beyond this to still higher frequencies, the scene suddenly dims and darkens. A being able to see only in the far ultraviolet would be in a very unfortunate position. To him, it would always be night, whether or not the Sun was above the horizon.

What has happened? Doesn't the Sun radiate in the far ultraviolet? Certainly it does, but this radiation is all blocked by the atmosphere, miles above our head. In the far ultraviolet, a few inches of ordinary air are as opaque as a sheet of metal.

Only with the development of rocket-borne instruments has it become possible to study this unknown region of the solar spectrum – a region, incidentally, which contains vital information about the Sun and the processes which power it by the atmosphere, miles above our head. In the far ultraviolet, if you started off from ground level on a bright, sunny day, this is what you would see.

At first, you would be in utter darkness, even though you were looking straight at the Sun. Then, about twenty miles up, you would notice a slow brightening, as you climbed through the opaque fog of the atmosphere. Beyond this, between twenty and thirty miles high, the ultraviolet Sun would break through in its awful glory.

I use that word 'awful' with deliberate intent. These rays can kill, and swiftly. They do not bother astronauts, because they can be easily filtered out by special glass. But if they reached the surface of the Earth – if they were not blocked by the upper atmosphere – most existing forms of life would be wiped out.

If you regard the existence of this invisible ultraviolet umbrella as in any way providential, you are confusing cause and effect. The screen was not put in the atmosphere to protect terrestrial life: it was put there by life itself, hundreds of millions of years before man appeared on Earth.

The Sun's raw ultraviolet rays, in all probability, did reach the surface of the primeval Earth; the earliest forms of life were adapted to it, perhaps even thrived upon it. In those days, there was no oxygen in the atmosphere; it is a by-product of plant life, and over geological aeons its amount slowly increased, until at last those oxygen-burning creatures called animals had a chance to thrive.

That filter in the sky is made of oxygen – or, rather, the grouping of three oxygen atoms known as ozone. Not until Earth's protective ozone layer was formed, and the short ultraviolet rays were blocked twenty miles up, did the present types of terrestrial life evolve. If there had been no ozone layer, they would doubtless have evolved into different forms. Perhaps we might still be here, but our skins would be very, very black.*

Life on Mars must face this problem, for that planet has no oxygen in its atmosphere and, therefore, no ozone layer. The far ultraviolet rays reach the Martian surface unhindered, and must profoundly affect all living matter there. It has been suggested that these rays are responsible for the colour changes which astronomers have observed on the planet. Whether or not this is true, we can predict that one of the occupational hazards of Martian explorers will be severe sunburn.

Just as ultraviolet lies beyond the violet, so still shorter rays lie beyond it. These are X-rays, which are roughly a thousand times shorter than visible light. Like the ultraviolet, these even more dangerous rays are blocked by the atmosphere; few of them come to within a hundred miles of Earth, and they have been detected by rocket instruments only during the last few years. The solar X-rays are quite feeble – only a millionth of the intensity of visible light – but their importance is much greater than this figure would indicate. We know now that blasts of X-rays from the Sun, impinging upon the upper atmosphere, can produce violent changes in radio communications, even to the extent of complete blackouts.

Men have lost their lives because the Sun has disrupted radio; nations are equally vulnerable, in this age of the ICBM.

You will recall that though the Sun shines with remarkable steadiness in the visible spectrum, it flares and sparkles furiously on

*I never imagined that, thirty years later, the ozone layer would be headline news!

the long (radio) waves. Exactly the same thing happens with its X-ray emission, even though these waves are a billion times shorter. Moreover, both the Sun's radio waves and its X-rays appear to come from the same localized areas of the solar surface – disturbed regions in the neighbourhood of sunspots, where clouds of incandescent gas larger than the Earth erupt into space at hundreds of miles a second.

For reasons not yet understood (there is not much about the Sun that we do thoroughly understand) solar activity rises and falls in an eleven-year cycle. The Sun was most active around 1957, which is why that date was chosen for the International Geophysical Year. In the 1960s it headed for a minimum but unfortunatley threatened to come back to the boil at around the time the first major space expeditions were being planned. The astronauts might have run into some heavy weather, since the Sun by then was shooting out not only vast quantities of ultraviolet, X-rays and radio waves, but other radiations which cannot be so easily blocked. (As it turned out, however, the risks were far less than had at one time been feared.)

We see, then, how complicated and how variable sunlight is, if we use that word in the widest sense to describe all the waves emitted by the Sun. Nevertheless, when we accept the evidence of our unaided eyes and describe the Sun as a yellow star, we have summed up the most important single fact about it – at *this* moment in time. It appears probable, however, that sunlight will be the colour we know for only a negligibly small part of the Sun's history.

For stars, like individuals, age and change. As we look out into space, we see around us stars at all stages of evolution. There are faint blood-red dwarfs so cool that their surface temperature is a mere 4,000 degrees Fahrenheit; there are searing ghosts blazing at 100,000 degrees, and almost too hot to be seen, for the greater part of their radiation is in the invisible ultraviolet. Obviously, the 'daylight' produced by any star depends upon its temperature; today (and for ages past, as for ages to come) our Sun is at about 10,000 degrees Fahrenheit, and this means that most of its light is concentrated in the yellow band of the spectrum, falling slowly in intensity towards both the longer and the shorter waves.

That yellow 'bump' will shift as the Sun evolves, and the light of day will change accordingly. It is natural to assume that as the Sun grows older and uses up its hydrogen fuel – which it is now doing at the spanking rate of half a billion tons *a second* – it will become steadily colder and redder.

But the evolution of a star is a highly complex matter, involving chains of interlocking nuclear reactions. According to one theory, the Sun is still growing hotter and will continue to do so for several billion years. Probably life will be able to adapt itself to these changes, unless they occur catastrophically, as would be the case if the Sun exploded into a nova. In any event, whatever the vicissitudes of the next five or ten billion years, at long last the Sun will settle down to the white dwarf stage.

It will be a tiny thing, not much bigger than the Earth, and therefore too small to show a disc to the naked eye. At first, it will be hotter than it is today, but because of its minute size it will radiate very little heat to its surviving planets. The daylight of that distant age will be as cold as moonlight, but much bluer, and the temperature of Earth will have fallen to 300 degrees below zero. If you think of mercury lamps on a freezing winter night, you have a faint mental picture of high noon in the year AD 7,000 million.

Yet that does not mean that life – even life as we know it today – will be impossible in the Solar System; it will simply have to move in towards the shrunken Sun. The construction of artificial planets would be child's play to the intelligences we can expect at this date; indeed, it will be child's play to us in a few hundred years' time.

Around the year 10,000 million the dwarf Sun will have cooled back to its present temperature, and hence to the yellow colour that we know today. From a body that was sufficiently close to it – say only a million miles away – it would look exactly like our present Sun, and would give just as much heat. There would be no way of telling, by eye alone, that it was actually a hundred times smaller, and a hundred times closer.

So matters may continue for another five billion years; but at last the inevitable will happen. Very slowly, the Sun will begin to cool, dropping from yellow down to red. Perhaps by the year 15,000 million it will become a red dwarf, with a surface temperature of a mere 4,000 degrees. It will be nearing the end of the evolutionary track, but reports of its death will be greatly exaggerated. For now comes one of the most remarkable, and certainly least appreciated, results of modern astrophysical theories.

When the Sun shrinks to a dull red dwarf, it will not be dying. It will just be starting to live – and *everything that has gone before will be merely a fleeting prelude to its real history.*

For a red dwarf, because it is so small and so cool, loses energy at such an incredibly slow rate that it can stay in business for *thousands* of times longer than a normal-sized white or yellow star. We must no longer talk in billions but of trillions of years if we are to measure its life span. Such figures are, of course, inconceivable. (For that matter, who can think of a thousand years?) But we can nevertheless put them into their right perspective if we relate the life of a star to the life of a man.

On this scale, the Sun is but a week old. Its flaming youth will continue for another month; then it will settle down to a sedate adult existence which may last at least eighty years.

Life has existed on this planet for two or three days of the week that has passed; the whole of human history lies within the last second, and there are eighty years to come.

In the wonderful closing pages of *The Time Machine*, the young H. G. Wells described the world of the far future, with a blood-red Sun hanging over a freezing sea. It is a sombre picture that chills the blood, but our reaction to it is wholly irrelevant and misleading. For we are creatures of the dawn, with eyes and senses adapted to the hot light of today's primeval Sun. Though we should miss beyond measure the blues and greens and violets which are the fading afterglow of Creation, they are all doomed to pass with the brief billion-year infancy of the stars.

But the eyes that will look upon that all-but-eternal crimson twilight will respond to the colours that we cannot see, because evolution will have moved their sensitivity away from the yellow, somewhere out beyond the visible red. The world of rainbow-hued heat they see will be as rich and colourful as ours – and as beautiful; for a melody is not lost if it is merely transposed an octave down into the bass.

So now we know that Shelley, who was right in so many things, was wrong when he wrote:

> Life, like a dome of many-coloured glass,
> Stains the white radiance of eternity.

For the radiance of eternity is not white: it is infra-red.

Source: Arthur C. Clarke, *By Space Possessed*, London, Gollancz, 1993.

Can We Know the Universe? Reflections on a Grain of Salt

Carl Sagan, Professor of Astronomy and Space Sciences at Cornell University, played a leading role in the Mariner, Viking and Voyager space programmes. Deeply interested in the possibility of life on other planets, he compiled the Voyager interstellar record – a message from earthdwellers to other civilizations in space. One of the most distinguished popular science writers, he won the Pulitzer Prize for *The Dragons of Eden: Speculations on the Evolution of Human Intelligence*. The following extract is from *Broca's Brain* (1979).

But to what extent can we *really* know the universe around us? Sometimes this question is posed by people who hope the answer will be in the negative, who are fearful of a universe in which everything might one day be known. And sometimes we hear pronouncements from scientists who confidently state that everything worth knowing will soon be known – or even is already known – and who paint pictures of a Dionysian or Polynesian age in which the zest for intellectual discovery has withered, to be replaced by a kind of subdued languor, the lotus eaters drinking fermented coconut milk or some other mild hallucinogen. In addition to maligning both the Polynesians, who were intrepid explorers (and whose brief respite in paradise is now sadly ending), as well as the inducements to intellectual discovery provided by some hallucinogens, this contention turns out to be trivially mistaken.

Let us approach a much more modest question: not whether we can know the universe or the Milky Way Galaxy or a star or a world. Can we know, ultimately and in detail, a grain of salt? Consider one microgram of table salt, a speck just barely large enough for someone with keen eyesight to make out without a microscope. In that grain of salt there are about 10^{16} sodium and chlorine atoms. This is a 1 followed by 16 zeros, 10 million billion atoms. If we wish to know a grain of salt, we must know at least the three-dimensional positions of each of these atoms. (In fact, there is much more to be known – for example, the nature of the forces between the atoms – but we are making only a modest calculation.) Now, is this number more or less than the number of things which the brain can know?

How much *can* the brain know? There are perhaps 10^{11} neurons in the brain, the circuit elements and switches that are responsible in their electrical and chemical activity for the functioning of our minds. A typical brain neuron has perhaps a thousand little wires, called dendrites, which connect it with its fellows. If, as seems likely, every bit of information in the brain corresponds to one of these connections, the total number of things knowable by the brain is no more than 10^{14}, one hundred trillion. But this number is only one percent of the number of atoms in our speck of salt.

So in this sense the universe is intractable, astonishingly immune to any human attempt at full knowledge. We cannot on this level understand a grain of salt, much less the universe.

But let us look a little more deeply at our microgram of salt. Salt happens to be a crystal in which, except for defects in the structure of the crystal lattice, the position of every sodium and chlorine atom is predetermined. If we could shrink ourselves into this crystalline world, we would see rank upon rank of atoms in an ordered array, a regularly alternating structure – sodium, chlorine, sodium, chlorine, specifying the sheet of atoms we are standing on and all the sheets above us and below us. An absolutely pure crystal of salt could have the position of every atom specified by something like 10 bits of information.* This would not strain the information-carrying capacity of the brain.

If the universe had natural laws that governed its behavior to the same degree of regularity that determines a crystal of salt, then, of course, the universe would be knowable. Even if there were many such laws, each of considerable complexity, human beings might have the capability to understand them all. Even if such knowledge exceeded the information-carrying capacity of the brain, we might store the additional information outside our bodies – in books, for example, or in computer memories – and still, in some sense, know the universe.

*Chlorine is a deadly poison gas employed on European battlefields in World War I. Sodium is a corrosive metal which burns upon contact with water. Together they make a placid and unpoisonous material, table salt. Why each of these substances has the properties it does is a subject called chemistry, which requires more than 10 bits of information to understand.

Human beings are, understandably, highly motivated to find regularities, natural laws. The search for rules, the only possible way to understand such a vast and complex universe, is called science. The universe forces those who live in it to understand it. Those creatures who find everyday experience a muddled jumble of events with no predictability, no regularity, are in grave peril. The universe belongs to those who, at least to some degree, have figured it out.

It is an astonishing fact that there *are* laws of nature, rules that summarize conveniently – not just qualitatively but quantitatively – how the world works. We might imagine a universe in which there are no such laws, in which the 10^{80} elementary particles that make up a universe like our own behave with utter and uncompromising abandon. To understand such a universe we would need a brain at least as massive as the universe. It seems unlikely that such a universe could have life and intelligence, because beings and brains require some degree of internal stability and order. But even if in a much more random universe there were such beings with an intelligence much greater than our own, there could not be much knowledge, passion or joy.

Fortunately for us, we live in a universe that has at least important parts that are knowable. Our common-sense experience and our evolutionary history have prepared us to understand something of the workaday world. When we go into other realms, however, common sense and ordinary intuition turn out to be highly unreliable guides. It is stunning that as we go close to the speed of light our mass increases indefinitely, we shrink toward zero thickness in the direction of motion, and time for us comes as near to stopping as we would like. Many people think that this is silly, and every week or two I get a letter from someone who complains to me about it. But it is a virtually certain consequence not just of experiment but also of Albert Einstein's brilliant analysis of space and time called the Special Theory of Relativity. It does not matter that these effects seem unreasonable to us. We are not in the habit of traveling close to the speed of light. The testimony of our common sense is suspect at high velocities.

Or consider an isolated molecule composed of two atoms shaped something like a dumbbell – a molecule of salt, it might be. Such a molecule rotates about an axis through the line connecting the two atoms. But in the world of quantum mechanics, the realm of the very small, not all orientations of our dumbbell molecule are possible. It might be that the molecule could be oriented in a horizontal position, say, or in a vertical position, but not at many angles in between. Some rotational positions are forbidden. Forbidden by what? By the laws of nature. The universe is built in such a way as to limit, or quantize, rotation. We do not experience this directly in everyday life; we would find it startling as well as awkward in sitting-up exercises, to find arms outstretched from the sides or pointed up to the skies permitted but many intermediate positions forbidden. We do not live in the world of the small, on the scale of 10^{-13} centimeters, in the realm where there are twelve zeros between the decimal place and the one. Our common-sense intuitions do not count. What does count is experiment – in this case observations from the far infrared spectra of molecules. They show molecular rotation to be quantized.

The idea that the world places restrictions on what humans might do is frustrating. Why *shouldn't* we be able to have intermediate rotational positions? Why *can't* we travel faster than the speed of light? But so far as we can tell, this is the way the universe is constructed. Such prohibitions not only press us toward a little humility; they also make the world more knowable. Every restriction corresponds to a law of nature, a regularization of the universe. The more restrictions there are on what matter and energy can do, the more knowledge human beings can attain. Whether in some sense the universe is ultimately knowable depends not only on how many natural laws there are that encompass widely divergent phenomena, but also on whether we have the openness and the intellectual capacity to understand such laws. Our formulations of the regularities of nature are surely dependent on how the brain is built, but also, and to a significant degree, on how the universe is built.

For myself, I like a universe that includes much that is unknown and, at the same time, much that is knowable. A universe in which everything is known would be static and dull, as boring as the heaven of some weak-minded theologians. A universe that is unknowable is no fit place for a thinking being. The ideal universe for us is one very much like the universe we inhabit. And I would guess that this is not really much of a coincidence.

Source: Carl Sagan, *Broca's Brain: The Romance of Science*, London, Hodder & Stoughton, 1979.

Brain Size

Trained as a zoologist at Oxford, Anthony Smith has ballooned over East African herds, discovered the world's first blind loach (*Noemacheilus smithi*) and explored wildernesses from the Arctic to the Antarctic. His book *The Body* (1968) sold 400,000 hardback in the US alone, and was translated into twelve languages. This is from the follow-up, *The Mind* (1984).

The human brain consists of ten to fifteen thousand million nerve cells. (The anatomy books are always more precise, each opting for fourteen or eleven or fifteen billion as if its choice is the true and unassailable figure.) If that kind of number is bewildering, being three times as many as there are human brains alive on this planet, the number of synapses (nerve cell connections) is a thousand times more so, there being about one hundred million million of them, or more than the total number of humans who have ever lived since we acquired this fantastic brain in full measure those thousand centuries ago. Coupled with the nerve cells, supporting and nourishing them, are the glial cells whose number is on a par with the nerve cells that they sustain. By way of comparison, appreciating that such figures can put normal minds in turmoil, the clever little honey bee has about seven thousand nerve cells.

The whole human brain weighs some three pounds in the male and about 10 per cent less in females (or 1,400 grams as against 1,250 grams). This disproportion can seem unfair to women but their brains are relatively bigger, being $2\frac{1}{2}$ per cent of total body weight as against 2 per cent for men. The three pound weight makes our brain among the lightest of our organs, being very much less than muscle (42 per cent of the total weight for males, 36 per cent for females), much less than the combined total for the 206 bones of the human skeleton, less than the twenty-plus square feet of skin, less than the twenty-eight feet of intestine, less than the eleven pounds or so of blood, and just less than the four pounds of liver. However, it weighs more than the heart (which is one pound), the kidneys (a mere five ounces each), the spleen (six ounces), the pancreas (three ounces), and the lungs (two and a half pounds). A foetus, with its relatively huge head plus brain and its small everything else, is arguably a more accurate representation of *Homo sapiens*, but that big head wanes proportionately as the child grows wiser. There is undoubted paradox about our most remarkable property, all three pounds of it.

There is also conflict over its abilities. On the one hand, and for many a normal day, a particular brain may exhibit precious little intelligence. Its owner may eat what has been set before him, walk to a bus stop, reach work, perform the same repetitive task, return home, eat again and sleep. An animal could do the same. On the other hand there is the musician Hans von Bülow travelling by train from Hamburg to Berlin, reading Stanford's *Irish Symphony*, previously unknown to him, and then conducting it that evening without a score. Some musicians prefer reading a piece of music to hearing the work, claiming that the experience is without the blemishes of an actual performance. Wolfgang Mozart confided that a whole new composition would suddenly arise in his mind. At convenient moments he would translate this entire fabric of rhythm, melody, harmony, counterpoint and tone into the written symbols of a score. For those who have trouble with a telephone number or with a name to fit a face, it is even problematical contemplating the gap between us and them, the normal and the genius. Someone once asked A. C. Aitken, professor at Edinburgh University, to make 4 divided by 47 into a decimal. After four seconds he started and gave another digit every three-quarters of a second: 'point 08510638297872340425531914'. He stopped (after twenty-four seconds), discussed the problem for one minute, and then restarted: '191489' – five-second pause – '361702127659574468. Now that's the repeating point. It starts again with 085. So if that's forty-six places, I'm right.' To many of us such a man is from another planet, particularly in his final comment.

The bizarre fact is that the brains of von Bülow, Mozart and Aitken were inherited from a long line of hunter-gatherers. Why on earth, or even how on earth, did a brain system evolve that could remember symphonies or perform advanced mental arithmetic when its palaeolithic requirements were assuredly less demanding? And why, as the second major conundrum, did the process stop at least 100,000 years ago? Only since then, via population increase, larger and more settled communities, division of labour and a subjugation of nature, has the brain of man begun to realize its potential. Yet it is a prehistoric brain, there being no detectable difference (so far as can be judged from fossils) between then and now, theirs and ours, extremely primitive and very modern man.

The solar system is vast, incomprehensible to most of us, and staggering in its distances, and to mention it in the same breath as our three pounds of brain is apparently to relate like with unlike, a thing colossal with a thing minute. But the bracketing together is fairer than might be imagined. The dimensions that astronomers talk about, and seem to understand, have their parallel in the numbers that neuroanatomists relate, almost in passing, as if these too are understood. Already mentioned are the fifteen billion nerve cells, which is also the numeric total (more or less) of stars in our galaxy. Also mentioned are the synapses, a thousandfold greater, and therefore as plentiful as the stars of a thousand galaxies. Astronomers do use such figures, being more aware than most of the thousands of millions of light years existing between us and the furthest parts of the known universe; but there must be a limit even to their comprehension.

The human brain, I suspect, can confound them, not in its neurons but in the range of its possibilities. Nerve cells are the basic units, but their synapses create a framework for interconnections, for a variety of ways in which one nerve cell may be linked with another, and for that other to be connected with others yet again. The figure of possible connections within our modern brain is as good as infinite. It is certainly larger than the number of atoms presumed to exist in the entire universe and no one, I warrant, can begin to grapple with that thought. Somehow or other a bipedal, fairly hairless, hunting, scavenging ape did acquire this incredible possession and then handed it on to us. Why it did no one knows, or can even surmise. 'I haven't the foggiest notion,' replied Richard Leakey, anthropologist and skilful finder of early hominids, when asked why or how such a swelling of brain power could have occurred among early, primitive, and scattered tribes of men.

Growth

The speed of that swelling was considerable. From about five hundred cubic centimetres – and therefore comparable in size with gorilla brains – it leaped to the human size of fourteen hundred cubic centimetres in about three million years. Assuming the brain cells of earliest man to be as compressed as in a modern brain this means that some nine billion cells were added during those years, or approximately one hundred and fifty thousand per generation. That seems like a big increase for every single leap from parent to offspring, particularly when it is remembered that many invertebrates, all quite astute, have far less than that number, but in size it is not very large. In the brain there are about ten million cells in every cubic centimetre, and therefore that generation gap of one hundred and fifty thousand occupies a sixtieth of one cubic centimetre or just fifteen cubic millimetres. Such an increment is modest if viewed simply as bulk, and many another animal has increased its body size by much more than that per generation, the weight increase being only 0.015 grams or one two-thousandth of an ounce.

If elephants had only achieved their seven-ton weight from their, say, one-ton ancestors at this increment of 0.015 grams a generation it would have taken about four hundred million generations, or roughly eight thousand million years. However, it is tempting to regard brain-tissue weight-gain as more problematical in evolution than elephant weight-gain. The brain-gain seems more so because brains are of more significance – at least from our *sapiens* point of view – than mere bulk, a thicker skin or larger trunk. It is easier to be impressed with a tripling of nerve cells in three million years.

The brain growth seems less remarkable if thought of solely in terms of cell division. To achieve 15,000 million nerve cells it is necessary to have just 33 doublings of the parent cell. To achieve half that number only 32 doublings are necessary. In this light the difference between primitive and modern man seems less marked – scarcely more than one extra doubling in three million years. As the adult complement of brain cells is made during the first three months of pregnancy the 33 divisions therefore take place at an average rate of about one every three days. Bacteria double their number every twenty minutes or so, and the foetal brain increase is therefore not particularly rapid. In fact it is equivalent to all sorts of other increases going on in the embryonic human at the same time: liver growth, skin growth and so on. Brain growth just seems more remarkable, particularly when it is brought down to the level of brain cells. To possess 15,000 million neurons at the end of three months' gestational activity means growing them at the rate of 2,000 a second. Knowing that many small animals lead quite complex lives with that number of nerve cells, it is arguable that we should be far more intelligent than is actually the case; but it is obviously wrong to compare insect ability, however complex and admirable, with human capability. We are not equivalent to seven million insects. We just happen to have as many brain cells as they possess.

These paragraphs of figures, with so many noughts in them, may confuse rather than enlighten. Their purpose was to show that the acquisition of our three-pound brain is full of contradiction. It was a tremendous increase; yet would have been of little note had it occurred with some other kind of tissue. The three pounds are only beginning to achieve their potential for a very few people in modern times; yet they were developed for all our ancestors in relatively simple times. The brain's cells and synapses are merely numerous; the quantity of interconnections is about as infinite as anything we know. The brain's size is plainly crucial; and yet those individuals with twice the brain of others are none the wiser for it. Its growth was undoubtedly critical for the emergence of *Homo sapiens*, and for the development of this species; yet its size was probably curtailed by the practical demands of a relatively minor portion of anatomy, namely the elasticity at birth of the pelvic canal. It was easy for evolution to permit a steady growing of the foetal head; but birth must have been an increasing problem. Teleologically speaking, it was a good time for the mammals to introduce live birth in place of the egg birth that had ruled, more or less, since life began; but viviparity meant, in time, a limitation to head size. (Even so, we now have a brain more than suited to our needs. Perhaps it will teach us one day how to tap its real potential.)

Source: Anthony Smith, *The Mind*, London, Hodder & Stoughton, 1984.

On Not Discovering

Ruth Benedict (1887–1948), the author of this extract, was an American anthropologist and poet who did fieldwork among the Pueblo, Apache and Blackfoot Indians. Her most famous book was *Patterns of Culture* (1934).

History is full of examples of apparently simple discoveries that were not made even when they would be surpassingly useful in that culture. Necessity is not necessarily the mother of invention. Men in most of Europe and Asia had adopted the wheel during the Bronze Age. It was used for chariots, as a pulley wheel for raising weights, and as a potter's wheel for making clay vessels. But in the two Americas it was not known except as a toy in any pre-Columbian civilization. Even in Peru, where immense temples were built with blocks of stone that weighed up to ten tons, these huge weights were excavated, transported, and placed in buildings without any use of wheels.

The invention of the zero is another seemingly simple discovery which was not made even by classic Greek mathematicians or Roman engineers. Only by the use of some symbol for nothingness can the symbol 1 be used so that it can have the value either of 1 or 10 or 100 or 1000. It makes it possible to use a small number of symbols to represent such different values as 129 and 921. Without such inventions figures cannot be added or subtracted by writing them one above another, and multiplication and division are even more difficult. The Romans had to try to divide CCCLVIII by XXIV and the difficulty was immense. It was not the Egyptians or the Greeks or the Romans who first invented the zero, but the Maya Indians of Yucatán. It is known that they had a zero sign and positional values of numbers by the time of the birth of Christ. Quite independently the Hindus made these inventions in India some five to seven centuries later. Only gradually was it adopted in medieval Europe, where it was known as Arabic notation because it was introduced there by the Arabs.

Source: Ruth Benedict's essay in *Man, Culture and Society*, ed. Harry L. Shapiro, New York, Oxford University Press, 1956.

Negative Predictions

Of all scientific Nobel Prize-winners from the English-speaking world, the British zoologist Sir Peter Medawar (1915–87) is perhaps the most remarkable for wit and panache, as evidenced in his autobiography, *Memoirs of a Thinking Radish* (1986). In *Pluto's Republic* (1982), from which this extract is taken, he does battle with several of his pet hates – psychoanalysts, mystical theologians, believers in 'rhapsodical intellection' and peddlers of paradoxes ('a paradox', wrote Medawar, 'has the same significance for the logician as the smell of burning rubber has for the electronics engineer'). He (and Macfarlane Burnet) were awarded the Nobel Prize in 1961 for their work on immunological tolerance in mice, which showed for the first time that the problem of transplanting tissues from one individual to another was soluble, and so opened the way for transplant surgery.

No kind of prediction is more obviously mistaken or more dramatically falsified than that which declares that something which is possible in principle (that is, which does not flout some estabished scientific law) will never or can never happen. I shall choose now from my own subject, medical science, a bouquet of negative predictions chosen not so much for their absurdity as for the way in which they illustrate something interesting in the history of science or medicine.

My favourite prediction of this kind was made by J. S. Haldane (the distinguished physiologist father of the geneticist J. B. S. Haldane), who in a book published in 1930 titled *The Philosophy of Biology* declared it to be 'inconceivable' that there should exist a chemical compound having exactly the properties since shown to be possessed by deoxyribonucleic acid (DNA). DNA is the giant molecule that encodes the genetic message which passes from one generation to the next – the message that prescribes how development is to proceed. The famous paper in the scientific journal *Nature* in which Francis Crick and James Watson described the structure of DNA and how that structure qualifies it to fulfil its genetic functions was published not so many years after Haldane's unlucky prediction. The possibility that such a compound as DNA might exist had been clearly envisaged by the German nature-philosopher Richard Semon in a book *The Mneme*, a reading of which prompted Haldane to dismiss the whole idea as nonsense.

In the days before the introduction of antisepsis and asepsis, wound infection was so regular and so grave an accompaniment of surgical operations that we can hardly wonder at the declaration of a well-known surgeon working in London, Sir John Erichsen (1818–96), that 'The abdomen is forever shut from the intrusions of the wise and humane surgeon.' Of course, the coming of aseptic surgery to which I refer below, combined with the improvement of anaesthesia, soon made nonsense of this and opened the door to the great achievements of gastrointestinal surgery in the first decade of our century.

One of the very greatest surgeons of this period was Berkeley George Moynihan of Leeds (1865–1936), a man whose track-record for erroneous predictions puts him in a class entirely by himself.

Around 1900 the famous British periodical, the *Strand* magazine (the first to publish the case records of Sherlock Holmes), thought that at the turn of the century its readers would be interested to know what was in store for them in the century to come; 'a Harley Street surgeon' (unmistakably Moynihan) was accordingly invited to tell them what the future of surgery was to be. Evidently not spectacular, for Moynihan opined that surgery had reached its zenith and that no great advances were to be looked for in the future – nothing as dramatic, for example, as the opening of the abdomen, an event regarded with as much awe as the opening of Japan.

Moynihan's forecast was not the hasty, ill-considered opinion of a busy man: it represented a firmly held conviction. In a Leeds University Medical School magazine in 1930 he wrote: 'We can surely never hope to see the craft of surgery made much more perfect than it is today. We are at the end of a chapter.' Moynihan repeated this almost word for word when he delivered Oxford University's most prestigious lecture, the Romanes Lecture, in 1932. He was a vain and arrogant man, and if these quotations are anything to go by a rather silly one too, but surgery is indebted to him nevertheless, for he introduced the delicacy and fastidiousness of technique that did away for ever with the image of the surgeon as a brusque, over-confident and rough-and-ready sawbones. Moreover Moynihan, along with William Stewart Halsted of Johns Hopkins (1852–1922), introduced into modern surgery the *aseptic* technique with all the rituals and drills that go with it: the scrupulous scrub-up, the gown, cap and rubber gloves, and the facial mask over the top of which the pretty young theatre nurse gazes with smouldering eyes at the handsome young intern who is planning to wrong her. These innovations may be said to have made possible the hospital soap opera and thus in turn TV itself – for what would TV be without the hospital drama, and what would the hospital drama be without cap and masks and those long, meaningful stares?

The full regalia of the surgical operation did not escape a certain amount of gentle ridicule – in which we may hear the voice of those older, coarser surgeons whom Moynihan supplanted. Moynihan was once described as 'the pyloric pierrot', and upon seeing Moynihan's rubber shoes a French surgeon is said to have remarked 'Surely he does not intend to stand in the abdomen?'

Source: Peter Medawar, *Pluto's Republic*, London, Oxford University Press, 1982.

Clever Animals

Lewis Thomas, a distinguished American research pathologist, won best-sellerdom with *The Lives of a Cell* (1974), a collection of his scientific journalism. This extract is from *Late Night Thoughts* (1983).

Scientists who work on animal behavior are occupationally obliged to live chancier lives than most of their colleagues, always at risk of being fooled by the animals they are studying or, worse, fooling themselves. Whether their experiments involve domesticated laboratory animals or wild creatures in the field, there is no end to the surprises that an animal can think up in the presence of an investigator. Sometimes it seems as if animals are genetically programmed to puzzle human beings, especially psychologists.

The risks are especially high when the scientist is engaged in training the animal to do something or other and must bank his professional reputation on the integrity of his experimental subject. The most famous case in point is that of Clever Hans, the turn-of-the-century German horse now immortalized in the lexicon of behavioral science by the technical term, the 'Clever Hans Error.' The horse, owned and trained by Herr von Osten, could not only solve complex arithmetical problems, but even read the instructions on a blackboard and tap out infallibly, with one hoof, the right answer. What is more, he could perform the same computations when total strangers posed questions to him, with his trainer nowhere nearby. For several years Clever Hans was studied intensively by groups of puzzled scientists and taken seriously as a horse with something very like a human brain, quite possibly even better than human. But finally in 1911, it was discovered by Professor O. Pfungst that Hans was not really doing arithmetic at all; he was simply observing the behavior of the human experimenter. Subtle, unconscious gestures – nods of the head, the holding of breath, the cessation of nodding when the correct count was reached – were accurately read by the horse as cues to stop tapping.

Whenever I read about that phenomenon, usually recounted as the exposure of a sort of unconscious fraud on the part of either the experimenter or the horse or both, I wish Clever Hans would be given more credit than he generally gets. To be sure, the horse couldn't really do arithmetic, but the record shows that he was considerably better at observing human beings and interpreting their behavior than humans are at comprehending horses or, for that matter, other humans.

Cats are a standing rebuke to behavioral scientists wanting to know how the minds of animals work. The mind of a cat is an inscrutable mystery, beyond human reach, the least human of all creatures and at the same time, as any cat owner will attest, the most intelligent. In 1979, a paper was published in *Science* by B. R. Moore and S. Stuttard entitled 'Dr. Guthrie and Felis domesticus or: tripping over the cat,' a wonderful account of the kind of scientific mischief native to this species. Thirty-five years ago, E. R. Guthrie and G. P. Horton described an experiment in which cats were placed in a glass-fronted puzzle box and trained to find their way out by jostling a slender vertical rod at the front of the box, thereby causing a door to open. What interested these investigators was not so much that the cats could learn to bump into the vertical rod, but that before doing so each animal performed a long ritual of highly stereotyped movements, rubbing their heads and backs against the front of the box, turning in circles, and finally touching the rod. The experiment has ranked as something of a classic in experimental psychology, even raising in some minds the notion of a ceremony of superstition on the part of cats: before the rod will open the door, it is necessary to go through a magical sequence of motions.

Moore and Stuttard repeated the Guthrie experiment, observed the same complex 'learning' behavior, but then discovered that it occurred only when a human being was visible to the cat. If no one was in the room with the box, the cat did nothing but take naps. The sight of a human being was all that was needed to launch the animal on the series of sinuous movements, rod or no rod, door or no door. It was not a learned pattern of behavior, it was a cat greeting a person.

Source: Lewis Thomas, *Late Night Thoughts*, London, Oxford University Press, 1985. First published in the USA as *Late Night Thoughts on Listening to Mahler's Ninth Symphony*, New York, Viking, 1983.

Great Fakes of Science

A celebrated popularizer of science and mathematics, Martin Gardner has devoted much of his life to debunking ESP, 'psychic' phenomena, metal-bending and other paranormality. This excerpt is from *Science Good, Bad and Bogus* (1983).

Politicians, real-estate agents, used-car salesmen, and advertising copy-writers are expected to stretch facts in self-serving directions, but scientists who falsify their results are regarded by their peers as committing an inexcusable crime. Yet the sad fact is that the history of science swarms with cases of outright fakery and instances of scientists who unconsciously distorted their work by seeing it through lenses of passionately held beliefs.

Gregor Johann Mendel, whose experiments with garden peas first revealed the basic laws of heredity, was such a hero of modern science that scientists in the thirties were shocked to learn that this pious monk probably doctored his data. R. A. Fisher, a famous British statistician, checked Mendel's reports carefully. The odds, he concluded, are about 10,000 to 1 that Mendel gave an inaccurate account of his experiments.

Brother Mendel was a Roman Catholic priest who lived in an abbey in Brünn, now part of Czechoslovakia. More than a century ago, working alone in a monastery garden, he found that his plants were breeding according to precise laws of probability. Later, these laws were explained by the theory of genes (now known to be sections along a helical DNA molecule), but it was Brother Mendel who laid the foundations for what later was called Mendelian genetics. His great work was totally ignored by the botanists of his time, and he died without knowing he would become famous.

Most of the monk's work was with garden peas. Seeds from dwarf pea plants always grow into dwarfs, but tall pea plants are of two kinds. Seeds from one kind produce only talls. Seeds from the other kind produce both talls and dwarfs. Mendel found that when he crossed true-breeding talls with dwarfs he got only talls. When he self-pollinated these tall hybrids he got a mixture of $\frac{1}{4}$ true-breeding talls, $\frac{1}{4}$ dwarfs, and $\frac{1}{2}$ talls that did not breed true.

Today one says that tallness in garden peas is dominant, dwarfness is recessive. Mendel's breeding experiment is like shaking an even mixture of red and blue beads in a hat, then taking out a pair. The probability is $\frac{1}{4}$ you will get red-red, $\frac{1}{4}$ you will get blue-blue, and $\frac{1}{2}$ you will get red-blue. These, however, are 'long-run' probabilities. Make such a test just once, with (say) 200 evenly mixed beads, and the odds are strongly against your getting *exactly* 25 red pairs, 25 blue, and 50 mixed. Statisticians would be deeply suspicious if you reported results that precise.

Mendel's figures are suspect for just this reason. They are too good to be true. Did the priest consciously fudge his data? Let us be charitable. Perhaps he was guilty only of 'wishful seeing' when he classified and counted his talls and dwarfs.

Geologists find strange things in the ground, but none so strange as the 'fossils' unearthed by Johann Beringer, a learned professor of science at the University of Würzburg. German Protestants of the early eighteenth century, like so many American fundamentalists today, could not believe that fossils were the relics of life that flourished millions of years ago. Professor Beringer had an unusual theory. Some fossils, he admitted, might be the remains of life that perished in the great flood of Noah, but most of them were 'peculiar stones' carved by God himself as he experimented with the kinds of life he intended to create.

Beringer was ecstatic when his teen-age helpers began to dig up hundreds of stones that supported his hypothesis. They bore images of the bodies of strange insects, birds, and fishes never seen on earth. One bird had a fish's head – an idea God had apparently discarded. Other stones showed the sun, moon, five-pointed stars, and comets with blazing tails. He began to find stones with Hebrew letters. One had 'Jehovah' carved on it.

In 1726 Beringer published a huge treatise on these marvelous discoveries. It was written in Latin and impressively illustrated with engraved plates. Colleagues tried to convince Beringer he was being bamboozled, but he dismissed this as 'vicious raillery' by stubborn, establishment enemies.

No one knows what finally changed the professor's mind. It was said that he found a stone with his own name on it! An inquiry was held. One of his assistants confessed. It turned out that the peculiar stones had been carved by two peculiar colleagues, one the university's librarian, the other a professor of geography.

Poor trusting, stupid Beringer, his career shattered, spent his life's savings buying up copies of his idiotic book and burning them. But the work became such a famous monument to geological gullery that twenty-seven years after Beringer's death a new edition was published in Germany. In 1963 a handsome translation was issued by the University of California Press. Beringer has become immortal only as the victim of a cruel hoax.

Was Paul Kammerer the victim of a similar hoax, or was he himself the perpetrator? In any case, when someone applied India ink to (or perhaps injected it into) the feet of several of Kammerer's frogs, the career of one of the most respected of Viennese biologists was brought to an inglorious end.

Kammerer was the last great champion of a theory of evolution called Lamarckism. In this view, named for the French naturalist Jean Lamarck [see p. 58], acquired traits are somehow passed on to descendants: when giraffes stretched their necks to nibble high leaves, their offspring were born with longer necks. Darwin himself was a Lamarckian. Modern genetics discards this theory, replacing it with the Mendelian view that natural selection operates on variations produced by random mutations.

In 1910 Lamarckism was still the 'establishment' view, but the new Mendelian theory was rapidly gaining ground. Eager to defend the older theory (he had written a book about it called *The Inheritance of Acquired Characteristics*), Kammerer devised a simple experiment with a species of frog known as the 'midwife toad.'

Most toads mate in water. To keep a firm grip on the female's slippery body, the male toad develops dark 'nuptial pads' on his feet. The male midwife toad, which mates on land, lacks such pads. Kammerer's scheme was to force midwife toads to copulate under water for several generations, then see if they develop nuptial pads. It was a stupid experiment, because, had it succeeded, Mendelians would have explained it as no more than a revival of a genetic blueprint. Nothing so complicated as a nuptial pad could have developed in just a few generations.

But Kammerer went ahead with his plan and soon reported it to be a huge success. The black pads had indeed appeared. The news was sensational, especially in Russia where Lamarckism then completely dominated biology. Russian scientists were so impressed that they offered Kammerer a post at the University of Moscow.

No sooner had Kammerer accepted this offer than it was discovered that his toad specimens had been crudely faked. It was the biggest science scandal of the decade. Kammerer blamed it all on an assistant, but nobody believed him. In 1926, at age 46, he took a pistol and shot himself through the head.

Kammerer continued to be a great hero in the Soviet Union throughout the period when Joseph Stalin and the plant-breeder Trofim Lysenko, both enthusiastic Lamarckians, saw to it that Mendelian geneticists were banished to Siberia. Now that Lysenko is dead and Soviet genetics has gone Mendelian, it is hard to find a biologist anywhere in the world who takes Lamarckism seriously.

Source: Martin Gardner, *Science Good, Bad and Bogus*, London, Oxford University Press, 1983.

Unnatural Nature

Lewis Wolpert is Professor of Biology as Applied to Medicine at University College, London. In *The Unnatural Nature of Science* (1992), from which this extract is taken, he argues that scientific ideas almost always run counter to common sense.

The physics of motion provides one of the clearest examples of the counter-intuitive and unexpected nature of science. Most people not trained in physics have some sort of vague ideas about motion and use these to predict how an object will move. For example, when students are presented with problems requiring them to predict where an object – a bomb, say – will land if dropped from an aircraft, they often get the answer wrong. The correct answer – that the bomb will hit that point on the ground more or less directly below the point at which the aircraft has arrived at the moment of impact – is often rejected. The underlying confusion partly comes from not recognizing that the bomb continues to move forward when released and this is not affected by its downwards fall. This point is made even more dramatically by another example. Imagine being in the centre of a very large flat field. If one bullet is dropped from your hand and another is fired horizontally from a gun at exactly the same time, which will hit the ground first? They will, in fact, hit the ground at the same time, because the bullet's rate of fall is quite independent of its horizontal motion. That the bullet which is fired is travelling horizontally has no effect on how fast it falls under the action of gravity . . .

Science also deals with enormous differences in scale and time compared with everyday experience. Molecules, for example, are so small that it is not easy to imagine them. If one took a glass of water, each of whose molecules were tagged in some way, went down to the sea, completely emptied the glass, allowed the water to disperse through all the oceans, and then filled the glass from the sea, then almost certainly some of the original water molecules would be found in the glass. What this means is that there are many more molecules in a glass of water than there are glasses of water in the sea. There are also, to give another example, more cells in one finger than there are people in the world. Again, geological time is so vast – millions and millions of years – that it was one of the triumphs of nineteenth-century geology to recognize that the great mountain ranges, deep ravines and valleys could be accounted for by the operation of forces no different from those operating at present but operating over enormous periods of time. It was not necessary to postulate catastrophes.

A further example of where intuition usually fails, probably because of the scale, is provided by imagining a smooth globe as big as the earth, round whose equator – 25,000 miles long – is a string that just fits. If the length of the string is increased by 36 inches, how far from the surface of the globe will the string stand out? The answer is about 6 inches, and is independent of whether the globe's equator is 25,000 or 25 million miles long.

Source: Lewis Wolpert, *The Unnatural Nature of Science,* Cambridge, Mass., Harvard University Press, 1993.

Rags, Dolls and Teddy Bears

The English psychiatrist D. W. Winnicott (1896–1971) is best known for his theory of 'transitional objects'. These are often bits of rag or soft toys to which the child becomes attached and which, Winnicott argues, play a vital role in reconciling it to the outside world – a role later taken over by art, religion and other sources of 'illusion'.

In common experience one of the following occurs, complicating an auto-erotic experience such as thumb-sucking:

(i) with the other hand the baby takes an external object, say a part of a sheet or blanket, into the mouth along with the fingers; or

(ii) somehow or other the bit of cloth is held and sucked, or not actually sucked; the objects used naturally include napkins and (later) handkerchiefs, and this depends on what is readily and reliably available; or

(iii) the baby starts from early months to pluck wool and to collect it and to use it for the caressing part of the activity; less commonly, the wool is swallowed, even causing trouble; or

(iv) mouthing occurs, accompanied by sounds of 'mum-mum', babbling, anal noises, the first musical notes, and so on.

One may suppose that thinking, or fantasying, gets linked up with these functional experiences.

All these things I am calling *transitional phenomena*. Also, out of all this (if we study one infant) there may emerge some thing or some phenomenon – perhaps a bundle of wool or the corner of a blanket or eiderdown, or a word or tone, or a mannerism – that becomes vitally important to the infant for use at the time of going to sleep, and is a defence against anxiety, especially anxiety of depressive type. Perhaps some soft object or other type of object has been found and used by the infant, and this then becomes what I am calling a *transitional object*. This object goes on being important. The parents get to know its value and carry it round when travelling. The mother lets it get dirty and even smelly, knowing that by washing it she introduces a break in continuity in the infant's experience, a break that may destroy the meaning and value of the object to the infant.

I suggest that the pattern of transitional phenomena begins to show at about four to six to eight to twelve months. Purposely I leave room for wide variations.

Patterns set in infancy may persist into childhood, so that the original soft object continues to be absolutely necessary at bed-time or at time of loneliness or when a depressed mood threatens. In health, however, there is a gradual extension of range and interest, and eventually the extended range is maintained, even when depressive anxiety is near. A need for a specific object or a behaviour pattern that started at a very early date may reappear at a later age when deprivation threatens.

This first possession is used in conjunction with special techniques derived from very early infancy, which can include or exist apart from the more direct auto-erotic activities. Gradually in the life of an infant teddies and dolls and hard toys are acquired. Boys to some extent tend to go over to use hard objects, whereas girls tend to proceed right ahead to the acquisition of a family. It is important to note, however, that *there is no noticeable difference between boy and girl in their use of the original 'not-me' possession*, which I am calling the transitional object.

As the infant starts to use organized sounds ('mum', 'ta', 'da') there may appear a 'word' for the transitional object. The name given by the infant to these earliest objects is often significant, and it usually has a word used by the adults partly incorporated in it. For instance, 'baa' may be the name, and the 'b' may have come from the adult's use of the word 'baby' or 'bear'.

The object is affectionately cuddled as well as excitedly loved and mutilated.

It must never change, unless changed by the infant . . .

It is not forgotten and it is not mourned. It loses meaning, and this is because the transitional phenomena have become diffused, have become spread out over the whole intermediate territory between 'inner psychic reality' and 'the external world as perceived by two persons in common', that is to say, over the whole cultural field.

At this point my subject widens out into that of play, and of artistic creativity and appreciation, and of religious feeling, and of dreaming, and also of fetishism, lying and stealing, the origin and loss of affectionate feeling, drug addiction, the talisman of obsessional rituals, etc.

Winnicott's caution to mothers against washing the transitional object, and his tracing of adult personality defects to breaks in the child's relationship with the object, may be relevant to the case of the poet Philip Larkin, who recalls in a letter (13 July 1959):

> My earliest toys were teddy bear, dog ('Rags') & rabbit, but only the last named meant anything to me. It sat on the table at meals, until one day it fell with its ears in the mint sauce. It was hung out many days to sweeten, & washed & scented, but I never felt the same about it.

Source: D. W. Winnicott, *Playing and Reality*, London, Tavistock Publications, 1971 (copyright, D. W. Winnicott).

The Man Who Mistook his Wife for a Hat

Dr Oliver Sacks was born in London in 1933, and educated in Oxford, California and New York, where he is Professor of Neurology at the Albert Einstein College of Medicine. His studies of bizarre neurological disorders are modern classics, raising profound questions about the mind and self-identity. *Awakenings*, which recounted the 'time-machine' effect of L-Dopa on patients who had for years been locked in a trance-like state following sleeping-sickness (*encephalitis lethargica*), was made into a feature film. This extract is from the title essay of his 1985 collection.

Dr P. was a musician of distinction, well-known for many years as a singer, and then, at the local School of Music, as a teacher. It was here, in relation to his students, that certain strange problems were first observed. Sometimes a student would present himself, and Dr P. would not recognize him; or, specifically, would not recognize his face. The moment the student spoke, he would be recognized by his voice. Such incidents multiplied, causing embarrassment, perplexity, fear – and, sometimes, comedy. For not only did Dr P. increasingly fail to see faces, but he saw faces when there were no faces to see: genially, Magoo-like, when in the street, he might pat the heads of water-hydrants and parking-meters, taking these to be the heads of children; he would amiably address carved knobs on the furniture, and be astounded when they did not reply. At first these odd mistakes were laughed off as jokes, not least by Dr P. himself. Had he not always had a quirky sense of humour, and been given to Zen-like paradoxes and jests? His musical powers were as dazzling as ever; he did not feel ill – he had never felt better; and the mistakes were so ludicrous – and so ingenious – that they could hardly be serious or betoken anything serious. The notion of there being 'something the matter' did not emerge until some three years later, when diabetes developed. Well aware that diabetes could affect his eyes, Dr P. consulted an ophthalmologist, who took a careful history, and examined his eyes closely. 'There's nothing the matter with your eyes,' the doctor concluded. 'But there is trouble with the visual parts of your brain. You don't need my help, you must see a neurologist.' And so, as a result of this referral, Dr P. came to me.

It was obvious within a few seconds of meeting him that there was no trace of dementia in the ordinary sense. He was a man of great cultivation and charm, who talked well and fluently, with imagination and humour. I couldn't think why he had been referred to our clinic.

And yet there *was* something a bit odd. He faced me as he spoke, was oriented towards me, and yet there was something the matter – it was difficult to formulate. He faced me with his *ears*, I came to think, but not with his eyes. These, instead of looking, gazing, at me, 'taking me in', in the normal way, made sudden strange fixations – on my nose, on my right ear, down to my chin, up to my right eye – as if noting (even studying) these individual features, but not seeing my whole face, its changing expressions, 'me', as a whole. I am not sure that I fully realized this at the time – there was just a teasing strangeness, some failure in the normal interplay of gaze and expression. He saw me, he *scanned* me, and yet . . .

'What seems to be the matter?' I asked him at length.

'Nothing that I know of,' he replied with a smile, 'but people seem to think there's something wrong with my eyes.'

'But *you* don't recognize any visual problems?'

'No, not directly, but I occasionally make mistakes.'

I left the room briefly, to talk to his wife. When I came back Dr P. was sitting placidly by the window, attentive, listening rather than looking out. 'Traffic,' he said, 'street sounds, distant trains – they make a sort of symphony, do they not? You know Honegger's *Pacific 234*?'

What a lovely man, I thought to myself. How can there be anything seriously the matter? Would he permit me to examine him?

'Yes, of course, Dr Sacks.'

I stilled my disquiet, his perhaps too, in the soothing routine of a neurological exam – muscle strength, co-ordination, reflexes, tone . . . It was while examining his reflexes – a trifle abnormal on the left side – that the first bizarre experience occurred. I had taken off his left shoe and scratched the sole of his foot with a key – a frivolous-seeming but essential test of a reflex – and then, excusing myself to screw my ophthalmoscope together, left him to put on the shoe himself. To my surprise, a minute later, he had not done this.

'Can I help?' I asked.

'Help what? Help whom?'

'Help you put on your shoe.'

'Ach,' he said, 'I had forgotten the shoe', adding, *sotto voce*, 'The shoe? The shoe?' He seemed baffled.

'Your shoe,' I repeated. 'Perhaps you'd put it on.'

He continued to look downwards, though not at the shoe, with an intense but misplaced concentration. Finally his gaze settled on his foot: 'That is my shoe, yes?'

Did I mis-hear? Did he mis-see?

'My eyes,' he explained, and put a hand to his foot. '*This* is my shoe, no?'

'No, it is not. That is your foot. *There* is your shoe.'

'Ah! I thought that was my foot.'

Was he joking? Was he mad? Was he blind? If this was one of his 'strange mistakes', it was the strangest mistake I had ever come across.

I helped him on with his shoe (his foot), to avoid further complication. Dr P. himself seemed untroubled, indifferent, maybe amused. I resumed my examination. His visual acuity was good: he had no difficulty seeing a pin on the floor, though sometimes he missed it if it was placed to his left.

He saw all right, but what did he see? I opened out a copy of the *National Geographic Magazine*, and asked him to describe some pictures in it.

His responses here were very curious. His eyes would dart from one thing to another, picking up tiny features, individual features, as they had done with my face. A striking brightness, a colour, a shape would arrest his attention and elicit comment – but in no case did he get the scene-as-a-whole. He failed to see the whole, seeing only details, which he spotted like blips on a radar screen. He never entered into relation with the picture as a whole – never faced, so to speak, *its* physiognomy. He had no sense whatever of a landscape or scene.

I showed him the cover, an unbroken expanse of Sahara dunes.

'What do you see here?' I asked.

'I see a river,' he said. 'And a little guest-house with its terrace on the water. People are dining out on the terrace. I see coloured parasols here and there.' He was looking, if it was 'looking', right off the cover, into mid-air and confabulating non-existent features, as if the absence of features in the actual picture had driven him to imagine the river and the terrace and the coloured parasols.

I must have looked aghast, but he seemed to think he had done rather well. There was a hint of a smile on his face. He also appeared to have decided that the examination was over, and started to look round for his hat. He reached out his hand, and took hold of his wife's head, tried to lift it off, to put it on. He had apparently mistaken his wife for a hat! His wife looked as if she was used to such things.

I could make no sense of what had occurred, in terms of conventional neurology (or neuropsychology). In some ways he seemed perfectly preserved, and in others absolutely, incomprehensibly devastated. How could he, on the one hand, mistake his wife for a hat and, on the other, function, as apparently he still did, as a teacher at the Music School?

I had to think, to see him again – and to see him in his own familiar habitat, at home.

A few days later I called on Dr P. and his wife at home, with the score of the *Dichterliebe* in my briefcase (I knew he liked Schumann), and a variety of odd objects for the testing of perception. Mrs P. showed me into a lofty apartment, which recalled fin-de-siècle Berlin. A magnificent old Bösendorfer stood in state in the centre of the room, and all round it were music-stands, instruments, scores . . . There were books, there were paintings, but the music was central. Dr P. came in and, distracted, advanced with outstretched hand to the grandfather clock, but, hearing my voice, corrected himself, and shook hands with me. We exchanged greetings, and chatted a little of current concerts and performances. Diffidently, I asked him if he would sing.

'The *Dichterliebe*!' he exlaimed. 'But I can no longer read music. You will play them, yes?'

I said I would try. On that wonderful old piano even my playing sounded right, and Dr P. was an aged, but infinitely mellow Fischer-Dieskau, combining a perfect ear and voice with the most incisive musical intelligence. It was clear that the Music School was not keeping him on out of charity . . .

I had stopped at a florist on my way to his apartment and bought myself an extravagant red rose for my buttonhole. Now I removed this and handed it to him. He took it like a botanist or morphologist given a specimen, not like a person given a flower.

'About six inches in length,' he commented. 'A convoluted red form with a linear green attachment.'

'Yes,' I said encouragingly, 'and what do you think it *is*, Dr P.?'

'Not easy to say.' He seemed perplexed. 'It lacks the simple symmetery of the Platonic solids, although it may have a higher symmetry of its own . . . I think this could be an inflorescence or flower.'

'Could be?' I queried.

'Could be,' he confirmed.

'Smell it,' I suggested, and he again looked somewhat puzzled, as if I had asked him to smell a higher symmetry. But he complied courteously, and took it to his nose. Now, suddenly, he came to life.

'Beautiful!' he exlaimed. 'An early rose. What a heavenly smell!' He started to hum 'Die Rose, die Lillie . . .' Reality, it seemed, might be conveyed by smell, not by sight.

I tried one final test. It was still a cold day, in early spring, and I had thrown my coat and gloves on the sofa.

'What is this?' I asked, holding up a glove.

'May I examine it?' he asked, and, taking it from me, he proceeded to examine it as he had examined the geometrical shapes.

'A continuous surface,' he announced at last, 'infolded on itself. It appears to have' – he hesitated – 'five outpouchings, if this is the word.'

'Yes,' I said cautiously. 'You have given me a description. Now tell me what it is.'

'A container of some sort?'

'Yes,' I said, 'and what would it contain?'

'It would contain its contents!' said Dr P., with a laugh. 'There are many possibilities. It could be a change-purse, for example, for coins of five sizes. It could . . .'

I interrupted the barmy flow. 'Does it not look familiar? Do you think it might contain, might fit, a part of your body?'

No light of recognition dawned on his face.*

No child would have the power to see and speak of 'a continuous surface . . . infolded on itself', but any child, any infant, would

*Later, by accident, he got it on, and exclaimed, 'My God, it's a glove!' This was reminiscent of Kurt Goldstein's patient 'Lanuti', who could only recognize objects by trying to use them in action.

immediately know a glove as a glove, see it as familiar, as going with a hand. Dr P. didn't. He saw nothing as familiar. Visually, he was lost in a world of lifeless abstractions. Indeed he did not have a real visual world, as he did not have a real visual self. He could speak about things, but did not see them face-to-face. Hughlings Jackson, discussing patients with aphasia and left-hemisphere lesions, says they have lost 'abstract' and 'propositional' thought – and compares them with dogs (or, rather, he compares dogs to patients with aphasia). Dr P., on the other hand, functioned precisely as a machine functions. It wasn't merely that he displayed the same indifference to the visual world as a computer but – even more strikingly – he construed the world as a computer construes it, by means of key features and schematic relationships. The scheme might be identified – in an 'identiti-kit' way – without the reality being grasped at all . . .

When the examination was over, Mrs P. called us to the table, where there was coffee and a delicious spread of little cakes. Hungrily, hummingly, Dr P. started on the cakes. Swiftly, fluently, unthinkingly, melodiously, he pulled the plates towards him, and took this and that, in a great gurgling stream, an edible song of food, until, suddenly, there came an interruption: a loud, peremptory rat-tat-tat at the door. Startled, taken aback, arrested, by the interruption, Dr P. stopped eating, and sat frozen, motionless, at the table, with an indifferent, blind, bewilderment on his face. He saw, but no longer saw, the table; no longer perceived it as a table laden with cakes. His wife poured him some coffee: the smell titillated his nose, and brought him back to reality. The melody of eating resumed.

How does he do anything, I wondered to myself? What happens when he's dressing, goes to the lavatory, has a bath? I followed his wife into the kitchen and asked her how, for instance, he managed to dress himself. 'It's just like the eating,' she explained. 'I put his usual clothes out, in all the usual places, and he dresses without difficulty, singing to himself. He does everything singing to himself. But if he is interrupted and loses the thread, he comes to a complete stop, doesn't know his clothes – or his own body. He sings all the time – eating songs, dressing songs, bathing songs, everything. He can't do anything unless he makes it a song.'

While we were talking my attention was caught by the pictures on the walls.

'Yes,' Mrs P. said, 'he was a gifted painter as well as a singer. The School exhibited his pictures every year.'

I strolled past them curiously – they were in chronological order. All his earlier work was naturalistic and realistic, with vivid mood and atmosphere, but finely detailed and concrete. Then, years later, they became less vivid, less concrete, less realistic and naturalistic; but far more abstract, even geometrical and cubist. Finally, in the last paintings, the canvases became nonsense, or nonsense to me – mere chaotic lines and blotches of paint. I commented on this to Mrs P.

'Ach, you doctors, you're such philistines!' she exclaimed, 'Can you not see *artistic development* – how he renounced the realism of his earlier years, and advanced into abstract, non-representational art?'

'No, that's not it,' I said to myself (but forbore to say it to poor Mrs P.). He had indeed moved from realism to non-representation to the abstract, but this was not the artist, but the pathology, advancing – advancing towards a profound visual àgnosia, in which all powers of representation and imagery, all sense of the concrete, all sense of reality, were being destroyed. This wall of paintings was a tragic pathological exhibit, which belonged to neurology, not art.

Source: Oliver Sacks, *The Man Who Mistook his Wife for a Hat*, London, Picador, Pan Book, 1986.

Seeing the Atoms in Crystals

This is part of an interview from *A Passion for Science* (1988), a book based on a BBC Radio 3 series produced by Alison Richards. The interviewer is Lewis Wolpert, the interviewee Dorothy Hodgkin.

In 1964 the *Daily Mail* carried a headline, 'Nobel Prize for British Wife'. The prize was for chemistry, and had been awarded to Dorothy Crowfoot Hodgkin for research on the structure of biologically important molecules including penicillin and vitamin B12. Using the technique of X-ray crystallography she had coaxed from these molecules the minute details of their three-dimensional structure, to the extent that the exact position of each atom was known. In 1969, she went on to solve the structure of insulin.

Professor Dorothy Hodgkin, Nobel Laureate, Fellow of the Royal Society, Emeritus Professor at the University of Oxford and the first woman since Florence Nightingale to have the Order of Merit conferred upon her, was born in 1910, in Cairo, where her father, Dr J. W. Crowfoot, was in the Egyptian Ministry of Education. Both her own family and that of her husband, Thomas Hodgkin, had a long tradition of intellectual achievement and social responsibility, and she has combined her distinguished scientific career with an active commitment to the cause of world peace. She has now officially retired and I went to talk to her at her home in a small rural village about 30 miles from Oxford. She looks just like the famous portrait by Brian Organ which hangs in the Royal Society, and her hands, crippled by arthritis since she was a child, are familiar too, from the drawings by Henry Moore. We sat by the fire in the cluttered sitting room, with its faded rugs on the floor and books and pictures everywhere. She is still very active, and divides her time between writing up research papers and Pugwash, an international movement of scientists working for peace. The delight she takes in her chosen field is infectious.

X-ray crystallography is a way of studying the structure of

molecules by shining X-rays through them. The beam is scattered, or diffracted, by the atoms in the molecule and registers on a photographic plate as a pattern of spots of varying intensities. That this pattern of spots could be used to determine how the atoms were arranged in the molecule was the brilliant insight of a 22-year-old Cambridge student, Lawrence Bragg, who three years later, in 1915, became the youngest person ever to receive a Nobel prize. The technique brought about a revolution in physics and chemistry, and also, more dramatically, in biology. Not only did it make Dorothy Hodgkin's achievement possible, but it also led directly to the solution of the structure of the genetic material, DNA.

Bragg worked with relatively simple, inorganic crystals such as salt. When the technique was applied to biological molecules, which are larger and more complicated, the diffraction patterns were, not surprisingly, also more complicated. Their interpretation became a highly skilled and immensely time-consuming occupation involving painstaking measurement, complex and lengthy calculations, and no small measure of intuition. When Dorothy Hodgkin began her research career, the ground rules for interpreting the X-ray data had still to be worked out.

Among the leaders in the field was Desmond Bernal, later to become Professor of Physics in the University of London. He was pioneering the application of X-ray crystallography to proteins, the most diverse and important chemical components of cells, at the Cavendish Laboratory in Cambridge. In 1932, after graduating from Somerville College, Oxford, Dorothy Hodgkin went to work under him. Together they obtained the first successful X-ray photograph of a protein single crystal. This was a major achievement, for not only does the skill lie in taking the actual photographs, but also in growing the crystals in the first place. It's not a matter of following rules, but of almost alchemical skill. Dorothy Hodgkin's future achievements were to depend both on her talent for growing suitable crystals, and on the intuitive and dogged brilliance she brought to the study of the impenetrable spots.

Two years later, she returned to Oxford. Here, just before the Second World War, Howard Florey and Ernst Chain began trying to isolate the actual antibacterial agent from the mould studied by Fleming. This fortunate set of circumstances gave her the opportunity to start work on penicillin as soon as sufficiently pure samples were available. It was a major undertaking, but by 1945 the structure was solved. Soon afterwards, in 1948, vitamin B12, the factor which prevents pernicious anaemia, was isolated, almost simultaneously, by two British and American groups. This was a more complex molecule than penicillin, and even with the aid of one of the first electronic computers, its structure was to occupy her for the next six years. Then came insulin, a still more complicated compound, which had preoccupied her since the beginning of her career. Its chemical structure – the number and order of the chemical units from which it is composed – had been worked out by Fred Sanger, in Cambridge, in the early 1950s, and won him the Nobel prize for chemistry in 1958. But the monumental task of determining the exact configuration of the constituent atoms still lay ahead. Dorothy Hodgkin took it on.

DH: I'm really an experimentalist. I used to say, I think with my hands. I just like manipulation. I began to like it as a child and it's continued to be a pleasure. I don't do very much experimental work now, but I get something of the same pleasure from going through the maps indicating the position of the atoms that result from the calculations that are carried out.

LW: I hadn't thought of crystallography as being an experimental subject.

DH: Well, it does involve experiments, usually, because you often have to modify the crystal in order to get understandable results from the intensities of the reflection of the X-rays.

LW: Now, to a non-X-ray crystallographer the reflections that you get from the X-rays look a little ordered, but it's very hard to see any structure in them. Is it just a logical process to interpret them or is there a great deal of intuitive skill?

DH: Of course now you can just interpret them by putting the photographs through a machine, and letting the machine place the reflections and measure their intensities and pass them into a tape full of numbers which you can put into your computer. It wasn't like that when I was young and it isn't what I think about. I'd start off any crystal structure operation by taking the photograph myself and looking at it and seeing straight away what there is about the structure that I can tell immediately from the distribution of the reflections on the photograph. I admit that I don't like some modern improvements which cut out photographs almost altogether and put everything through a counter. I got a lot of pleasure myself out of just looking at the photographs and guessing the answers even if one guessed imperfectly and wrong. Also some photographs are really very beautiful you know.

LW: So you had a great skill in being able to go from those two-dimensional, ghost-like pictures to a three-dimensional object. Why were you so successful?

DH: Well I don't know that it requires all that skill if you know the lines on which these things work. It was a great advantage to start early. I mean one gets a certain amount of notoriety from being the first person to do things which anybody else really could have done. What I find difficult to know is why more people didn't take up this particular method of attacking problems at the same stage as we did. It seems to me that once W. L. Bragg had taken the first step, the chemists and physicists should have realized much more than they did that this was a tremendous opportunity. But for those who came in at the early stages there was so much gold lying about that we couldn't help finding some of it.

LW: Your pleasure, you said, comes partly from handling crystals. Is this something that developed very early?

DH: When I was quite young, I think I was ten at the time, I went to a small PNEU class in Beccles. PNEU stands for Parents National Educational Union, and it was founded by a Miss Mason of Ambleside to improve the education given by governesses, in a private way, all over the country. They produced small books that would enable the governesses to introduce their pupils to the different sciences in turn. So the small book on chemistry began with growing crystals, which I think is quite a common way to begin chemistry, growing crystals of copper sulphate and alum. I found this fascinating and repeated the experiments at home, when we had a home, which was the following year. My father and mother had been abroad most of the war and came home to look for a house for us to live in so that we could settle down near the local secondary school for our further education.

LW: So you made a little laboratory at home?

DH: Yes. I did go on crystal growing, and then, when I knew about the elements of analytical chemistry, I also used to carry out analyses on a collection of minerals. Now how I came by this is quite a nice story which perhaps illustrates the situation. My father and mother, as I said, worked abroad in the Sudan, and when I was thirteen they were just about to retire. They thought it would be interesting for us children to see how they lived out there and so they took the two eldest of us away from school for a term to stay in Khartoum with them. We didn't do very much in the way of lessons but my mother took us about with her to see the different things she was interested in. One of the visits that we paid was to the Wellcome Laboratories and we first of all went to the medical one. Then, next door, was geology. The geologists there had just brought some little tiny pellets of gold back, and to amuse us children, they showed us how they got these by panning the sand from the bottom of streams. Of course this started me off thinking why shouldn't we find gold. So we went and panned the sand at the bottom of the little water channel running through our garden, and found a black shiny mineral. Now, I had already made friends with the chemical section of the Wellcome Laboratories. Its head was a particular friend of my father, Dr A. F. Joseph, whom we called Uncle Joseph. I went across to him and said 'Please can I analyse this mineral and find out what it is?' I guessed and told him I thought it might be manganese dioxide because it was black and shiny like manganese dioxide. So he helped me try the tests and, of course, it wasn't manganese dioxide. It was ilmenite, which is a mixed ore of iron and titanium. After that, he gave me the proper sort of surveyor's box with little bottles of reagents. One could carry it about the country and test for different elements in the minerals one found. It had a sample set of minerals in little tubes, so when I got home I used quite often to try out experiments and see whether I found the things in these little tubes that were supposed to be in them, according to the books. Then on his advice I bought a very large and serious text book of analytical chemistry, and continued this interest all through the years that I was at school.

LW: So you really were a crystallographer at heart from the beginning?

DH: Well, again, there's a quite interesting thing about that period. You see, my mother was very much interested in my choice of subject. She approved of it. She and my father were not scientists at all, they were both archaeologists as far as they had a profession, though they were at that time working mainly in education. She bought me, amongst other things, W. H. Bragg's books based on the subjects of his Christmas Lectures to children at the Royal Institution. If you come across them, they're very good and still perfectly readable. One is called *Concerning the Nature of Things* and the other one is called *Old Trades and New Knowledge*. My mother was particularly interested in *Old Trades and New Knowledge*. I think she got it because she was interested in weaving and potting and things of that kind. But the one *Concerning the Nature of Things* describes the X-ray diffraction of crystals and has in it the words: 'by this means you can "see" the atoms in the crystals'. And so I really decided then that this was what I would do. It was very exciting.

LW: Now when you started work you began with insulin, but you didn't solve it straight away.

DH: No, no, good heavens no. The first measurements on insulin, like Bernal's first measurements on pepsin which I was a little involved in a year before, were wholly ahead of their time as far as there being any conceivable chance that we could work out the structure. We were both really totally inexperienced in even simple structure analysis at the time. We were faced with an enormously complex problem and though, right at the very beginning, Bernal suggested the way in which it could be solved, it seemed to me that no way was I, at the age of twenty-four, going to set out on that path without trying the proposed method on very much simpler problems first.

LW: So did you abandon insulin then?

DH: I never wholly abandoned it. I left it, yes, but in a curious way I didn't even really leave it. I went on doing the sort of things that would eventually have to be done, but in rather imperfect ways. Very slowly and gradually during the war, doing the measurements out in this house where I brought my little child to be safe away from possible bombing. I knew they weren't really good enough measurements to solve the structure, yet I couldn't help going on doing them somehow. But I put my real effort, when I was back in the lab again, into soluble problems. The first one was a carryover from the work that I was doing in Cambridge with Bernal, which was concerned with finding the structure of the sterols, and particularly of cholesterol. The next one was penicillin.

LW: Why did you choose penicillin?

DH: Penicillin was just historical accident. The work on penicillin began in Oxford just before the war and one of my friends at that time was Ernst Chain. In Oxford we go up and down South Parks Road, and going along South Parks Road one morning I met Ernst Chain in a state of great excitement having just been carrying out the experiment that is now famous. They had four mice which they injected with streptococcus and penicillin and four mice which they had injected with streptococcus alone. One group, the last group, died and the other group lived. And, as they were trying to isolate penicillin, Ernst was extremely excited, and said 'Some day we'll have crystals for you.'

LW: Now, when you chose to work on penicillin, was it that you really cared about penicillin, or was it that here was an important molecule which offered you the pleasure of finding out its structure?

DH: I think that both elements went into that particular operation. I mean nobody who lived through the first year or two of the trials of penicillin in Oxford could possibly not care about what it was. But also it's difficult not to enjoy just growing the crystals.

LW: Now, the structure of penicillin was soluble, unlike insulin which you were still holding in the background. Was that because it was simpler?

DH: Yes, there's all the difference in the world between working out the arrangement of atoms in the space of a small molecule in which you've got 16 or 17 atoms in the assymetric unit and one in which you've got a thousand.

LW: How complicated, then, was vitamin B12?

DH: That's intermediate. That's of the order of 100. We didn't know it was intermediate when we started. Vitamin B12 came about through the fact that I got to know the people in the pharmaceutical industry rather well through the penicillin work, and Dr Lester Smith of Glaxo was working on the isolation of vitamin B12 just after the war. He got crystals first in 1948, within a week or two of crystals being obtained in America by the Merck Group. But to get its structure was again a long process because of the number of atoms in the molecule. The fact was, of course, at that moment we knew nothing, but nothing, about the structure, and what really held us up was just the state of computing in the world. You see electronic computers were being built, but when we started the beginning of the B12 work they hadn't been used at all in X-ray crystallography. They weren't really in a fit state. We did the end of the penicillin calculations on an old punched card machine and we brought this back again for the beginning of the B12 calculations. But it was very slow. A calculation which at the end of the story was taking, well, still a few hours on one of the early electronic computers, took us three months on punched cards . . .

LW: At some stage you must have realized that you had talents or abilities that set you apart from your friends and colleagues. When was that?

DH: I don't think it was so very obvious, you know, because, in a curious way, of the sketchiness of my education. In the early period, before the age of eleven, when my parents came home and I started secondary education, I had moved from one school, one little sort of private school, to another, and one year we'd spent actually being taught entirely by my mother, which was a very fascinating time. So when I first went to secondary school, I was rather behind, if anything. I was *terribly* behind in arithmetic, and it was only at the end of my time there, at the very end of the last year, that I was first in the form. One of the other girls was also very good indeed, and she was generally first. Actually, she did better in chemistry in school certificate than I did. She was very, very good.

LW: So what were your special qualities? Why didn't she go on and do all the things that you did?

DH: I've always thought that her case is really a case history for the problems of girls' education in this country. You see, a girl's future has depended a good deal on ambition and the advice that the young get given. She just didn't think in terms of going to a university, although there's no reason at all why she shouldn't have. I think in terms of the present organization, she would have done so, and would very likely have ended up in research.

LW: Are these same attitudes reflected in the headline in the *Daily Mail* which read: 'British Wife gets Nobel Prize?'

DH: Oh, I thought it said 'Grandmother'!

LW: But do you mind that sort of thing?

DH: I didn't mind that one, no. The one I was slightly worried about concerned the penicillin work. The headline on my election to the Royal Society read 'Mother was first'. I wasn't quite sure that my chemical colleagues would have really appreciated that.

LW: Have you felt strongly about the position of women in science?

DH: No. I think it's because I didn't really notice it very much, that I was a woman amongst so many men. And the other thing is, of course, that I'm a little conscious that there were moments when it was to my advantage. My men colleagues at Oxford were very often particularly nice and helpful to me as a lone girl. And at the time just after the war, when there was an air of liberalism abroad and the first elections of women to the Royal Society were made, that probably got me in earlier than one might have as a man, just because one was a woman.

Source: Lewis Wolpert and Alison Richards, *A Passion for Science*, Oxford, Oxford University Press, 1988.

The Plan of Living Things

The most famous breakthrough in modern science was the discovery of the structure of DNA – the genetic material of all organisms in nature – by Francis Crick and James Watson. The term 'gene' was coined in 1909 by Wilhelm Johannsen, and by the mid-1930s it had been established that genes were physical entities. By the early 1950s it was known that the chemical material of the gene was DNA (deoxyribonucleic acid). Watson and Crick, in 1953, proposed a model for the DNA molecule in the form of a double helix, with two distinct chains wound round one another about a common axis. They also suggested that to replicate DNA the cell unwinds the two chains and uses each as a template to guide the formation of a new companion chain – thus producing two double helices, each with one new and one old chain. In this extract from his autobiography, *What Mad Pursuit* (1989), Crick explains the process of earlier discovery that made possible his and Watson's advance. RNA, mentioned in the extract, is ribonucleic acid.

At the time I started in biology – the late 1940s – there was already some rather indirect evidence suggesting that a single gene was perhaps no bigger than a very large molecule – that is, a macromolecule. Curiously enough, a simple, suggestive argument based on common knowledge also points in this direction.

Genetics tells us that, roughly speaking, we get half of all our genes from our mother, in the egg, and the other half from our father, in the sperm. Now, the head of a human sperm, which contains these genes, is quite small. A single sperm is far too tiny to be seen clearly by the naked eye, though it can be observed fairly easily using a high-powered microscope. Yet in this small space must be housed an almost complete set of instructions for building an entire human being (the egg providing a duplicate set). Working through the figures, the conclusion is inescapable that a gene must be, by everyday standards, very, very small, about the size of a very large chemical molecule. This alone does not tell us what a gene does, but it does hint that it might be sensible to look first at the chemistry of macromolecules.

It was also known at that time that each chemical reaction in the cell was catalysed by a special type of large molecule. Such molecules were called enzymes. Enzymes are the machine tools of the living cell. They were first discovered in 1897 by Edouard Buchner, who received a Nobel Prize ten years later for his discovery. In the course of his experiments, he crushed yeast cells in a hydraulic press and obtained a rich mixture of yeast juices. He wondered whether such fragments of a living cell could carry out any of its chemical reactions, since at that time most people thought that the cell must be intact for such reactions to occur. Because he wanted to preserve the juice, he adopted a stratagem used in the kitchen: he added a lot of sugar. To his astonishment, the juice fermented the sugar solution! Thus were enzymes discovered. (The word enzyme means 'in yeast'.) It was soon found that enzymes could be obtained from many other types of cell, including our own, and that each cell contained very many distinct kinds of enzymes. Even a simple bacterial cell may contain more than a thousand different *types* of enzymes. There may be hundreds or thousands of molecules of any one type.

In favourable circumstances an enzyme could be purified away from all the others and its action studied by itself in solution. Such studies showed that each enzyme was very specific, and catalysed only one particular chemical reaction or, at most, a few related ones. Without that particular enzyme the chemical reaction, under the mild conditions of temperature and acidity usually found in living cells, would proceed only very, very slowly. Add the enzyme and the reaction goes at a good pace. If you make a well-dispersed solution of starch in water, very little will happen. Spit into it and the enzyme amylase in your saliva will start to digest the starch and release sugars.

The next major discovery was that each of the enzymes studied was a macromolecule and that they all belonged to the same family of macromolecules called proteins. The key discovery was made in 1926 by a one-armed American chemist called James Sumner. It is not all that easy to do chemistry when you have only one arm (he had lost the other in a shooting accident when he was a boy) but Sumner, who was a very determined man, decided he would nevertheless demonstrate that enzymes were proteins. Though he showed that one particular enzyme, urease, was a protein and obtained crystals of it, his results were not immediately accepted. In fact, a group of German workers hotly contested the idea, which somewhat embittered Sumner, but it turned out that he was correct. In 1946 he was awarded part of the Nobel Prize in Chemistry for his discovery. Though very recently a few significant exceptions to this rule have turned up, it is still true that almost all enzymes are proteins.

Proteins are thus a family of subtle and versatile molecules. As soon as I learned about them I realized that one of the key problems was to explain how they were synthesized.

There was a third important generalization, though in the 1940s this was sufficiently new that not everybody was inclined to accept it. This idea was due to George Beadle and Ed Tatum. (They too were to receive a Nobel Prize, in 1958, for their discovery.) Working with the little bread-mould *Neurospora*, they had found that each mutant of it they studied appeared to lack just a single enzyme. They coined the famous slogan 'One gene – one enzyme'.

Thus the general plan of living things seemed almost obvious. Each gene determines a particular protein. Some of these proteins are used to form structures or to carry signals, while many of them are the catalysts that decide what chemical reactions should and should not take place in each cell. Almost every cell in our bodies has a complete set of genes within it, and this chemical programme directs how each cell metabolizes, grows, and interacts with its neighbours. Armed with all this (to me) new knowledge, it did not take much to recognize the key questions. What are genes made of? How are they copied exactly? And how do they control, or at least influence, the synthesis of proteins?

It had been known for some time that most of a cell's genes are located on its chromosomes and that chromosomes were probably made of nucleoprotein – that is, of protein and DNA, with perhaps some RNA as well. In the early 1940s it was thought, quite erroneously, that DNA molecules were small and, even more erroneously, simple. Phoebus Levene, the leading expert on nucleic acid in the 1930s, had proposed that they had a regular repeating structure [the so-called tetranucleotide hypothesis]. This hardly suggested that they could easily carry genetic information. Surely, it was thought, if genes had to have such remarkable properties, they must be made of proteins, since proteins as a class were known to be capable of such remarkable functions. Perhaps the DNA there had some associated function, such as acting as a scaffold for the more sophisticated proteins.

It was also known that each protein was a polymer. That is, it consisted of a long chain, known as a polypeptide chain, constructed by stringing together, end to end, small organic molecules, called monomers since they are the elements of a polymer. In a homopolymer, such as nylon, the small monomers are usually all the same. Proteins are not as simple as that. Each protein is a heteropolymer, its chains being strung together from a selection of somewhat different small molecules, in this instance amino acids. The net result is that, chemically speaking, each polypeptide chain has a completely regular backbone, with little side-chains attached at regular intervals. It was believed that there were about twenty different possible side-chains (the exact number was not known at that time). The amino acids (the monomers) are just like the letters in a font of type. The base of each kind of letter from the font is always the same, so that it can fit into the grooves that hold the assembled type, but the top of each letter is different, so that a particular letter will be printed from it. Each protein has a characteristic number of amino acids, usually several hundred of them, so any particular protein could be thought of crudely as a paragraph written in a special language having about twenty (chemical) letters. It was not then known for certain, as it is now, that for each protein the letters have to be in a particular order (as indeed they have to be in a particular paragraph). This was first shown a little later by the biochemist Fred Sanger, but it was easy enough to guess that this was likely to be true.

Of course each paragraph in our language is really one long line of letters. For convenience this is split up into a series of lines, written one under the other, but this is only a secondary matter, since the meaning is exactly the same whether the lines are long or short, few or many, provided we take care about splitting the words at the end of each line. Proteins were known to be very different. Although the polypeptide backbone is chemically regular, it contains flexible links, so that in principle many different three-dimensional shapes are possible. Nevertheless, each protein appeared to have its own shape, and in many cases this shape was known to be fairly compact (the word used was 'globular') rather than very extended (or 'fibrous'). A number of proteins had been crystallized, and these crystals gave detailed X-ray diffraction patterns, suggesting that the three-dimensional structure of each molecule of a particular kind of protein was exactly (or almost exactly) the same. Moroever many proteins, if heated briefly to the boiling point of water, or even to some temperature below this, became denatured, as if they had unfolded so that their three-dimensional structure had been partly destroyed. When this happened the denatured protein usually lost its catalytic or other function, strongly suggesting that the function of such a protein *depended on its exact three-dimensional structure.*

And now we can approach the baffling problem that appeared to face us. If genes are made of protein, it seemed likely that each gene had to have a special three-dimensional, somewhat compact structure. Now, a vital property of a gene was that it could be copied exactly for generation after generation, with only occasional mistakes. What we were trying to guess was the general nature of this copying mechanism. Surely the way to copy something was to make a complementary structure – a mould – and then to make a further complementary structure of the mould, to produce in this way an exact copy of the original. This, after all, is how, broadly speaking, sculpture is copied. But then the dilemma arose: it is easy to copy the *outside* of a three-dimensional structure in this way, but how on earth could one copy the *inside*? The whole process seemed so utterly mysterious that one hardly knew how to begin thinking about it.

Of course, now that we know the answer, it all seems so completely obvious that no one nowadays remembers just how puzzling the problem seemed then. If by chance you do *not* know the answer, I ask you to pause a moment and reflect on what the answer might be. There is no need, at this stage, to bother about the details of the chemistry. It is the principle of the idea that matters. The problem was not made easier by the fact that many of the properties of proteins and genes just outlined were not known for certain. All of them were plausible and most of them seemed very probable but, as in most problems near the frontiers of research, there were always nagging doubts that one or more of these assumptions might be dangerously misleading. In research the front line is almost always in a fog.

So what was the answer? Curiously enough, I had arrived at the correct solution before Jim Watson and I discovered the double-helical structure of DNA. The basic idea (which was not entirely new) was this: all a gene had to do was to get the *sequence* of the amino acids correct in that protein. Once the correct polypeptide chain had been synthesized, with all its side-chains in the right order, then, following the laws of chemistry, the protein *would fold itself up correctly into a unique three-dimensional structure.* (What the exact three-dimensional structure of each protein was remained to be determined.) By this bold assumption the problem was changed from a three-dimensional one to a one-dimensional one, and the original dilemma largely disappeared.

Of course, this had not *solved* the problem. It had merely transformed it from an intractable one to a manageable one. For the problem still remained: how to make an exact copy of a one-dimensional sequence. To approach that we must return to what was known about DNA.

By the late 1940s our knowledge of DNA had improved in several important respects. It had been discovered that DNA molecules were not, after all, very short. Exactly how long they were was not clear. We know now that they appeared to be short because, being long molecules (in the sense that a piece of string is long), they could easily be broken in the process of getting them out of the cell and manipulating them in the test tube. Just stirring a DNA solution is enough to break the longer molecules. Their chemistry was now known more correctly, and moreover the tetranucleotide hypothesis was dead, killed by some very beautiful work by a chemist at Columbia, the Austrian refugee Erwin Chargaff. DNA was known to be a polymer, but with a very different backbone and with only four letters in its alphabet, rather than twenty. Chargaff showed that DNA from different sources had rather different amounts of those four bases (as they were called). Perhaps DNA was not such a dumb molecule after all. It might conceivably be long enough and varied enough to carry some genetic information.

Even before I left the Admiralty there had been some quite unexpected evidence pointing to DNA as near the center of the mystery. In 1944 Avery, MacLeod, and McCarty, who worked at the Rockefeller Institute in New York, had published a paper claiming that the 'transforming factor' of pneumococcus consisted of pure DNA. The transforming factor was a chemical extracted from a strain of bacteria having a smooth coat. When added to a related strain lacking such a coat it 'transformed' it, so that some of the recipient bacteria acquired the smooth coat. More important, all the descendants of such cells had the same smooth coat. In the paper the authors were rather cautious in interpreting their results, but in the now-famous letter to his brother Avery expressed himself more freely. 'Sounds like a virus – may be a gene,' he wrote.

This conclusion was not immediately accepted. An influential biochemist, Alfred Mirsky, also at the Rockefeller, was convinced that it was an impurity of the DNA that was causing the transformation. Subsequently more careful work by Rollin Hotchkiss at the Rockefeller showed that this was highly unlikely. It was argued that Avery, MacLeod, and McCarty's evidence was flimsy, in that only one character had been transformed. Hotchkiss showed that another character could also be transformed. The fact that these transformations were often unreliable, tricky to perform, and only altered a minority of cells did not help matters. Another objection was that the process had been shown to occur just in these particular bacteria. Moreover, at that time no bacterium of any sort had been shown to have genes, though this was discovered not long afterward by Joshua Lederberg and Ed Tatum. In short, it was feared that transformations might be a freak case and misleading as far as higher organisms were concerned. This was not a wholly unreasonable point of view. A single isolated bit of evidence, however striking, is always open to doubt. It is the accumulation of several different lines of evidence that is compelling.

It is sometimes claimed that the work of Avery and his colleagues was ignored and neglected. Naturally there was a mixed spectrum of reactions to their results, but one can hardly say no one knew about it. For example, that august and somewhat conservative body, the Royal Society of London, awarded the Copley Medal to Avery in 1945, specifically citing his work on the transforming factor. I would dearly love to know who wrote the citation for them.

Nevertheless, even if all the objections and reservations are brushed aside, the fact that the transforming factor was pure DNA does not in itself prove that DNA alone is the genetic material in pneumococcus. One could quite logically claim that a gene there was made of DNA *and* protein, each carrying part of the genetic information, and it was just an accident of the system that in transformation the altered DNA part was carrying the information to change the polysaccharide coat. Perhaps in another experiment a protein component might be found that would also produce a heritable change in the coat or in other cell properties.

Whatever the interpretation, because of this experiment and because of the increased knowledge of the chemistry of DNA, it was now possible that genes might be made of DNA alone.

Source: Francis Crick, *What Mad Pursuit: A Personal View of Scientific Discovery*, London, Weidenfeld & Nicolson, 1989.

Willow Seeds and the *Encyclopaedia Britannica*

Richard Dawkins (b. 1941) lectures in Zoology at Oxford University. His first book, *The Selfish Gene* (1976) was an international bestseller, translated into eleven languages. This extract is from *The Blind Watchmaker* (1986), which defends Darwin's theory of evolution by natural selection against some modern attacks.

It is raining DNA outside. On the bank of the Oxford canal at the bottom of my garden is a large willow tree, and it is pumping downy seeds into the air. There is no consistent air movement, and the seeds are drifting outwards in all directions from the tree. Up and down the canal, as far as my binoculars can reach, the water is white with floating cottony flecks, and we can be sure that they have carpeted the ground to much the same radius in other directions too. The cotton wool is mostly made of cellulose, and it dwarfs the tiny capsule that contains the DNA, the genetic information. The DNA content must be a small proportion of the total, so why did I say that it was raining DNA rather than cellulose? The answer is that it is the DNA that matters. The cellulose fluff, although more bulky, is just a parachute, to be discarded. The whole performance, cotton wool, catkins, tree and all, is in aid of one thing and one thing only, the spreading of DNA around the countryside. Not just any DNA, but DNA whose coded characters spell out specific instructions for building willow trees that will shed a new generation of downy seeds. Those fluffy specks are, literally, spreading instructions for making themselves. They are there because their ancestors succeeded in doing the same. It is raining instructions out there; it's raining programs; it's raining tree-growing, fluff-spreading, algorithms. That is not a metaphor, it is the plain truth. It couldn't be any plainer if it were raining floppy disks.

It is plain and it is true, but it hasn't long been understood. A few years ago, if you had asked almost any biologist what was special about living things as opposed to nonliving things, he would have told you about a special substance called protoplasm. Protoplasm wasn't like any other substance; it was vital, vibrant, throbbing, pulsating, 'irritable' (a schoolmarmish way of saying responsive). If you look at a living body and cut it up into ever smaller pieces, you would eventually come down to specks of pure protoplasm. At one time in the last century, a real-life counterpart of Arthur Conan Doyle's Professor Challenger thought that the 'globigerina ooze' at the bottom of the sea was pure protoplasm. When I was a schoolboy, elderly textbook authors still wrote about protoplasm although, by then, they really should have known better. Nowadays you never hear or see the word. It is as dead as phlogiston and the universal aether. There is nothing special about the substances from which living things are made. Living things are collections of molecules, like everything else.

What is special is that these molecules are put together in much more complicated patterns than the molecules of nonliving things, and this putting together is done by following programs, sets of instructions for how to develop, which the organisms carry around inside themselves. Maybe they do vibrate and throb and pulsate with 'irritability', and glow with 'living' warmth, but these properties all emerge incidentally. What lies at the heart of every living thing is not a fire, not warm breath, not a 'spark of life'. It is information, words, instructions. If you want a metaphor, don't think of fires and sparks and breath. Think, instead, of a billion discrete, digital characters carved in tablets of crystal. If you want to understand life, don't think about vibrant, throbbing gels and oozes, think about information technology . . .

The basic requirement for an advanced information technology is some kind of storage medium with a large number of memory locations. Each location must be capable of being in one of a discrete number of states. This is true, anyway, of the *digital* information technology that now dominates our world of artifice. There is an alternative kind of information technology based upon *analogue* information. The information on an ordinary gramophone record is analogue. It is stored in a wavy groove. The information on a modern laser disk (often called 'compact disk', which is a pity, because the name is uninformative and also usually mispronounced with the stress on the first syllable) is digital, stored in a series of tiny pits, each of which is either definitely there or definitely not there: there are no half measures. That is the diagnostic feature of a digital system: its fundamental elements are either definitely in one state or definitely in another state, with no half measures and no intermediates or compromises.

The information technology of the genes is digital. This fact was discovered by Gregor Mendel in the last century, although he wouldn't have put it like that. Mendel showed that we don't blend our inheritance from our two parents. We receive our inheritance in discrete particles. As far as each particle is concerned, we either inherit it or we don't. Actually, as R. A. Fisher, one of the founding fathers of what is now called neo-Darwinism, has pointed out, this fact of particulate inheritance has always been staring us in the face, every time we think about sex. We inherit attributes from a male and a female parent, but each of us is either male or female, not hermaphrodite. Each new baby born has an approximately equal *probability* of inheriting maleness or femaleness, but any one baby inherits only one of these, and doesn't combine the two. We now know that the same goes for all our particles of inheritance. They don't blend, but remain discrete and separate as they shuffle and reshuffle their way down the generations. Of course there is often a powerful appearance of blending in the effects that the genetic units have on bodies. If a tall person mates with a short person, or a black person with a white person, their offspring are often intermediate. But the appearance of blending applies only to effects on bodies, and is due to the summed small effects of large numbers of particles. The particles themselves remain separate and discrete when it comes to being passed on to the next generation.

The distinction between blending inheritance and particulate inheritance has been of great importance in the history of evolutionary ideas. In Darwin's time everybody (except Mendel who, tucked away in his monastery, was unfortunately ignored until after his death) thought that inheritance was blending. A Scottish engineer called Fleeming Jenkin pointed out that the fact (as it was thought to be) of blending inheritance all but ruled out natural selection as a plausible theory of evolution. Ernst Mayr rather unkindly remarks that Jenkin's article 'is based on all the usual prejudices and misunderstandings of the physical scientists'. Nevertheless, Darwin was deeply worried by Jenkin's argument. It was most colourfully embodied in a parable of a white man shipwrecked on an island inhabited by 'negroes':

> . . . grant him every advantage which we can conceive a white to possess over the native; concede that in the struggle for existence his chance of a long life will be much superior to that of the native chiefs; yet from all these admissions, there does not follow the conclusion that, after a limited or unlimited number of generations, the inhabitants of the island will be white. Our shipwrecked hero would probably become king; he would kill a great many blacks in the struggle for existence; he would have a great many wives and children, while many of his subjects would live and die as bachelors . . . Our white's qualities would certainly tend very much to preserve him to a good old age, and yet he would not suffice in any number of generations to turn his subjects' descendants white . . . In the first generation there will be some dozens of intelligent young mulattoes, much superior in average intelligence to the negroes. We might expect the throne for some generations to be occupied by a more or less yellow king; but can any one believe that the whole island will gradually acquire a white, or even a yellow population, or that the islanders would acquire the energy, courage, ingenuity, patience, self-control, endurance, in virtue of which qualities our hero killed so many of their ancestors, and begot so many children; these qualities, in fact, which the struggle for existence would select, if it could select anything?

Don't be distracted by the racist assumptions of white superiority. These were as unquestioned in the time of Jenkin and Darwin as our speciesist assumptions of *human* rights, *human* dignity, and the sacredness of *human* life are unquestioned today. We can rephrase Jenkin's argument in a more neutral analogy. If you mix white paint and black paint together, what you get is grey paint. If you mix grey paint and grey paint together, you can't reconstruct either the original white or the original black. Mixing paints is not so far from the pre-Mendelian vision of heredity, and even today popular culture frequently expresses heredity in terms of a mixing of 'bloods'. Jenkin's argument is an argument about swamping. As the generations go by, under the assumption of blending inheritance, variation is bound to become swamped. Greater and greater uniformity will prevail. Eventually there will be no variation left for natural selection to work upon.

Plausible as this argument must have sounded, it is not only an argument against natural selection. It is more an argument against inescapable facts about heredity itself! It manifestly isn't *true* that variation disappears as the generations go by. People are *not* more similar to each other today than they were in their grandparents' time. Variation is maintained. There is a pool of variation for selection to work on. That was pointed out mathematically in 1908 by W. Weinberg, and independently by the eccentric mathematician G. H. Hardy, who incidentally, as the betting book of his (and my) college records, once took a bet from a colleague of 'One half penny to his fortune till death, that the sun will rise tomorrow'. But it took R. A. Fisher and his colleagues, the founders of modern population genetics, to develop the full answer to Fleeming Jenkin in terms of Mendel's theory of *particle* genetics. Fisher and his colleagues showed that Darwinian selection made sense, and Jenkin's problem was elegantly solved, if what changed in evolution was the relative *frequency* of discrete hereditary particles, or genes, each of which was either there or not there in any particular individual body. Darwinism post-Fisher is called neo-Darwinism. Its digital nature is not an incidental fact that happens to be true of genetic information technology. Digitalness is probably a necessary precondition for Darwinism itself to work.

In our electronic technology the discrete, digital locations have only two states, conventionally represented as 0 and 1 although you can think of them as high and low, on and off, up and down: all that matters is that they should be distinct from one another, and that the pattern of their states can be 'read out' so that it can have some influence on something. Electronic technology uses various physical media for storing 1s and 0s, including magnetic discs, magnetic tape, punched cards and tape, and integrated 'chips' with lots of little semiconductor units inside them.

The main storage medium inside willow seeds, ants and all other living cells is not electronic but chemical. It exploits the fact that certain kinds of molecule are capable of 'polymerizing', that is joining up in long chains of indefinite length. There are lots of different kinds of polymer. For example, 'polythene' is made of long chains of the small molecule called ethylene – polymerized ethylene. Starch and cellulose are polymerized sugars. Some polymers, instead of being uniform chains of one small molecule like ethylene, are chains of two or more different kinds of small molecule. As soon as such heterogeneity enters into a polymer chain, information technology becomes a theoretical possibility. If there are two kinds of small molecule in the chain, the two can be thought of as 1 and 0 respectively, and immediately any amount of information, of any kind, can be stored, provided only that the chain is long enough. The particular polymers used by living cells are called polynucleotides. There are two main families of polynucleotides in living cells, called DNA and RNA for short. Both are chains of small molecules called nucleotides. Both DNA and RNA are heterogeneous chains, with four different kinds of nucleotides. This, of course, is where the opportunity for information storage lies. Instead of just the two states 1 and 0, the information technology of living cells uses four states, which we may conventionally represent as A, T, C and G. There is very little difference, in principle, between a two-state binary information technology like ours, and a four-state information technology like that of the living cell.

There is enough information capacity in a single human cell to store the *Encyclopaedia Britannica*, all 30 volumes of it, three or four times over. I don't know the comparable figure for a willow seed or an ant, but it will be of the same order of staggeringness. There is enough storage capacity in the DNA of a single lily seed or a single salamander sperm to store the *Encyclopaedia Britannica* 60 times over. Some species of the unjustly called 'primitive' amoebas have as much information in their DNA as 1,000 *Encyclopaedia Britannicas*.

Source: Richard Dawkins, *The Blind Watchmaker*, London, Longman Scientific and Technical, 1986.

Shedding Life

Miroslav Holub, poet and immunologist, was born in western Bohemia in 1923. At school he specialized in Latin and Greek, and he worked during the war as a railway labourer. He started to write poetry as a student of science and medicine in Prague, and became editor of the scientific magazine *Vesmír*, the equivalent of *New Scientist*. In 1954 he joined the immunological section of the Biological Institute of the Czechoslovak Academy of Science, receiving his PhD in the same year as his first book of poems appeared. He has published over 140 scientific papers, 3 scientific monographs, and 14 books of poetry. This piece is from *The Dimension of the Present Moment and Other Essays* (1990).

A muskrat, also called musquash, or technically, *Ondatra zibethica zibethica* Linn. 1766 – the creature didn't give a hoot about nomenclature – fell into our swimming pool, which was empty except for a puddle of winter water. It huddled in a corner, wild frightened eyes, golden-brown fur, hairless muddied tail. Before I could find instruments suitable for catching and removing muskrats, a passing neighbour (unfamiliar with rodents *per se*, or even with rodents living in Czechoslovakia since 1905), deciding he'd come across a giant rat as bloodthirsty as a tiger and as full of infections as a plague hospital, ran home, got his shotgun, and fired at the muskrat until all that was left was a shapeless soggy ball of fur with webbed hind feet and bared teeth. There was blood all over the sides and bottom of the pool, all over the ball of fur, and the puddle of water was a little red sea. The hunting episode was over, and I was left to cope with the consequences. Humankind can generally be divided into hunters and people who cope with consequences.

I buried the deceased intruder under the spruces in the backyard, and, armed with a bundle of rags, I went to clean up the shooting gallery. The swimming pool doesn't have a drain, so the operation looked more like an exercise in rag technology, chasing the blood north, south, east, west, up, and down. Chasing blood around an empty swimming pool is as inspirational as listening to a record of Haydn's 'Farewell Symphony' with the needle stuck in the same groove, I became very intimate with the blood in that hour, and I began to daydream about it. The blood wasn't just that unpleasant stuff that under proper and normal conditions belonged inside the muskrat. It was the muskrat's secret life forced out. The puddle of red sea was, in fact, a vestige of an ancient Silurian sea. It was kept as an inner environment when life came ashore. Kept so that even – though it's changed to a radically different concentration of ions, a different osmotic pressure, and different salts – the old metabolism hasn't needed too much reshuffling.

In any case, the muskrat was cast ashore from its own little red sea. Billions of red blood cells were coagulating and disintegrating, their haemoglobin molecules puzzled as to how and where to pass their four molecules of oxygen.

The blood corpuscles were caught in tender, massive nets of fibres formed from fibrinogen, stimulated by thrombin that was formed from prothrombin. A long sequence of events occurred one after the other in the presence of calcium ions, phospholipids from blood platelets, and thromboplastin, through which the shot arteries were trying to show that the bleeding should be stopped because it was bad for the muskrat (though in the long run it didn't matter). And in the serum around the blood cells, the muskrat's inner-life signals were probably still flickering, dimming, and fading out: instructions from the pituitary gland to the liver and adrenals, from the thyroid gland to all kinds of cells, from the adrenal glands to sugars and salts, from the pancreas to the liver and fat tissues – the dying debate of an organism whose trillions of cells coexist thanks to unified information.

And, especially because of the final chase, the adrenalin and the stress hormone corticotrophin were still sounding their alarms. Alarms were rushing to the liver to mobilize sugar reserves, alarms were sounding to distend the coronary and skeletal muscle arteries, to increase heart activity, to dilate bronchioles, to contract skin arteries and make the hair stand up, to dilate the pupils. And all that militant inner tumult was abandoned by what should obey it. Then there were endorphins, which lessen the pain and anxiety of a warrior's final struggle, and substances to sharpen the memory, because the struggle for life should be remembered well.

So there was this muskrattish courage, an elemental bravery transcending life.

But mainly, among the denaturing proteins and the disintegrating peptide chains, the white blood cells lived, really lived, as anyone knows who has ever peeked into a microscope, or anyone knows who remembers how live tissue cells were grown from a sausage in a Cambridge laboratory (the sausage having certainly gone through a longer funereal procedure than blood that is still flowing). There were these shipwrecked white blood cells in the cooling ocean, millions and billions of them on the concrete, on the rags, in the wrung-out murkiness. Bewildered by the unusual temperature and salt concentration, lacking unified signals and gentle ripples of the vascular endothelium, they were nevertheless alive and searching for whatever they were destined to search for. The T lymphocytes were using their receptors to distinguish the muskrat's self-markers from non-self bodies. The B lymphocytes were using their antibody molecules to pick up everything the muskrat had learned about the outer world in the course of its evolution. Plasma cells were dropping antibodies in various places. Phagocyte cells were creeping like amoebas on the bottom of the pool, releasing their digestive granules in an attempt to devour its infinite surface. And here and there a blast cell divided, creating two new, last cells.

In spite of the escalating losses, these huge home-defence battalions were still protecting the muskrat from the sand, cement, lime, cotton and grass; they recognized, reacted, signalled, immobilized, died to the last unknown soldier in the last battle beneath the banner of an identity already buried under the spruces.

Multicellular life is complicated, as is multicellular death. What is known as the death of an individual and defined as the stoppage of the heart – or, more accurately, as the loss of brain functions – is not, however, the death of the system that guards and assures its individuality. Because of this system's cells – phagocytes and lymphocytes – the muskrat was still, in a sense, running around the pool in search of itself.

Not to mention the possibility that a captured lymphocyte, when exposed to certain viruses or chemicals, readily fuses with a cell of even another species, forgetting about its previous self but retaining in its hybrid state both self and non-self information; it can last more or less forever there, providing the tissue culture is technically sound.

Not to mention the theoretical possibility that the nucleus of any live cell could be inserted into an ovum cell of the same species whose nucleus has been removed, and after implantation into the surrogate mother's uterus, the egg cell will produce new offspring with the genetic makeup of the inserted nucleus.

The shed blood shows that there is not one death, but a whole stream of little deaths of varying degrees and significances. The dark act of the end is as special and prolonged as the dark act of the beginning, when one male and one female cell start the flow of divisions and differentiations of cells and tissues, the activation of some hereditary information and the repression of some other, the billions of cellular origins, endings, arrivals and departures.

So in a way the great observer William Harvey was at least a little right when he called blood the main element of the four basic Greek elements of the world and body. In 1651 he wrote: 'We conclude that blood lives of itself and that it depends in no ways upon any parts of the body. Blood is the cause not only of life in general, but also of longer or shorter life, of sleep and waking, of genius, aptitude and strength. It is the first to live and the last to die.'

Blood will have its way, I thought, wringing out another rag.

It is the colour of blood that makes death so horrible. People and other creatures (unless they happen to be the likes of a shark, hyena or wolf) have a fear of shed blood for this reason. It is a fear that hinders further violence when mere immobility, spiritlessness and breathlessness can't. A fear that keeps the published photographs of a killing or slaughter from being true to life. The human reaction to the colour of blood is a faithful counterpart to the microscopic reality, the lethal cascade we so decently provoke by the final shot in the right place. There are an extraordinary number of last things in anyone's bloodbath. Including a muskrat's. And if any tiny bit of soul can be found there, there is not one tiny bit of salvation.

They say you can't see into blood. But I think you can, if only through that instinctive fear.

Lucky for the Keres, the goddesses of bloodshed, that no one concerns himself with the microscopy of battlefields; lucky for the living that molecular farewell symphonies can't be heard; lucky for hunters that they don't have to clean up the mess.

Source: Miroslav Holub, *The Dimension of the Present Moment and Other Essays*, ed. David Young, London, Faber and Faber, 1990.

The Greenhouse Effect: An Alternative View

Freeman Dyson is a Member of the Institute for Advanced Studies at Princeton who has made a second career writing science for non-scientists. He published his first book at the age of 55. Born in England in 1923 he was sent to a prep school where no science was taught. The curriculum was based on Latin and the headmaster was a sadist who punished mistakes in Latin grammar with a riding whip. Clever boys, like Dyson, who generally escaped this, were tortured by the other boys, their favourite instrument being sandpaper applied to the face or other tender areas of skin. As a refuge from brutality, Dyson and other members of the persecuted minority started a science society, where they learned 'that science is a conspiracy of brains against ignorance, that science is a revenge of victims against oppressors, that science is a territory of freedom and friendship in the midst of tyranny and hatred'.

Among the problems discussed in Dyson's *From Eros to Gaia* (1992) is that of 'the missing carbon dioxide'. The amount of carbon dioxide annually pumped into the air from the burning of fossil fuels is much greater than the annual increase of carbon dioxide in the atmosphere. In 1990, for example, the quantity of carbon dioxide released into the atmosphere was 6 gigatons (where one gigaton equals a billion metric tons of carbon), but the increase of carbon dioxide in the atmosphere was only 3.5 gigatons. Where does the rest go? The answer, it seems, must be into the ocean or into the biosphere (trees and topsoil). Dyson favours the latter solution. The experts have been misguided, he suggests, in concentrating on the possible climatic effects of increased carbon dioxide, and paying little attention to how the biosphere adapts to the new conditions.

To indicate the crucial nature of the nonclimatic effects of carbon dioxide, it is sufficient to mention the fact that a field of corn plants growing in full sunshine will completely deplete the carbon dioxide from the air within one meter of ground in a time of the order of five minutes . . .

Owners of commercial greenhouses discovered long ago that seedlings grow faster when the air in the greenhouses is enriched with carbon dioxide. Many experiments have been done in growth chambers designed so that the carbon dioxide can be accurately controlled and the response of plants accurately measured. A typical experiment was done in 1975, measuring the effects of carbon dioxide on the growth and transpiration of leaves of the American poplar, *Populus deltoides*. Experiments on other plant species usually give similar results. The poplar experiment used atmospheres ranging from one-tenth to three times the present outdoor concentration of carbon dioxide. The growth rate is zero at one-tenth of the outdoor level, rises rapidly as the carbon dioxide is increased up to the outdoor level, then rises more slowly as the carbon dioxide is increased to twice the outdoor level, then becomes constant as the carbon dioxide increases beyond twice the outdoor level. The saturation value of growth rate is one and a half times the rate at the outdoor level, and is reached at three times the outdoor level. So far, the results are unsurprising.

More surprising and of greater practical importance are the measurements of water transpiration in the poplar experiment. Transpiration means the loss of water by evaporation from the leaves. The rate of transpiration falls steadily as carbon dioxide increases, and is reduced to about half its present value when the carbon dioxide is enriched threefold. How is this decrease of transpiration to be explained? The essential point is that carbon dioxide molecules are rare in the atmosphere. They are hard for a plant to catch. The only way a plant can catch a carbon dioxide molecule is to keep open the little stomata or pores on the surface of its leaves, and wait for the occasional carbon dioxide molecule to blunder in. But the air inside the stomata is saturated with water vapor. On the average, about two hundred water molecules will stumble out of the hole for every one carbon dioxide molecule that stumbles in. The poplar experiment measured the water loss and the carbon gain of the leaves simultaneously, and found that the loss is a hundred times the gain when the carbon dioxide is at the outdoor level. The plant is forced to lose a lot of water in order to collect a little carbon dioxide. When the carbon dioxide in the atmosphere is enriched, the plant has a choice. The plant may either keep its stomata open and lose water as rapidly as before while increasing photosynthesis. Or it may partially close its stomata and reduce the loss of water while keeping photosynthesis constant. Or it may make a compromise, closing the stomata only a little, so that water loss is decreased while photosynthesis is increased.

The poplar leaves in the experiment chose the compromise strategy. Plants will in general choose whatever strategy they find optimum, depending on local conditions of temperature, humidity, and sunlight.

The moral of this story is that for plants growing under dry conditions, enriched carbon dioxide in the atmosphere is a substitute for water. Give a plant more carbon dioxide, and it can make do with less water. Since the growth of plants, both in agriculture and in the wild, is frequently limited by lack of water, the effect of carbon dioxide in reducing transpiration may be of greater practical importance than the direct effect in increasing photosynthesis. It is easy to measure both these effects of carbon dioxide in greenhouses and growth chambers. It is difficult to measure them in the real world out-of-doors. Here then is the crucial task for understanding the human dimensions of the carbon-dioxide problem. Our research programs must come to grips with the responses of crop plants and grasses and trees all over the world to increased carbon dioxide. To measure these responses, experiments in growth chambers are inadequate and computer simulations are useless. There is no substitute for field observations, widely distributed in place and extended in time.

If we can establish research programs, putting as much money and time and talent into the measurement of ecological responses to carbon dioxide as we are now spending on climatic effects, we may be able within a few years to answer the politically crucial questions. Is the direct effect of increasing carbon dioxide on food production and forests more important than the effect on climate? Is the human species already hooked on carbon dioxide, needing a continued increase of fossil fuel burning to fertilize our crops? When the coal and oil are all gone, shall we be burning limestone to keep the atmosphere enriched in carbon dioxide at the level to which the biosphere will have become addicted? I am not saying that the answers to these questions should be yes. But we must be aware that we do not have the knowledge to answer them with a confident no.

Source: Freeman Dyson, *From Eros to Gaia*, London, Penguin Books, 1993.

Fractals, Chaos and Strange Attractors

Benoit Mandelbrot (b. 1924), a maverick mathematician of Lithuanian extraction, coined the word 'fractal' (from the Latin *fractus*, broken) in 1975, to describe irregular geometrical shapes that repeat themselves endlessly on smaller and smaller scales – a feature known as 'self-similarity'. Fractals, he found, were everywhere in nature – in coastlines, trees, mountains, forked lightning, the cardio-vascular system, or cauliflowers (where each floret of the edible head resembles the whole). Fractal pictures, produced by repeatedly feeding equations into computers, became popular in T-shirt and poster design in the 1980s. The connection between fractals and chaos theory is explained here by Caroline Series, reader in mathematics at Warwick University.

Most people have become familiar in recent years with pictures of fractals, those elusive shapes that, no matter how you magnify them, still look infinitely crinkled. The pictures you saw were probably drawn by computer, but examples abound in nature – the edge of a leaf, the outline of a tree, or the course of a river. Fractal curves differ from those studied in normal geometry. The curve of a circle, for instance, if magnified sufficiently, just about becomes a straight line. A fractal curve, on the other hand, when viewed on many different scales, from macroscopic to microscopic, reveals the same intricate pattern of convolutions.

In recent years, there has been a revolution of interest in fractals. Previously, only a few people had appreciated the significance and beauty of these strange shapes. Benoit Mandelbrot drew much attention to their potential use in describing the natural world. At the same time, the development of high-speed computing and computer graphics has made them easily accessible and this has drawn many people to study them more closely.

Another reason for the interest in fractals is that they are connected with chaos. In mathematics, chaos has a specialized meaning. The easiest way to understand chaos is by some examples.

Suppose a particle is moving in a confined region of space according to a definite deterministic law. Following the path traced out by the particle, we are likely to observe that it settles down to one of three possible behaviours – the geometrical description of which is called an attractor. The particle may be attracted to a final resting position (like, for example, the bob on a pendulum as it gradually settles down to rest). In this case, the attractor is just a point – the final resting position of the bob. Or the particle may settle down in a periodic cycle (like the planets in their orbits around the Sun). Here the attractor is an ellipse and the future motion can be predicted with astonishingly high accuracy as far ahead as we want. The last possibility is that the particle may continue to move wildly and erratically while, nevertheless, remaining in some bounded region of space. The motion of some of the asteroids, for example, appears to exhibit exactly this phenomenon. Tiny inaccuracies in measuring the position and speed of the asteroid quickly lead to enormous errors in predicting its future path. This phenomenon is the signal of chaotic motion. The regions of space traced out by such motions are called strange attractors.

Once a particle is attracted to a strange attractor there is no escaping. Almost anywhere you start inside the attractor, the point moves, on the average, in the same way, just as no matter how you start off a pendulum, it always eventually comes to rest at the same point. Although the motion is specified by precise laws, for all practical purposes, the particle behaves as if it were moving randomly. The interesting point here is that strange attractors are very frequently fractals.

The meteorologist Edward Lorenz pioneered chaos theory in a 1963 paper (though he did not call it chaos theory – the name was invented in 1972 by the mathematician James Yorke). Studying weather systems, Lorenz attributed their unpredictability to the fact that a very small initial difference could enormously change the future state of the system. This became known as 'the butterfly effect' from the title of Lorenz's 1979 paper 'Predictability: Does the Flap of a Butterfly's Wings in Brazil Set Off a Tornado in Texas?' Lorenz at first used a seagull as his example, but the butterfly was more dramatic.

Fractal pictures, and their link with chaos, have inspired science-writers to make claims about the underlying 'beauty' of nature or mathematics. Ian Stewart's rhapsody is typical.

> Chaos is beautiful. This is no accident. It is visible evidence of the beauty of mathematics, a beauty normally confined within the inner eye of the mathematician but which here spills over into the everyday world of human senses. The striking computer graphics of chaos have resonated with the global consciousness; the walls of the planet are papered with the famous Mandelbrot sets.

This seems to assume that beauty is an absolute, rather than a subjective value. But of course it is not. In reality any of us might protest that fractal pictures are not beautiful, but as kitsch as a luminous cauliflower, and no one could prove us wrong. To any arts undergraduate this would be an obvious point, and Stewart's neglect of it illustrates the gap that still exists between the two cultures. So (in the reverse direction) does the confusion among arts students about chaos theory, which is widely interpreted as meaning both that causation does not exist ('nature is chaotic') and, simultaneously, that chains of causation are so rigid that a butterfly's wing can cause a tornado. Paul Davies's is one of the most lucid attempts to resolve these difficulties in terms intelligible to the semi-numerate.

All science is founded on the assumption that the physical world is ordered. The most powerful expression of this order is found in the laws of physics. Nobody knows where these laws come from, nor why they apparently operate universally and unfailingly, but we see them at work all around us: in the rhythm of night and day, the pattern of planetary motions, the regular ticking of a clock.

The ordered dependability of nature is not, however, ubiquitous. The vagaries of the weather, the devastation of an earthquake, or the fall of a meteorite seem to be arbitrary and fortuitous. Small wonder that our ancestors attributed these events to the moodiness of the gods. But how are we to reconcile these apparently random 'acts of God' with the supposed underlying lawfulness of the Universe?

The ancient Greek philosophers regarded the world as a battleground between the forces of order, producing cosmos, and those of disorder, which led to chaos. They believed that random or disordering processes were negative, evil influences. Today, we don't regard the role of chance in nature as malicious, merely as blind. A chance event may act constructively, as in biological evolution, or destructively, such as when an aircraft fails from metal fatigue.

Though individual chance events may give the impression of lawlessness, disorderly processes, as a whole, may still display statistical regularities. Indeed, casino managers put as much faith in the laws of chance as engineers put in the laws of physics. But this raises something of a paradox. How can the same physical processes obey both the laws of physics and the laws of chance?

Following the formulation of the laws of mechanics by Isaac Newton in the 17th century, scientists became accustomed to thinking of the Universe as a gigantic mechanism. The most extreme form of this doctrine was strikingly expounded by Pierre Simon de Laplace in the 19th century. He envisaged every particle of matter as unswervingly locked in the embrace of strict mathematical laws of motion. These laws dictated the behaviour of even the smallest atom in the most minute detail. Laplace argued that, given the state of the Universe at any one instant, the entire cosmic future would be uniquely fixed, to infinite precision, by Newton's laws.

The concept of the Universe as a strictly deterministic machine governed by eternal laws profoundly influenced the scientific world view, standing as it did in stark contrast to the old Aristotelian picture of the cosmos as a living organism. A machine can have no 'free will'; its future is rigidly determined from the beginning of time. Indeed, time ceases to have much physical significance in this picture, for the future is already contained in the present. Time merely turns the pages of a cosmic history book that is already written.

Implicit in this somewhat bleak mechanistic picture was the belief that there are actually no truly chance processes in nature. Events may appear to us to be random but, it was reasoned, this could be attributed to human ignorance about the details of the processes concerned. Take, for example, Brownian motion. A tiny particle suspended in a fluid can be observed to execute a haphazard zigzag movement as a result of the slightly uneven buffeting it suffers at the hands of the fluid molecules that bombard it. Brownian motion is the archetypical random, unpredictable process. Yet, so the argument ran, if we could follow in detail the activities of all the individual molecules involved, Brownian motion would be every bit as predictable and deterministic as clockwork. The apparently random motion of the Brownian particle is attributed solely to the lack of information about the myriads of participating molecules, arising from the fact that our senses are too coarse to permit detailed observation at the molecular level.

For a while, it was commonly believed that apparently 'chance' events were always the result of our ignoring, or effectively averaging over, vast numbers of hidden variables, or degrees of freedom. The toss of a coin or a die, the spin of a roulette wheel – these would no longer appear random if we could observe the world at the molecular level. The slavish conformity of the cosmic machine ensured that lawfulness was folded up in even the most haphazard events, albeit in an awesomely convoluted tangle.

Two major developments of the 20th century have, however, put paid to the idea of a clockwork universe. First there was quantum mechanics. At the heart of quantum physics lies Heisenberg's uncertainty principle, which states that everything we can measure is subject to truly random fluctuations. Quantum fluctuations are not the result of human limitations or hidden degrees of freedom; they are inherent in the workings of nature on an atomic scale. For example, the exact moment of decay of a particular radioactive nucleus is intrinsically uncertain. An element of genuine unpredictability is thus injected into nature.

Despite the uncertainty principle, there remains a sense in which quantum mechanics is still a deterministic theory. Although the outcome of a particular quantum process might be undetermined, the relative probabilities of different outcomes evolve in a deterministic manner. What this means is that you cannot know in any particular case what will be the outcome of the 'throw of the quantum dice', but you can know completely accurately how the betting odds vary from moment to moment. As a statistical theory, quantum mechanics remains deterministic. Quantum physics thus builds chance into the very fabric of reality, but a vestige of the Newtonian-Laplacian world view remains.

Then along came chaos. The essential ideas of chaos were already present in the work of the mathematician Henri Poincaré at the turn of the century, but it is only in recent years, especially with the advent of fast electronic computers, that people have appreciated the full significance of chaos theory.

The key feature of a chaotic process concerns the way that predictive errors evolve with time. Let me first give an example of a non-chaotic system: the motion of a simple pendulum. Imagine two identical pendulums swinging in exact synchronism. Suppose that one pendulum is slightly disturbed so that its motion gets a little out of step with the other pendulum. This discrepancy, or phase shift, remains small as the pendulums go on swinging.

Faced with the task of predicting the motion of a simple pendulum, one could measure the position and velocity of the bob at some instant, and use Newton's laws to compute the subsequent behaviour. Any error in the initial measurement propagates through the calculation and appears as an error in the prediction. For the simple pendulum, a small input error implies a small output error in the predictive computation. In a typical non-chaotic system, errors accumulate with time. Crucially, though, the errors grow only in proportion to the time (or perhaps a small power thereof), so they remain relatively manageable.

Now let me contrast this property with that of a chaotic system. Here a small starting difference between two identical systems will rapidly grow. In fact, the hallmark of chaos is that the motions diverge exponentially fast. Translated into a prediction problem, this means that any input error multiples itself at an escalating rate as a function of prediction time, so that before long it engulfs the calculation, and all predictive power is lost. Small input errors thus swell to calculation-wrecking size in very short order.

The distinction between chaotic and non-chaotic behaviour is well illustrated by the case of the spherical pendulum, this being a pendulum free to swing in two directions. In practice, this could be a ball suspended on the end of a string. If the system is driven in a plane by a periodic motion applied at the pivot, it will start to swing about. After a while, it may settle into a stable and entirely predictable pattern of motion, in which the bob traces out an elliptical path with the driving frequency. However, if you alter the driving frequency slightly, this regular motion may give way to chaos, with the bob swinging first this way and then that, doing a few clockwise turns, then a few anticlockwise turns in an apparently random manner.

The randomness of this system does not arise from the effect of myriads of hidden degrees of freedom. Indeed, by modelling mathematically only the three observed degrees of freedom (the three possible directions of motion), one may show that the behaviour of the pendulum is nonetheless random. And this is in spite of the fact that the mathematical model concerned is strictly deterministic . . .

Chaos evidently provides us with a bridge between the laws of physics and the laws of chance. In a sense, chance or random events can indeed always be traced to ignorance about details, but whereas Brownian motion appears random because of the enormous number of degrees of freedom we are voluntarily overlooking, deterministic chaos appears random because we are necessarily ignorant of the utlra-fine detail of just a few degrees of freedom. And whereas Brownian chaos is complicated because the molecular bombardment is itself a complicated process, the motion of, say, the spherical pendulum is complicated even though the system itself is very simple. Thus, complicated behaviour does not necessarily imply complicated forces or laws. So the study of chaos has revealed how it is possible to reconcile the complexity of a physical world displaying haphazard and capricious behaviour with the order and simplicity of underlying laws of nature.

Though the existence of deterministic chaos comes as a surprise, we should not forget that nature is not, in fact, deterministic anyway. The indeterminism associated with quantum effects will intrude into the dynamics of all systems, chaotic or otherwise, at the atomic level. It might be supposed that quantum uncertainty would combine with chaos to amplify the unpredictability of the Universe. Curiously, however, quantum mechanics seems to have a subduing effect on chaos. A number of model systems that are chaotic at the classical level are found to be non-chaotic when quantized. At this stage, the experts are divided about whether quantum chaos is possible, or how it would show itself if it did exist. Though the topic will undoubtedly prove important for atomic and molecular physics, it is of little relevance to the behaviour of macroscopic objects, or to the Universe as a whole.

What can we conclude about Laplace's image of a clockwork universe? The physical world contains a wide range of both chaotic and non-chaotic systems. Those that are chaotic have severely limited predictability, and even one such system would rapidly exhaust the entire Universe's capacity to compute its behaviour. It seems, then, that the Universe is incapable of digitally computing the future behaviour of even a small part of itself, let alone all of itself. Expressed more dramatically, the Universe is its own fastest simulator.

This conclusion is surely profound. It means that, even accepting a strictly deterministic account of nature, the future states of the Universe are in some sense 'open'. Some people have seized on this openness to argue for the reality of human free will. Others claim that it bestows upon nature an element of creativity, an ability to bring forth that which is genuinely new, something not already implicit in earlier states of the Universe, save in the idealized fiction of the real numbers. Whatever the merits of such sweeping claims, it seems safe to conclude from the study of chaos that the future of the Universe is not irredeemably fixed. The final chapter of the great cosmic book has yet to be written.

Chaos theory reached the West End stage with Tom Stoppard's play *Arcadia*, where Valentine explains it to Chloë:

CHLOË: The future is all programmed like a computer – that's a proper theory, isn't it?
VALENTINE: The deterministic universe, yes.
CHLOË: Right. Because everything including us is just a lot of atoms bouncing off each other like billiard balls.
VALENTINE: Yes. There was someone, forget his name, nineteenth century, who pointed out that from Newton's laws you could predict everything to come – I mean, you'd need a computer as big as the universe but the formula would exist.
CHLOË: But it doesn't work, does it?
VALENTINE: No. It turns out the maths is different.

Stoppard's Valentine rejoices in unpredictability, because he sees it as restoring mystery to ordinary life:

The unpredictable and the predetermined unfold together to make everything the way it is. It's how nature creates itself, on every scale, the snowflake and the snowstorm. It makes me so happy. To be at the beginning again, knowing almost nothing. People were talking about the end of physics. Relativity and quantum looked as if they were going to clean out the whole problem between them. A theory of everything. But they only explained the very big and the very small. The universe, the elementary particles. The ordinary-sized stuff which is our lives, the things people write poetry about – clouds – daffodils – waterfalls – and what happens in a cup of coffee when the cream goes in – these things are full of mystery, as mysterious to us as the heavens were to the Greeks. We're better at predicting events at the edge of the galaxy or inside the nucleus of an atom than whether it'll rain on auntie's garden party three Sundays from now. Because the problem turns out to be different. We can't even predict the next drip from a dripping tap when it gets irregular. Each drip sets up the conditions for the next, the smallest variation blows prediction apart, and the weather is unpredictable the same way, will always be unpredictable. When you push the numbers through the computer you can see it on the screen. The future is disorder. A door like this has cracked open five or six times since we got up on our hind legs. It's the best possible time to be alive, when almost everything you thought you knew is wrong.

Arcadia's scientific adviser was Robert May, a Royal Society Research Professor at Oxford and Imperial College London, and one of the pioneers of chaos theory. His programme note for the play is the best succinct summary:

The vision given to us by Newton and by those who followed in the Age of Enlightenment is of an orderly and predictable world, governed by laws and rules – laws and rules which can best be expressed in mathematical form. If the circumstances are simple enough (for instance, a planet moving around a sun, bound by the inverse square law of gravitational attraction), then the system behaves in a simple and predictable way. Effectively unpredictable situations (for instance, a roulette ball whose fate – the winning number – is governed by a complex concatenation of croupier's hand, spinning wheel, and so on) were thought to arise only because the rules were many and complicated.

Over the past 20 years or so, this Newtonian vision has splintered and blurred. It is now widely recognised that the simplest rules or algorithms or mathematical equations, containing no random elements whatsoever, can generate behaviour which is as complicated as anything we can imagine.

This mathematics which 'is different' is the mathematics of 'deterministic chaos'. What it says is that a situation can be both deterministic *and* unpredictable; that is, unpredictable without being random (on the one hand) or (on the other hand) attributable to very complicated causes.

'Simple' as they may be in themselves, these chaos-generating equations have the property of being 'nonlinear'. In a *linear* equation you can 'guess ahead'. Imagine a road lined with telegraph poles in a perspective drawing. Given two or three poles, you can easily draw in the rest for yourself. But nature often draws itself differently, using *nonlinear* equations. Imagine a river running alongside the road. The water has flat bits and bumpy bits. But however many I draw in for you, there is no way for you to tell (with a real river) where the next flat bit or bumpy bit is going to be. This holds true on every scale.

Look down from a balloon and you'll see that parts of the bumpy bits look relatively flat. Put your face close to the water and you'll see that the flat bits contain relatively bumpy bits. The maths is the same for each case, and equally unpredictable.

In this sense, 'nonlinear' means two and two do not necessarily make four. Much of physics and other areas of science where so much progress has come, are linear (or at least decomposable into essentially linear bits). And so mathematical texts and courses have focused on linear problems. But increasingly it seems that most of what is interesting in the natural world, and especially in the biological world of living things, involves nonlinear mathematics. It was biologists – working on the ups and downs of animal population – who were among the first to see that not only can simple rules give rise to behaviour which looks very complicated, but the behaviour can be so sensitive to the starting conditions as to make long term prediction impossible (even when you know the rule).

There is a flip side to the chaos coin. Previously, if we saw complicated, irregular or fluctuating behaviour – weather patterns, marginal rates of Treasury Bonds, colour patterns of animals or shapes of leaves – we assumed the underlying causes were complicated. Now we realize that extraordinarily complex behaviour can be generated by the simplest of rules. It seems likely to me that much complexity and apparent irregularity seen in nature, from the development and behaviour of individual creatures to the structure of ecosystems, derives from simple – but chaotic – rules. (But, of course, a lot of what we see around us is very complicated because it is intrinsically so!)

I believe all this adds up to one of the real revolutions in the way we think about the world. Knowing the simple rule or equation that governs a system is not always sufficient to predict its behaviour. And, conversely, exceedingly complicated patterns or behaviour may derive not from exceedingly complex causes, but from the chaotic workings of some very simple algorithm. Ultimately, the mathematics of chaos offers new and deep insights into the structure of the world around us, and at the same time raises old questions about *why* abstract mathematics should be so unreasonably effective in describing this world.

Sources: (for Caroline Series's and Paul Davies's pieces) *The New Scientist Guide to Chaos*, ed. Nina Hall, London, Penguin, 1991; Tom Stoppard, *Arcadia*, Faber and Faber, 1993; Robert May, Programme note to *Arcadia*, 1993.

The Language of the Genes

Steve Jones is Professor of Genetics at University College, London. His book *The Language of the Genes*, based on his 1991 BBC Reith Lectures, is a model of how wit, learning and clear-headedness can make a complex subject intelligible to a huge audience.

The language of the genes has a simple alphabet, not with twenty-six letters, but just four. These are the four different DNA bases – adenine, guanine, cytosine and thymine (A, G, C and T for short). The bases are arranged in words of three letters such as CGA or TGG. Most of the words code for different amino acids, which themselves are joined together to make proteins, the building blocks of the body.

Just how economical the language of inheritance is can be illustrated with a rather odd quotation from a book called *Gadsby*, written in 1939 by one Ernest Wright: 'I am going to show you how a bunch of bright young folks did find a champion, a man with boys and girls of his own, a man of so dominating and happy individuality that youth was drawn to him as a fly to a sugar bowl.' This sounds rather peculiar, as does the rest of the fifty-thousand word book, and it is. The quotation, and the whole book, does not contain the letter 'e'. It is possible to write a meaningful sentence with twenty-five letters instead of twenty-six, but only just. Life manages with a mere four.

Although the inherited vocabulary is simple its message is very long. Each cell in the body contains about six feet of DNA. A useless but amusing fact is that if all the DNA in all the cells in a single human being were stretched out it would reach to the moon and back eight thousand times. There is now a scheme, the Human Genome Project, to read the whole of its three thousand million letters and to publish what may be the most boring book ever written; the equivalent of a dozen or so copies of the *Encyclopaedia Britannica*. There is much disagreement about how to set about reading the message and even about whether it is worth doing at all. It probably is. The Admiralty sent the *Beagle* to South America with Darwin on board not because they were interested in evolution but because they knew that the first step to understanding (and, with luck, controlling) the world was to make a map of it. The same is true of the genes. To make this map will be expensive – about as much as a single Trident nuclear submarine. The task will be stupefyingly tedious for those who have to do the work, but, before the end of the century, someone will publish the inherited lexicon of a human being. To be more precise, there will be a map of a sort of Mr Average – the chart is, of course, of a male – as the information will come from short bits of DNA from dozens of different people . . .

Human genetics was for most of its history more or less restricted to studying pedigrees which stood out because they contained an abnormality. This limited its ability to trace patterns of descent to those few families – like the Hapsburgs – who appeared to deviate from the perfect form. Biology has now shown that this perfect form does not exist. Instead there is a huge amount of inherited variation. Thousands of inherited characters – perfectly normal diversity, not diseases – distinguish each person. There is so much variety that everyone alive today is different, not only from everyone else, but from everyone who ever has lived or ever will live. The mass of diversity can be used to look at patterns of shared ancestry in any family, aristocratic or plebeian; healthy or ill. Because all modern genes are copies of those in earlier generations each can be used as a message from the past. They bring clues from the beginnings of humanity more than a hundred thousand years ago and from the origin of life three thousand million years before that . . .

There have been claims that we may soon find the gene that makes us human. The ancestral message will then at last allow us to understand what we really are. The idea seems to me ridiculous.

Just how ridiculous it is can be seen by looking at the search for another important gene, one which I inherited from my father, and he from his and so on back to a distant ancestor that lived long before the birth of our own species. This is the gene that makes me male. The maleness gene was tracked down recently and its message spelt out in the four DNA letters, A, G, C and T. It starts like this: GAT AGA GTG AAG CGA. There are 240 of these letters altogether and, between them, they contain the whole tedious biological story of being a man. This brief ancestral bulletin does nothing to tell that half of the population which is unfortunate enough not to have it what it is really like to be male rather than female. Being a man involves a lot more than a sequence of DNA bases; and the same is true for being a human.

Source: Steve Jones, *The Language of the Genes*, London, HarperCollins, 1993.

The Good Earth is Dying

First published in *Der Spiegel* in 1971, Isaac Asimov's warning fittingly concludes this book, since it takes a relentlessly scientific look at what is still humankind's most pressing problem – humankind.

How many people is the earth able to sustain?

The question is incomplete as it stands. One must modify the question by asking further: At what level of technology? And modify it still further by asking: At what level of human dignity?

As for technology, perhaps we can simply ask for the best. We can say that the more advanced technology is, the more people the earth can support, so let us not stint. After all, technology could give us the atomic bomb and put men on the moon and we should set no limits to it.

Let us accept, then, the dream that technology is infinitely capable and proceed from there. How many people can the earth sustain assuming that technology can solve all reasonable problems?

To begin with, it is estimated that there are twenty million million tons of living tissue on the earth, of which 10 percent, or two million million tons, is animal life. As a first approximation, this may be considered a maximum, since plant life cannot increase in mass without an increase in solar radiation or an increase in its own efficiency in handling sunlight. Animal life cannot increase in mass without an increase in the plant mass that serves as its ultimate food.

The mass of humanity has been increasing throughout history; and it is still increasing, but is doing so at the expense of other forms of animal life. Every additional kilogram of humanity has meant, as a matter of absolute necessity, one less kilogram of nonhuman animal life. We might argue, then, that the earth can support, as a maximum, a mass of mankind equal to the present mass of all animal life. At that point, the number of human beings on the earth would be forty million million, or over eleven thousand times the present number. And no other species of animal life would then exist.

What will this mean? The total surface of the earth is five hundred twenty million square kilometers, so that when human population reaches its ultimate number, the average density of population will be eighty thousand per square kilometer – twice the density of New York's island of Manhattan. Imagine such a density everywhere if the earth's population is spread out evenly – including over the polar regions, the deserts, and the oceans.

Can we imagine, then, a huge, world-girdling complex of high-rise apartments (over both land and sea) for housing, for offices, for industry? The roof of this complex will be given over entirely to plant growth; either algae, which are completely edible, or higher plants that must be treated appropriately to make all parts edible.

At frequent intervals, there will be conduits down which water and plant products will pour. The plant products will be filtered out, dried, treated, and prepared for food, while the water is returned to the tanks above. Other conduits, leading upward, will bring up the raw minerals needed for plant growth, consisting of (what else) human wastes and chopped-up human corpses. And at this point, of course, no further increase in human numbers is possible; so rigid population planning would then be necessary if it had not been before.

But if this number can be supported in theory, does it represent a kind of life – and this is for each of you to ponder – consonant with human dignity?

Can we buy space and time by transferring human beings to the moon? To Mars?

Consider – How long, under present conditions, will it take us to reach the global high-rise? At present, the earth's population is thirty-six hundred million and it is increasing at a rate that will double the figure in thirty-five years. Let us suppose that this rate of increase can be maintained. In that case, the ultimate population of forty million million will be reached in 465 years. The global high-rise will be in full splendor in AD 2436.

In that case, how many men do you think it will be possible to place, and support, on the moon, Mars, and elsewhere in the next 465 years? Be reasonable. Subtract your figure from forty million million and ask yourself if the contribution the other worlds can make is significant.

Can we buy further time by going beyond the sun? Can we make use of hydrogen fusion power to irradiate plant life? Or can we make food in the laboratory, with artificial systems and synthetic catalysts, and declare ourselves independent of the plant world altogether?

But that requires energy and here we come to another point. The sun pours down on the earth's day-side, some fifteen thousand times as much energy per day as mankind now uses. The earth's night-side must radiate exactly that much heat back into space if the earth's average temperature is to be maintained. If mankind adds to the heat on earth by burning coal, that additional energy must also be radiated out to space; and to accomplish this the earth's average temperature must rise slightly.

At present, man's addition to solar energy produces a terrestrial temperature-rise that is utterly insignificant; that addition, however, is doubling every twenty years. At this rate, in a hundred sixty-five years (by 2136) mankind's contribution to the heat that the earth must radiate away will amount to one percent of the sun's supply, and this will begin to produce unacceptable changes in the earth's temperature.

So, far from helping ourselves with further energy expenditures in the global high-rise world of AD 2436, we must accept a limitation of energy expenditure a full three centuries earlier, when man's population is less than a five-hundredth of its ultimate. We might improve matters by increasing the efficiency with which energy is used; but the efficiency cannot rise above a hundred percent, and that does *not* represent an enormous increase over present levels.

But, and this is a large 'but,' can we really depend on technology to make the necessary advances to bring us to energy-limits safely in a century and a half? By then the population, at the present level of increase, will be twenty times what it is today; and to bring man's level of nourishment to a desirable point, we would need a fortyfold increase in the food supply. We would also have to ask technology not only to arrange the necessary hundred-fifty-fold increase in energy utilization in a century and a half but to arrange to take care of what will be, very likely, a hundred-fifty-fold increase in environmental pollution and in waste production of all kinds.

How do things look at present?

Far from making strides to keep up with the population increase, technology is falling visibly behind. How can the global high-rise be a reasonable future vision when present-day housing is steadily deteriorating even in the most advanced nations? How can we reach our limit of energy expenditure when New York City finds itself, each year, with a growing deficit of power supply? Only yesterday, the third landing of men on the moon caused television viewing to go up, and a cutback in electrical voltage was immediately made necessary.

The earth's population will be at least six thousand million in the year 2000. Will the planet's technology be able to support that population even at present-day, wholly unsatisfactory levels? Will human dignity be compatible with such a population (let alone with forty million million), when in our cities *today* human dignity is disappearing; when it is impossible to walk safely by night (and often by day) in the largest city of the world's most technologically advanced nation.

Let us not look into the future at all, then. Let us gaze firmly at the present. The United States is the richest nation on earth and every other nation would like to be at least as rich. But the United States can live as it does only because it consumes slightly more than half of all the energy produced on earth for human consumption – although it has only a sixteenth of the earth's population.

What, then, if some wizard were to wave his magic wand and produce an earth on which every part of the population everywhere were able to live at the scale and the standard of Americans? In that case, the rate of energy expenditure would increase instantly to eight times the present world level and, inevitably, the production of waste and pollution would increase similarly – this with no increase in population at all.

And can present-day technology supply an eightfold increase in energy utilization (and that of other resources as well) and handle an eightfold increase in waste and pollution, when it is falling desperately short of supplying and handling present levels? Do you ask for time in which technology can arrange for such an eightfold increase? Very well, but in that time, population will increase, too, very likely more than eightfold.

In short, then, to the revised question, How many people is the earth able to sustain at a desirable level of technology and dignity? there can be only a short and horrifying answer –

Fewer than now exist!

The earth cannot support its present population at the average level of the American standard of living. Perhaps, at the moment, it can only support five hundred million people at that standard. Nor can technology improve itself to better this mark, with the present population clamoring for what it cannot have and with that population growing at a terrible rate.

What, then, will happen?

If matters continue as they are now going, there will be a continuing decline in the well-being of the individual human being on earth. Calories per mouth will decrease; available living quarters will dwindle; attainable comfort will diminish. What is more, in the increasing desperation to reverse this, man may well make wild attempts to race the technology engine at all costs and will then further pollute the environment and decrease its ability to support mankind. With all this taking place, there will be a struggle of man against man, with each striving to grasp an adequate share of a shrinking life-potential. And there cannot help but be an intensification of the human-jungle characteristics of our centers of population.

In not too long a time, the population increase will halt; but for the worst possible reason – there will be a catastrophic rise in the death rate. The famines will come, the pestilence will strike, civil disorder will intensify, and by AD 2000 some governmental leader may well be desperate enough to push the nuclear button.

How to prevent this, then?

We must stop living by the code of the past. We have, over man's history, developed a way of life that fits an empty planet and a short existence marked by high infant-mortality and brief life-expectancy. In such a world there was a virtue in having many children, in striving for growth in numbers and power, in expansion into endless space, in total commitment to that limited portion of mankind that could make up part of a viable society.

But none of this is so any longer. At the moment, child mortality is low, life expectancy high, the earth full. There are no empty spaces of worth, and so interdependent is man that it is no longer safe to confine loyalty to only a portion of mankind.

What was common sense in a world that once existed has become myth in the totally different world that now exists, and suicidal myth at that.

In our overpopulated world we can no longer behave as though woman's only function in life is to be a baby-producing machine. We can no longer believe that the greatest blessing a man can have is many children.

Motherhood is a privilege we must literally ration, for children, if produced indiscriminately, will be the death of the human race, and any woman who deliberately has more than two children is committing a crime against humanity.

We also have to alter our attitude toward sex. Through all the history of man it has been necessary to have as many children as possible, and sex has been made the handmaiden of that fact. Men and women have been taught that the only function of sex is to have children; that otherwise it is a bestial and wicked act. Men and women have been taught that only those forms of sex that make conception possible are tolerable, that everything else is perverse, unnatural, and criminal.

Yet we can no longer indulge in such views. Since sex cannot be suppressed, it must be divorced from conception. Birth control must become the norm and sex must become a social and interpersonal act rather than a child-centered act.

We also have to alter our attitude toward growth. The feeling of 'bigger and better' that bore up mankind through his millennia on this planet must be abandoned. We have reached the stage where bigger is no longer better. Although the notion of more people, more crops, more products, more machines, more gadgets – more, more, more – has worked, after a fashion, up to this generation, it will no longer work. If we attempt to force it to work, it will kill us rather quickly.

In our new and finite world, where for the first time in history we have reached, or are reaching, our limit, we must accept the fact of limit. We must limit our population, limit the strain we put on the earth's resources, limit the wastes we produce, limit the energy we use. We must preserve. We must preserve the environment, preserve the other forms of life that contribute to the fabric and viability of the biosphere, preserve beauty and comfort. And if we do limit and preserve, we will have room for deeper growth even so – growth in knowledge, in wisdom, and in love for one another.

We also have to alter our attitude toward localism. We can no longer expect to profit by another's misfortune. We can no longer settle quarrels by wholesale murder. The price has escalated to an unacceptable level. World War II was the last war that could be fought on this planet by major powers using maximum force. Since 1945, only limited wars have been conceivable, and even these have been monumental stupidities, as the situations in Southeast Asia and in the Middle East make clear.

The world is too small for the kind of localism that leads to wars. We can have special pride in our country, our language, and our literature, our customs and culture and tradition, but it has to be the abstract pride we have in our baseball team or our college – a pride that cannot and must not be backed by force of arms.

Localism doesn't even have the virtue of being useful in times of peace. The problems of the world today are planetary in scope. No one nation, not even if it is as rich as the United States, as centralized as the Soviet Union, or as populous as China, can solve its important problems today. No matter how a nation stabilizes population within its own borders, no matter how it rationalizes the use of its own resources, no matter how it conserves its own environment, all would come to nothing if the rest of the world continued its rabbit-multiplication and its poisoning of free will.

Even if every nation sincerely took measures, independent of each other, to correct the situation, the solutions one nation arrived at would not necessarily match those of its neighbors, and all might fall.

To put it bluntly, planetary problems require a planetary program and a planetary solution, and that means cooperation among nations, *real* cooperation. To put it still more bluntly, we need a world government that can come to logical and humane decisions and can then enforce them.

This does not mean a world government that will enforce conformity in every respect. The cultural diversity of mankind is surely a most valuable characteristic and it must be preserved – but not where it will threaten the species with suicide.

All these requirements for change go against the grain. Who really wants to downgrade motherhood and regard babies as enemies? Who is comfortable at the thought of dissociating sex and parenthood? Who is ready to submit his national pride to a truly effective world government? Who is willing to abandon the attempt to get as much as possible out of the world and settle instead for a controlled and limited exploitation?

Yet the logic of events is actually forcing us in that direction, willing or not. The birth rate is dropping in those nations that have access to birth-control methods. Sexual mores are loosening everywhere. The people are growing more concerned about the environment, and the clamour for cleaner air, water, and soil is becoming louder every day.

Most of all, and most heartening, localism is retreating. There is increasing social and economic cooperation among neighboring nations; a stronger drive in the direction of regionalism. More important still, there is a clear understanding that a major war, particularly one between the United States and the Soviet Union, is inadmissable. These two superpowers have quarrelled at levels of intensity that at any time up to the 1930s would have meant war – and now those quarrels do not even bring about a rupture in diplomatic relations. Not only must these nations not fight; they must not even snub each other.

But this motion in the right direction does not seem to be a matter of choice. Rather, stubborn humanity is inching forward to help itself only because the pressure of circumstance has closed all other passageways.

And this motion in the right direction is not fast enough. The population increase continues to outpace the education for birth control; the environment continues to deteriorate more rapidly than we can bring ourselves to correct matters; and, worse, nations still stubbornly quarrel, and continue to place local pride over the life and death of the species.

We must not only reorient our thinking toward motherhood, sex, growth and localism as we are beginning to do; but we must do it more *quickly.* Our society cannot survive another generation of the steadily intensifying stresses placed on it. If we continue as we are and change no faster, then by 2000 the technological structure of human society will almost certainly have been destroyed. Mankind, having been reduced to barbarism, may possibly be on the way to extinction. The planet itself may find its ability to support life seriously compromised.

The good earth is dying; so in the name of humanity let us move. Let us make our hard but necessary decisions. Let us do it quickly. Let us do it now.

Steve Connor, reporting from San Francisco in the *Independent* on 22 February 1994, brings Asimov's figures up to date.

Over-population threatens to become a global crisis unless drastic action to dramatically cut birth rates begins now, the American Association for the Advancement of Science was told yesterday. World population is already nearly 4 billion more than the 2 billion the planet can comfortably sustain, according to an ecological study of natural resources to be published later this year. Fertile soil for growing crops, unpolluted water, fossil fuels and the flora and fauna on which humanity depends are all being depleted at a rate that will lead to catastrophic natural, social and political disasters by the end of the next century, a leading ecologist told the meeting.

David Pimentel, Professor of Ecology at Cornell University at Ithaca, New York, released the results of a year-long study into the optimum human population – the number of people the planet can comfortably support with a reasonable standard of living for all. The study concludes the present population of 5.6 billion will have to shrink to 2 billion. However, the projected population for 2100 is expected to be between 12 and 15 billion. Professor Pimentel acknowledged that drastic adjustments to cut the population to 2 billion will cause serious difficulties. 'But continued rapid population growth will result in even more severe social, economic and political conflicts – plus catastrophic public health and environmental problems.'

Sources: Isaac Asimov, *The Roving Mind. A Panoramic View of Fringe Science, Technology, and the Society of the Future*. Oxford University Press, 1987. The *Independent*, 22 February 1994.

Acknowledgements

For some years I have pestered friends and colleagues for help and suggestions. I should like to thank them all for their kindness and forbearance. I am particularly conscious of my debt to Dinah Birch, David Bodanis, David Cairns, Richard Dawkins, Michael Dunnill, Artur Ekert, Xandra Hardie, Harriett Hawkins, Kevin Jackson, Peter MacDonald, Sir John Maddox, Robert May and Claire Preston. Chris Reid and Julian Loose at Fabers have been unfailingly helpful and supportive. All the mistakes, of course, are mine.

J.C.

For permission to reprint copyright material the publishers gratefully acknowledge the following:

DANNIE ABSE: to Sheil Land Associates for 'In the Theatre' from *Collected Poems* 1948–1976 (Hutchinson, 1977). NEIL ARMSTRONG et al: to Time Warner Inc for *First on the Moon: a Voyage with Neil Armstrong, Michael Collins, Edwin E. Aldrin*, with Grace Farmer and Dora Jane Hamblin (Little, Brown, 1970). ISAAC ASIMOV: to the Estate of Isaac Asimov, c/o Ralph M. Vicinanza Ltd for 'Black Holes' and 'The Good Earth is Dying', reprinted in *The Roving Mind: A Panoramic View of Fringe Science, Technology, and the Mind of the Future* (Prometheus Books, 1983). PETER ATKINS: to Penguin Books Ltd for *Creation Revisited: The Origin of Space, Time and The Universe* (Penguin Books, 1994; first published by W. H. Freeman). Copyright © Peter Atkins, 1992. W. H. AUDEN: to Faber & Faber Ltd and Random House Inc for 'In Memory of Sigmund Freud' from *Collected Poems*, ed. Edward Mendelson. Copyright © 1940 and renewed 1968 by W. H. Auden. WILLIAM BEEBE: to Prentice Hall Inc for *Half Mile Down* (Harcourt Brace, 1934). RUTH BENEDICT: to Oxford University Press for *Man, Culture and Society*, ed. Harry L.

Shapiro (1956). HECTOR BERLIOZ: to Victor Gollancz Ltd for *Memoirs*, trans. David Cairns (Cardinal, 1990). DAVID BODANIS: to June Hall Agency for *Web of Words: The Ideas Behind Politics* (Macmillan, 1988). LYNDON BOLTON: to Scientific American Inc for an essay in *Scientific American* (5 February 1921). Copyright 1921 by Scientific American Inc. All rights reserved. DANIEL J. BOORSTIN: to Random House Inc for *The Discoverers*. Copyright © 1983 by Daniel J. Boorstin. SERGE BRAMLEY: to HarperCollins Publishers Inc for *Leonardo: The Artist and the Man*, trans. Sian Reynolds (Burlinghame Books, 1991). P. W. BRIDGMAN: to Harper's Magazine for 'The New Vision of Science' in *Harper's Magazine* (March 1929). Copyright 1929 by Harper's Magazine. All rights reserved. NIGEL CALDER: to BBC Books for *The Key to the Universe: A Report on the New Physics* (1977). ITALO CALVINO: to Macmillan Ltd for *Mr Palomar*, trans. William Weaver (Pan, 1986). RACHEL CARSON: to Frances Collin Literary Agency for *The Sea Around Us* (1951). ARTHUR C. CLARKE: to the author and the author's agents, Scovil Chichak Galen Literary Agency Inc, New York, and David Higham Associates Ltd for *By Space Possessed* (Gollancz, 1993). STEVE CONNOR: to Newspaper Publishing plc for an article in *The Independent* (22 February 1994). FRANCIS CRICK: to Orion Publishing Group Ltd for *What Mad Pursuit* (Weidenfeld & Nicolson, 1989). EVE CURIE: to Reed Consumer Books and Curtis Brown Ltd, London for *Madame Curie*, trans. Vincent Sheean (Heinemann, 1938). PAUL DAVIES: to Orion Publishing Group for *Other World: Space, Superspace and the Quantum Universe* (Dent, 1980); to New Scientist for 'Chaos frees the universe' from *New Scientist* (6 October 1990). RICHARD DAWKINS: to Peters Fraser & Dunlop Group Ltd for *The Blind Watchmaker* (Longman, 1986). FREEMAN DYSON: to Penguin Books Ltd and Penguin Books USA Inc for *From Eros to Gaia* (Penguin, 1993). Copyright © Freeman Dyson, 1992. ALBERT EINSTEIN: to Reed Consumer Books for *Relativity: The Special and General Theory, a Popular Exposition*, trans. Robert W. Lawson (Methuen, 1920). WILLIAM EMPSON: to Random House UK Ltd for 'Camping Out' from *Collected Poems* (Chatto & Windus, 1955). LAURA FERMI: to University of Chicago Press for *Atoms in the Family: My Life with Enrico Fermi* (Allen & Unwin, 1955). SIGMUND FREUD: to Random House UK Ltd for *Three Essays on Sexuality* from *The Standard Edition of the Complete Psychological Works, Vol.* VII (1901–1905), trans. James Strachey (Hogarth Press, 1953). OTTO FRISCH: to Vista Publications for 'Atomic Particles' from *Science Survey, I*, eds Laslett & St John (1960). MARTIN GARDNER: to Oxford University Press and Prometheus Books for *Science: Good, Bad and Bogus* (1983). MURRAY GELL-MANN: to Little Brown & Company (UK) and W. H. Freeman & Co for *The Quark and the Jaguar: Adventures in the Simple and the Complex*. STEPHEN JAY GOULD: to Penguin Books Ltd and W. W. Norton & Company Inc for *Flamingo's Smile: Reflections in Natural History* (Penguin, 1987). Copyright © Stephen Jay Gould, 1985. LAVINIA GREENLAW: to Faber & Faber Ltd for 'The Innocence of Radium' from *Night Photograph* (1993). J. B. S. HALDANE: to Random House UK Ltd for 'On Being the Right Size' from *Possible Worlds and Other Essays* (Chatto & Windus, 1927). MIROSLAV HOLUB: to Faber & Faber Ltd for *The Dimension of the Present Moment and other Essays*, ed. David Young (1990). TED HUGHES: to Faber & Faber Ltd for 'Fire-Eater' from *Lupercal* (1960). BERNARD JAFFE: to Dover Publications Inc for *Crucibles: The Story of Chemistry from Ancient Alchemy to Nuclear Fission* (Dover, 1976). STEVE JONES: to Harper Collins Publishers Ltd for *The Language of the Genes* (1991). GEOFFREY KEYNES: to Oxford University Press for *The Life of William Harvey* (1966). JAMES KIRKUP: to the author for 'A Correct Compassion' from *A Correct Compassion and Other Poems* (Oxford University Press, 1952). NICHOLAS KURTI: to the author and Institute of Physics Publishing for 'Bird's Custard: The True Story' from *But the Crackling was Superb: An Anthology on Food and Drink by Fellows and Foreign Members of the Royal Society*, eds. Nicholas and Giana Kurti, Adam Hilger (1988). EDWARD LARRISSY: to the author for 'Galactic Diary of an Edwardian Lady' from *The Independent* (February 1994). PHILIP LARKIN: to Faber & Faber Ltd and Farrar, Straus & Girowe Inc for letter (13 July 1959). PRIMO LEVI: to Michael Joseph Ltd for *The Periodic Table*, trans. Raymond Rosenthal (1985). Copyright © Giulio Einaudi Editore s.p.a., Torino, 1975; translation copyright © Schocken Books Inc, 1984. J. E. LOVELOCK: to Oxford University Press for *Gaia: A New Look at Life on Earth* (1979). EDWARD MACCURDY: to Random House UK Ltd for *The Notebooks of Leonardo da Vinci*, Arranged, Rendered in English, with Introduction by Edward MacCurdy (Cape, 1938). HUGH MACDIARMID: to Martin Brian & O'Keefe Ltd for 'Two Scottish Boys' from *The Complete*

Poems 1920–1976 (1978). ANGUS MCLAREN: to Blackwell Publishers for *A History of Contraception: From Antiquity to the Present Day* (1990). THOMAS MANN: to Reed Consumer Books for *The Magic Mountain*, trans. H. T. Lowe-Porter (Secker & Warburg/Penguin, 1960). JOHN MASEFIELD: to the Society of Authors for commemorative poem to Ronald Ross from *The Times* (20 August 1957). ROBERT M. MAY: to the author for programme note ('From Newton to Chaos') for Tom Stoppard's play, *Arcadia* (Faber, 1993). PETER MEDAWAR: to Oxford University Press for *Pluto's Republic* (1982). VLADIMIR NABOKOV: to Orion Publishing Group Ltd and Smith/Skolnik Literary Management, New York for *Speak, Memory* (Penguin Books, 1969). BEAUMONT NEWHALL: to Reed Consumer Books for *The History of Photography: From 1839 to the Present*, trans. Beaumont Newhall (Secker & Warburg, 1982). JOHN FREDERICK NIMS: to New Directions Publishing Corporation for 'Klutz' from *The Six-cornered Snowflake*. Copyright © 1990 by John Frederick Nims. ALFRED NOYES: to Sheed & Ward Ltd for *The Torch-bearers* (1937). CHARLES OFFICER and JAKE PAGE: to Oxford University Press for *Tales of the Earth: Paroxysms and Perturbations of the Blue Planet* (1993). GEORGE ORWELL: to A. M. Heath & Co Ltd for 'Toads' from *Tribune* (April 1946), reprinted in *Collected Essays, Journalism and Letters: Volume 4* (Penguin, 1970). Copyright © The Estate of the late Sonia Brownell Orwell and Martin Secker & Warburg Ltd. ADAM PHILLIPS: to Faber & Faber Ltd and Harvard University Press for *On Kissing, Tickling and Being Bored* (1993). BERTRAND RUSSELL: to Bertrand Russell Peace Foundation for *ABC of Relativity* (Routledge & Kegan Paul, 1925). Copyright © The Bertrand Russell Peace Foundation Ltd. OLIVER SACKS: to Gerald Duckworth & Co Ltd for *The Man Who Mistook His Wife for a Hat* (1985). CARL SAGAN: to Hodder Headline plc for *Broca's Brain: The Romance of Science* (Hodder & Stoughton, 1979). GEORGE B. SCHALLER: to the author and University of Chicago Press for *The Year of the Gorilla* (1964). CAROLINE SERIES: to New Scientist for 'Fractals, reflections and distortions' from *New Scientist* (22 September 1990). GEORGE BERNARD SHAW: to Society of Authors for Preface to *Back to Methuselah* (1921). CHARLES SHERRINGTON: to U. M. Sherrington, literary executor, and Cambridge University Press for *Man on His Nature* (1951). ANTHONY SMITH: to Hodder Headline plc for *The Mind* (Hodder & Stoughton, 1984). C. P.

SNOW: to Curtis Brown Group Ltd, London, for *The Physicists* (Macmillan, 1981). Copyright © 1981 the Estate of C. P. Snow. JOHN STEINBECK: to Reed Consumer Books and Viking Penguin, a division of Penguin Books USA Inc for *Steinbeck: A Life in Letters*, eds. Elaine Steinbeck and Robert Wallsten (Heinemann, 1975). Copyright © 1952 by John Steinbeck, © 1969 by The Estate of John Steinbeck , © 1975 by Elaine A. Steinbeck and Robert Wallsten; and *The Log From the Sea of Cortez* (Heinemann, 1958). Copyright © 1941 by John Steinbeck and Edward F. Ricketts. Copyright renewed © 1969 by John Steinbeck and Edward F. Ricketts, Jr. TOM STOPPARD: to Faber & Faber Ltd and Peters, Fraser & Dunlop Group Ltd for *Arcadia* (1993). LEWIS THOMAS: to Viking Penguin, a division of Penguin Books USA Inc, for 'Clever Animals' from *Late Night Thoughts on Listening to Mahler's Ninth*. Copyright © 1982 by Lewis Thomas. MARK TWAIN: to Regents of the University of California and University of California Press for *What is Man? and other philosophical writings*, trans./ed. Paul Baender (1973). Copyright © 1973 The Mark Twain Company. JOHN UPDIKE: to Penguin Books Ltd for 'Ode to Entropy' from *Facing Nature* (Deutsch, 1984). Copyright © John Updike, 1984; and 'Cosmic Gall' from *Telephone Poles and other poems* (Deutsch, 1960). Copyright © John Updike, 1960. ANDREAS VESALIUS: to Atlantis Publishers for *Andreas Vesalius's First Public Anatomy at Bologna, 1540, An Eyewitness Report* (Almqvist & Wiksell International AB, 1959). RICHARD WILBUR: to Harcourt Brace & Company for 'Lamarck Elaborated' from *Things of this World*. Copyright © 1956 and renewed 1984 by Richard Wilbur. WILLIAM CARLOS WILLIAMS: to Carcanet Press and New Directions Publishing Corporation for 'St Francis Einstein of the Daffodils' from *Collected Poems*. EDWARD O. WILSON: to Penguin Books Ltd and Harvard University Press for *The Diversity of Life* (Harvard University Press, 1992). Copyright © Edward O. Wilson, 1992. D. W. WINNICOTT: to Routledge Ltd for *Playing and Reality* (Tavistock Publications, 1971). LEWIS WOLPERT: to Faber & Faber Ltd and Harvard University Press for *The Unnatural Nature of Science* (1992). LEWIS WOLPERT and ALISON RICHARDS: to Oxford University Press for *A Passion for Science* (1988).

Every effort has been made to trace or contact copyright holders. The publishers will be pleased to make good in future editions or reprints any omissions or corrections brought to their attention.

Index of Names

5006